The Role of Neurotransmitters in Brain Injury

The Role of Neurotransmitters in Brain Injury

Edited by

Mordecai Y.-T. Globus

and

W. Dalton Dietrich

University of Miami School of Medicine
Miami, Florida

SPRINGER SCIENCE+BUSINESS MEDIA, LLC

Library of Congress Cataloging-in-Publication Data

The Role of neurotransmitters in brain injury / edited by Mordecai Y.-T.
Globus. and W. Dalton Dietrich
     p.   cm.
  "Proceedings of an official satellite symposium of BRAIN-91: the
role of neurotransmitters in brain injury, held June 7-9, 1991, in
Key West, Florida."--T.p. verso.
  Includes bibliographical references and index.
  ISBN 978-1-4613-6528-0   ISBN 978-1-4615-3452-5 (eBook)
  DOI 10.1007/978-1-4615-3452-5
  1. Brain damage--Pathophysiology--Congresses.
2. Neurotransmitters--Pathophysiology--Congresses. 3. Excitatory
amino acids--Pathophysiology--Congresses.   I. Globus, Mordecai, Y.-T.
II. Dietrich, W. Dalton.   III. Brain-91 (1991 : Miami, Fla.)
  [DNLM: 1. Brain Injuries--congresses. 2. Neuroregulators-
-physiology--congresses.   WL 354 R745 1991]
RC387.5.R65 1992
616.8--dc20
DNLM/DLC
for Library of Congress   92-49215
                                                        CIP

Proceedings of an official satellite symposium of BRAIN-91:
The Role of Neurotransmitters in Brain Injury,
held June 7-9, 1991, in Key West, Florida

ISBN 978-1-4613-6528-0

© 1992 Springer Science+Business Media New York
Originally published by Plenum Press, New York in 1992
Softcover reprint of the hardcover 1st edition 1992

PREFACE

Neuroscientists from various disciplines have given extraordinary attention to the role of neurotransmitters in the field of neuronal injury. This volume summarizes the original oral and poster contributions which were presented at the symposium, "The Role of Neurotransmitters in Brain Injury," in Key West, Florida, between June 7-9, 1991. This symposium was the official Satellite of Brain-91, the Fifteenth International Symposium on Cerebral Blood Flow and Metabolism, held in Miami the previous week.

The two principal goals of the Key West satellite meeting were to document recent progress and, more importantly, to explore future directions for investigative studies of the role of neurotransmitters in brain injury. To achieve these goals we assembled participants from diverse scientific fields and specialties who brought their collective expertise to discussions on the importance of neurotransmitters in neuronal and vascular injury following brain ischemia, trauma, and epilepsy. Their contributions are reflected in this volume.

An important section of this volume is devoted to the role of glutamate and glutamatergic receptors in the development of ischemic neuronal damage. Topics covered include the mechanisms underlying injury-induced release of glutamate and other excitatory amino acids, and the role of different glutamatergic receptors in brain injury, including the NMDA and the non-NMDA receptors. The involvement of other neurotransmitters in the process of ischemic brain injury is well established. Recent studies which were presented at the satellite meeting confirm the role of dopamine in global and focal ischemia, and provide evidence for the role of the noradrenergic system in modulating excitotoxicity. Other neurotransmitters which were presented as modulators of brain injury include serotonin and adenosine. Manipulation of these neurotransmitter systems may provide an important route for the treatment of brain injury. An important section of this volume is focused on the role of second-messenger systems in brain injury. Topics include changes in gene expression, phospholipase A_2, and the effect of calpain inhibitors on excitotoxicity and ischemia. The role of the blood-brain barrier in brain injury is summarized, and the potential for blood-borne neurotransmitters contributing to neuronal injury also is discussed. Further presentations introduce the concept that neurotransmitter actions may be involved in disruption of the blood-brain barrier as their primary target, especially glutamate and NMDA.

We wish to acknowledge the following institutions for their generous support of this symposium: the International Society of Cerebral Blood Flow and Metabolism, the National Institute of Neurological Disorders and Stroke, the National Parkinson Foundation, the Department of Neurology of the University of Miami School of Medicine, the Warner-Lambert Company, the Upjohn Company, and the Titmus Foundation.

We believe this volume will enable those engaged in brain-injury and neurotransmitter research to inform themselves of the most recent developments in the field.

Mordecai Y.-T. Globus, M.D.
W. Dalton Dietrich, Ph.D.

CONTENTS

1. INJURY-INDUCED RELEASE OF EXCITATORY AMINO ACIDS

2. THE ROLE OF GLUTAMATERGIC RECEPTOR IN BRAIN INJURY

3. THE ROLE OF MONOAMINES IN BRAIN INJURY

4. THE ROLE OF STEROIDS, ADENOSINE, GABA, AND ACETYLCHOLINE IN BRAIN INJURY

5. INTRACELLULAR MESSENGERS AND BRAIN INJURY

6. BLOOD BRAIN BARRIER IN BRAIN INJURY

7. NEUROTRANSMITTER MODULATION OF CEREBRAL BLOOD FLOW AND METABOLISM

8. NEUROTRANSMITTERS AND FREE RADICAL MEDIATED INJURY

Chapter 1

Injury-Induced Release of Excitatory Amino Acids

EXCITATORY AMINO ACID RELEASE INDUCED BY INJURY

Brian Meldrum, Maria H. Millan and Tihomir P. Obrenovitch*

Department of Neurology, Institute of Psychiatry
London, SE5 8AF, U.K. and
Department of Neurological Surgery*, Institute of
Neurology, Queen Square, London, WC1N 3BG, U.K.

INTRODUCTION

The pioneering work of Van Harreveld introduced the two concepts
that concern us here, firstly that injury of the brain is associated
with an abnormal release of glutamate and secondly that glutamate
focally in the brain induces cytopathology (Van Harreveld & Fifkova,
1971). Van Harreveld studied two models of cerebral injury, spreading
depression (induced by focal injury or high K^+) and asphyxia. He
provided evidence that spreading depression in the retina and chick
brain was associated with release of glutamate (Fifkova & Van Harreveld,
1970).

In the last 7 years in vivo microdialysis studies have focussed
attention on the increase in the extracellular concentration of
glutamate and aspartate that occurs during global and focal cerebral
ischemia, profound hypoglycemia and spinal or hemispheric trauma
(Benveniste et al., 1984; Hagberg et al., 1985; Butcher et al., 1987;
Liu et al., 1991). These studies have only partially answered the
following crucial questions,- what mechanisms give rise to this increase
in extracellular glutamate and aspartate? how significant is this
increase in determining the pathological outcome? and what procedures
most effectively modify the release (and the pathological outcome)?
This chapter will review our present limited understanding of these
issues, outlining the basic concepts and the experimental approaches
available for resolving the present uncertainties.

EXTRACELLULAR GLUTAMATE CONCENTRATION

Whereas the cytoplasmic and intravesicular concentrations of
glutamate can be estimated at approximately 10 mM and 100 mM
respectively, more uncertainty attaches to estimates of the
concentration of glutamate in the extracellular fluid. The CSF
concentration is around 1 µM, but microdialysis studies, when correction
is made not only for recovery but also for limited in vivo diffusion
("tortuosity factor"), suggest that mean extracellular concentrations
around neuropil are higher than this (Benveniste & Huttemeier, 1990).
Na^+/K^+(gradient)-dependent carrier-mediated uptake processes account for

The Role of Neurotransmitters in Brain Injury, Edited by
M. Globus and W.D. Dietrich, Plenum Press, New York, 1992

the principle uptake of glutamate and aspartate into both neurons and glia (Nicholls & Attwell, 1990).

This carrier has a high affinity for glutamate (K_d 20-50µM), and apparently cotransports $3Na^+$ (or $2Na^+$ and H^+) and Glu^- inwards and K^+ outwards resulting in a net inward movement of one positive ion, i.e. it is electrogenic. This uptake is thus critically dependent on the transmembrane gradients of Na^+ and K^+.

Methods of Studying Injury-Induced EAA Release

Injury-induced EAA release can be studied in in vitro models or in vivo. In vitro models include synaptosomes, neuronal cultures, brain slice preparations, or whole retina, in which EAA release is induced by high $[K^+]$, veratrine, anoxia or metabolic poisons (Drejer et al., 1985). Release can be measured by analysis of glutamate or aspartate or sulphur containing amino acids in superfusate. In vivo the usual system is to study local changes in extracellular glutamate concentration using a push-pull system or microdialysis with a flexible fibre or a rigid probe. Clearly when studying the effects of cerebral trauma the flexible fibre system is more appropriate than rigid probes (which have to be removed before and re-inserted after the trauma) (Panter et al, 1990; Liu et al, 1991). Microdialysis probes may modify the processes being studied, exacerbating, for example, ischemic damage (Phebus et al., 1991) or altering drug-induced epileptic activity (Chastain et al., 1990).

It is also possible to visualise the movement of glutamate. This can be done in hippocampal slices utilising glutamate dehydrogenase and NAD^+ (Mitani et al., 1991) or by using immunocytochemical methods with in vivo fixation with glutaraldehyde and EM visualisation with immunogold procedures to permit ultrastructural localisation (Torp et al., 1991).

Increases in Extracellular EAA Concentrations; Differentiating the Mechanisms

The synaptic release of glutamate is Ca^{++} dependent, and less directly (because of the requirement for ATP to generate the proton gradient that enables vesicular uptake of glutamate) ATP-dependent. The simplest procedure for demonstrating synaptic release of glutamate in vitro and in vivo is the application of high K^+ solutions (50-100 mM). Such a K^+-induced glutamate release is largely but not entirely Ca^{++} dependent. Recent data suggest that the calcium-independent component of glutamate release may be non-vesicular and due to a reversal of glutamate uptake (Szatkowski et al., 1990, Sarantis & Attwell, 1990). Potassium-induced release can be decreased by the presynaptic inhibitors of glutamate release. Presynaptic receptors which can decrease glutamate release include those for the adenosine A1 receptor; the kappa opioid receptor (at which dynorphin appears to be the endogenous agonist and); the $GABA_B$ receptor; the glutamate, 2-amino-4-phosphonobutyric acid receptor; and a cholinergic/muscarinic receptor. These mostly seem to act by decreasing the Ca^{++} entry into nerve terminals. They are found on different types of nerve ending with very different densities or efficacies. Thus adenosine receptors are highly effective on excitatory terminals in the hippocampus but have little efficacy on inhibitory terminals (Scholz & Miller, 1991; Yoon & Rothman, 1991), whereas the reverse is true for $GABA_B$ receptors. Dynorphin receptors decrease both dopamine and glutamate release.

The increased glutamate release induced by kainate is largely synaptic. As reported by Millan et al. in this volume hippocampal microdialysis with pilocarpine applied via the probe gives rise to an increase in glutamate concentration in the dialysate that is not seen when the dialysis fluid contains no Ca^{++} but EDTA. Pilocarpine may be acting to depolarize glutamatergic neurons since muscarinic receptors are abundant on dendrites of pyramidal cells (Rotter et al., 1979). The question arises whether the injury induced increases in extracellular glutamate concentration can be suppressed either by the presynaptic receptor agonists or by the absence of Ca^{++}. Evidence on these matters remains controversial. Drejer et al., (1985) studying microdialysates from the rat hippocampus during ischemia found that the increase in glutamate produced by transient global ischemia was absent when the perfusion fluid contained 10 mM $CoCl_2$ and no Ca^{++}.

Of course at the extremes the mechanisms of injury induced release can be readily defined. Severe trauma sufficient to disrupt cell membranes causes a direct leakage of cytoplasmic fluid, and indeed this is seen upon insertion of a dialysis probe. The mildest stress is spreading depression where the ionic shifts and glutamate release are approximately synchronous. This process can occur in association with trauma or mild to moderate ischemia.

Carrier-Mediated Glutamate Uptake/Exchange

The plasma membrane carrier systems in neurons and glia appear closely similar although the proteins involved are different. An increase in extracellular glutamate that is attributed to "a failure of the glutamate reuptake mechanism" appears to imply that the primary cause of the increase is vesicular release. However if the uptake system is perceived as an equilibrium system dependent on the Na^+ and K^+ concentration gradients then an increase in $[Glut]_e$ can occur in the absence of vesicular glutamate release. This reversal of carrier-mediated glutamate uptake appears to be the basis of the glutamate release induced in vitro and in vivo by veratridine.

Thus the veratrine-induced release is dependent on an increased influx of Na^+, and tetrodotoxin completely prevents the neurotransmitter release. Thus the changes in $[Na^+]$ presumably lead to reversal of the Na^+-dependent glutamate carrier in the plasma membrane; under some circumstances depolarisation may also facilitate Ca^{++} entry through voltage sensitive channels, triggering vesicular release of glutamate (Villanueva et al., 1988). A significant proportion of the released glutamate and aspartate can arise from synaptic terminals, as is shown by the decrease in veratridine-induced release seen with striatal dialysis following cortical deafferentation,- which can not be presumed to derive from vesicular release (Butcher & Hamberger, 1987).

It also appears that the reverse-uptake mechanism makes a major contribution to the increase in $[Glut]_e$ seen in ischemia. This arises primarily as a consequence of the anoxic depolarisation of neurons and glia (Szatkowski et al., 1990; Sarantis & Attwell, 1990) that is associated with major shifts in Na^+ and K^+. There is evidence that the sudden DC potential shift that indicates anoxic depolarisation occurs synchronously with a massive release of monoamines, glutamate and aspartate (Scheller et al., 1989; Obrenovitch et al., 1990). Thus in the 4-vessel occlusion model of cerebral ischemia microdialysis with on-line detection of glutamate shows a massive increase in glutamate concentration associated with anoxic depolarisation (Obrenovitch et al., 1990). However, with penumbral ischemia where electrical activity is abolished but ionic homeostasis is preserved (no anoxic depolarization), a moderate rise in [Glut]$_e$ still remains detectable (Obrenovitch et al., 1990). It is interesting to note that, under such conditions, [K$^+$]$_e$ is known to increase up to 10-12 meg/1, a level that has been recently shown in vitro to produce the release of non-vesicular glutamate by reversed electrogenic glutamate uptake (Szatkowski et al., 1990; Sarantis & Attwell, 1990). That the release of glutamate is contributing in a causal way to the anoxic depolarisation is suggested by the observation that kynurenate (which antagonises NMDA and non-NMDA receptor mediated excitation) delays terminal depolarisation in cerebral ischemia (Katayama et al., 1989). This concept is supported by more decisive experiments in spreading depression where the acute depolarisation can be triggered by aspartate or glutamate release and can be totally blocked by low doses of NMDA receptor antagonists (Marrannes et al., 1988).

It is sometimes proposed that injury-induced glutamate release could be a non-specific phenomenon associated with breakdown of membrane permeability properties such that soluble cytoplasmic constituents would diffuse out freely. Available measurements of dialysate composition in a wide range of injuries do not support this interpretation. However the glutamate release is part of a more or less synchronous release of neuroactive compounds that includes aspartate, taurine, GABA, dopamine, serotonin, adenosine and various neuropeptides.

Strategies for Reducing Injury-Induced EAA Release

a) decreasing synthesis of glutamate

The metabolic route by which neurotransmitter glutamate is derived is still not well defined (see Torgner & Kvamma, 1990). The ultimate precursor is thought to be plasma glucose, but neurons are incapable of converting glucose to glutamate as they lack TCA cycle enzymes and pyruvate carboxylase. In vitro experiments have been interpreted as showing a predominant glutamate formation from glutamine (via glutaminase), with glutamine being supplied by astrocytes, who derive it from glucose or glutamate. The effect of glutaminase inhibitors (such as azaserine and DON) has not been assessed in in vivo models. Alternative approaches include ornithine transaminase inhibition (by L-canaline) and inhibition of the mitochondrial membrane carrier for ketodicarboxylic acids (by phenyl-succinate). Phenyl succinate applied via a dialysis tube blocks K$^+$-induced release of glutamate but not that induced by ischemia (see Diemer et al; this volume).

b) decreasing synaptic release

i) by decreasing neuronal activity. Barbiturates and other anaesthetics may act by this means (part of this may be a specific action on non-NMDA excitatory neurotransmission via non-NMDA receptors).

ii) by presynaptic receptor activation

Of the receptors mentioned above the greatest attention in terms of cerebroprotective potential has been paid to the adenosine A1 presynaptic receptor and the kappa-opioid or dynorphin receptor. As the evidence is strongest for the kappa opioid agonists (such as U 50488H and U69593) we shall restrict our discussion to them. A cerebroprotective action has been reported in cerebral ischemia (both global and focal) (Birch et al., 1991) and in cerebral trauma (Hall et al., 1987). Although there is clear evidence that kappa opioids do decrease the synaptic release of glutamate (and of dynorphin and dopamine) (Gannon & Terrian., 1991) it is nevertheless highly probable that the protective action of kappa opioids is related to their diuretic and anti-oedema actions (Silvia et al., 1987).

c) decreasing reverse transport by plasma membrane glutamate carrier

   i) by preventing ionic changes (Na^+ and K^+). A simple demonstration of the contribution of Na^+ entry to ischemic process is provided by the delaying effect of tetrodotoxin on anoxic depolarisation in the striatum (Prenen et al., 1988). Recent data of Ben Ari et al. (1990) sugggest that K^+ agonists may also be capable of reducing anoxic depolarisation.

   ii) by direct action on the carrier. Evidence for this mechanism is limited. Lamotrigine and the related compound BW 1003C (see Millan et al this volume) may act by this means, but they may also act on Na^+ entry.

Apart from such direct pharmacological appoaches, it remains the case that reducing the temperature of the brain is a highly effective method of reducing injury-induced glutamate release (Busto et al., 1989).

The Significance of Glutamate Release for Neuronal Cell Death

The role of glutamate in determining cell death following ischemia or other injury is extremely complex, and in vitro experiments may be a poor guide to the sequential events involving multiple cell to cell interactions that occur in vivo. Many experiments are consistent with the view that the early glutamate release is necessary as a trigger or primer for later events that determine cell death (e.g Benveniste et al., 1989). Such later events may also involve glutamate receptors and may be a better therapeutic target than the injury-induced release itself.

ACKNOWLEDGEMENTS

We thank the Wellcome Trust, the Leverhulme Trust and the Medical Research Council for financial support.

REFERENCES

Ben-Ari, Y., Krnjevic, K., and Crépel, V., 1990, Activators of ATP-sensitive K^+ channels reduce anoxic depolarization in CA3 hippocampal neurons. Neuroscience, 37:55-60.
Benveniste, H., Drejer, J., Schousboe, A., and Diemer, N.H., 1984, Elevation of the extracellular concentrations of glutamate and aspartate in rat hippocampus during transient cerebral ischemia monitored by intracerebral microdialysis, J. Neurochem., 43:1369-1374.
Benveniste, H., Jorgensen, B., Sandberg, M., Christensen, T., Hagberg, H., and Diemer, N.H., 1989, Ischaemic damage in hippocampal CA1 is dependent on glutamate release and intact innervation from CA3, J. Cereb. Blood Flow Metab., 9:629-639.
Benveniste, H., and Hüttemeier, P.C., 1990, Microdialysis-Theory and application, Prog. Neurobiol., 35:195-215.

Birch, P.J., Rogers, H., Hayes, A.G., Hayward, N.J., Tyers, M.B., Scopes, D.I.C., Naylor, A., and Judd, D.B., 1991, Neuroprotective actions of GR 89696, a highly potent and selective kappa-opioid receptor agonist, Brit. J. Pharmacol., 103:1819-1823.

Busto, R., Globus, M.Y.T., Dietrich, D., Martinez, E., Valdes, I., and Ginsberg, M.D., 1989, Effect of mild hypothermia on ischemia-induced release of neurotransmitters and free fatty acids in rat brain, Stroke, 20:904-10.

Butcher, S.P., and Hamberger, A., 1987, In vivo studies on the extracellular, and veratrine-releasable, pools of endogenous amino acids in the rat striatum: effects of corticostriatal deafferentation and kainic acid lesion, J. Neurochem., 48:713-721.

Butcher, S.P., Sandberg, M., Hagberg, H., and Hamberger, A., 1987, Cellular origins of endogenous amino acids released into the extracellular fluid of the rat striatum during severe insulin-induced hypoglycemia, J. Neurochem., 48:722-728.

Chastain, J.E., Jr., Samson, F., Neldon, S.R., and Pazdernik, T.L., 1990, Effects of microdialysis on brain metabolism in normal and seizure states, Neuroscience 37:155-161.

Drejer, J., Benveniste, H., Diemer, N.H., and Schousboe, A., 1985, Cellular origin of ischemia-induced glutamate release from brain tissue in vivo and in vitro, J. Neurochem., 45:145-151.

Fifkova, E. and Van Harreveld, A, 1970, Glutamate effects in the developing chicken. Exp. Neurol., 28:286-298.

Hagberg, H., Lehmann, A., Sandberg, M., Nyström, B., Jacobson, I., and Hamberger, A., 1985, Ischemia-induced shift of inhibitory and excitatory amino acids from intra- to extracellular compartments, J. Cereb. Blood Flow Metab., 5:413-419.

Hall, E.D., Wolf, D.L., Althaus, J.S., and Von Voigtlander, P.F., 1987, Beneficial effects of the kappa opioid receptor agonist U-50488H in experimental acute brain and spinal cord injury. Brain Res., 435:174-180.

Gannon, R.L., and Terrian, D.M., 1991, U-50,488H inhibits dynorphin and glutamate release from guinea pig hippocampal mossy fiber terminals. Brain Res. 548:242-247.

Katayama, Y., Becker, D.P., Tamura, T., Martin, N.A., Cheung, M.K., and Tsubokawa, T., 1989, Inhibition of massive ionic fluxes during cerebral ischemia (terminal depolarization) with an excitatory amino acid antagonist. J. Cereb. Blood Flow Metab., 9:S57. (Abstract).

Liu, D., Thangnipon, W., and McAdoo, D.J., 1991, Excitatory amino acids rise to toxic levels upon impact injury to the rat spinal cord. Brain Res., 547, 344-348.

Marrannes, R., Willems, R., De Prins, E., and Wauquier, A., 1988, Evidence for a role of the N-methyl-D-aspartate (NMDA) receptor in cortical spreading depression in the rat. Brain Res., 457:226-240.

Mitani, A., Kadoya, F., Nakamura, Y., and Kataoka, K., 1991, Visualization of hypoxia-induced glutamate release in gerbil hippocampal slice. Neurosci. Lett. 122:167-170.

Nicholls, D. and Attwell, D., 1990, The release and uptake of excitatory amino acids. TIPS., 11:462-468.

Obrenovitch, T.P., Sarna, G.S., and Symon, L., 1990, Ionic homeostasis and neurotransmitter changes in ischaemia, in: "Pharmacology of Cerebral Ischemia, 1990," J. Krieglstein and H. Oberpichler, eds., p.97, Wissenschaftliche Verlagsgesellschaft mbh, Stuttgart.

Panter, S.C., Yum, S.W., and Faden, A.I., 1990, Alteration in extracellular amino acids after trumatic spinal cord injury, Ann. Neurol., 27:96-99.

Phebus, L., Mincy, B.E., and Clemens, J.A., 1991, Microdialysis perfusion accelerates striatal ischemic damage. Communication to the 1991 Int. Symp. on Microdialysis and Allied Analytical Techniques, Indianapolis, IN, May 15-17, Abstract No.72.

Prenen, G.H.M., Go, K.G., Postema, F., Zuiderveen, F. and Korf, J. 1988, Cerebral cation shifts in hypoxic-ischemic brain damage are prevented by the sodium channel blocker tetrodotoxin. Exp. Neurol., 99:118-132.

Rotter, A., Birdsall, N.J., Burgen, A.S.V., Field, P.M., Hulme, E.C., and Raisman, G., 1979, Muscarinic receptors in the central nervous system of the rat. I. Technique for autoradiographic localization and of the binding of [³H]propylbenzilycholine mustard and the distribution in the forebrain. Brain Res. Dev. 1:141-165.

Sarantis, M., and Attwell, D., 1990, Glutamate uptake in mammalian retinal glia is voltage- and potassium-dependent, Brain Res., 516:322-325.

Scheller, D., Heister, U., Peters, U., and Holler, M., 1989, Glutamate and aspartate are released concomitantly with the terminalDC negativation after global cerebral ischemia, J. Cereb. Blood Flow Metab., 9:S372. (Abstract).

Scholz, K.P., and Miller, R.J., 1991, Analysis of adenosine actions on CA2+ currents and synaptic transmission in cultured rat hippocampal pyramidalneurones. J. Physiol. (Lond.), 435:373-393.

Szatkowski, M., Barbour, B., and Attwell, D., 1990, Non-vesicular release of glutamate from glial cells by reversed electrogenic glutamate uptake, Nature, 348:443-446.

Torgner, I., and Kvamma, E., 1990, Synthesis of transmitter glutamate and the glial-neuron interrelationship. Mol. Chem. Neuropathol., 12:11-17.

Torp, R., Andiné, P., Hagberg, H., Karagülle, T., Blackstad, T.W., and Ottersen, O.P., 1991, Cellular and subcellular redistribution of glutamate-, glutamine- and taurine-like immunoreactivities during forebrain ischemia: A semiquantitative electron microscopic study in rat hippocampus, Neuroscience, 41:433-447.

Van Harreveld, A., and Fifkova, E., 1971, Light- and electron-microscopic changes in central nervous tissue after electrophoretic injection of glutamate, Exp. Mol. Pathol. 15:61-81.

Villaneuva, S., Frenz, P., Dragnic, Y., and Orrego, F., 1988, Veratridine-induced release of endogenous glutamate from rat brain cortex slices: a reappraisal of the role of calcium, Brain Res. 461:377-380.

Yoon, K.-W., and Rothman, S.M., 1991, Adenosine inhibits excitatory but not inhibitory synaptic transmission in the hippocampus, J. Neurosci., 11:1375-1380.

EXCITATORY AMINO ACID RELEASE AFTER FOCAL CEREBRAL ISCHAEMIA: INFARCT VOLUME DETERMINES EAA RELEASE

Ross Bullock, Steven Butcher, David Graham,
Graham Teasdale

University Departments of Neurosurgery and
Neuropathology, Institute of Neurological Sciences
Glasgow, and Department of Pharmacology
University of Edinburgh

INTRODUCTION

The release of neuro-excitatory amino acids during cerebral ischaemia is a "corner-stone" of the excitotoxic hypothesis.[6] Recently, microdialysis has demonstrated release of excitatory amino acid neurotransmitters, (EAAs) particularly glutamate and aspartate after focal and global ischaemic events and traumatic brain injury, in various animal models.[1,2,5] The factors which affect the ECF concentration of EAAs, and inhibitory neurotransmitters during ischaemia are, however, poorly understood. Shimada et al. have shown that glutamate release in a cat global ischaemia model is a threshold phenomenon; glutamate increases rapidly below a blood flow level of 20 mls/100g/min.[8]

Other factors which may affect EAA levels after an ischaemic event include:

1. Release of EAAs from presynaptic vesicles; [6]
2. Release of non-transmitter metabolic amino acids from the cytosol.
3. Integrity of EAA uptake mechanisms, into presynaptic terminals.
4. Diffusion of EAAs away from the ischaemic focus into surrounding tissue. This may be influenced by residual blood flow, particularly in "penumbral regions" after a focal lesion.
5. The size of the ischaemic lesion, and thus the quantity of EAAs released.

We have recently used microdialysis to study release of EAAs in two animal models of focal cerebral ischaemia.[1,2] We have tested the hypothesis that the volume of the focal ischaemic lesion is a determinant of the magnitude of EAA release.

MATERIALS AND METHODS

General preparation

Twenty-two adult male Sprague-Dawley rats weighing 337-510 grams were anaesthetised with halothane (0.5-2%) and a nitrous oxide oxygen mixture. After tracheostomy and cannulation of femoral arteries and veins and the surgical procedures (see below) the animals were mounted in a stereotactic frame and physiologically monitored (blood pressure, blood gases, core temperature, blood glucose).

Surgical procedures - middle cerebral artery occlusion (MCAO)

In eight animals middle cerebral artery occlusion was carried out using the method of Tamura.[9] In four rats occlusion was carried out proximal to the olfactory tract and lenticulo striate arteries. In a further four animals, the MCA was coagulated distal to the olfactory tract to produce a smaller cortical infarct. In two control animals the MCA was exposed but not occluded. After the procedure the wounds were sutured, and dialysis probes inserted.

Subdural haematoma (SDH)

SDH was generated by injecting 400 microlitres of autologous blood into the subdural space via a parietal burr hole over seven minutes.[3] Subdural haematomas were performed in eight rats and in four control animals, sham surgical procedures were performed but no blood was injected.

Microdialysis

Precalibrated 3mm microdialysis probes were implanted into the parietal cortex and caudate nucleus (MCAO) and into the hippocampus and parietal cortex (SDH). The probes were perfused with Krebs bicarbonate buffer using a Carnegie CM100 infusion pump. After a one hour equilibration period, 20 microlitre dialysate fractions were collected before, and for four hours (MCAO group) and two hours (SDH group) after the surgical procedure. Dialysate amino acid content was determined by high performance liquid chromatography with fluorescence detection (aspartate, glutamate, glutamine, taurine, alanine, valine).

Volumetric histopathology

The MCAO animals were perfusion fixed and six micron serial forebrain sections were cut. The volume of ischaemic damage was traced at eight predetermined stereotactic levels throughout the brain. An image analysing computer was used to determine the volume of ischaemic damage for each brain.[4]

Cerebral blood flow measurement

In SDH animals the volume of ischaemic tissue was determined from iodoantipyrene autoradiograms. CBF was measured by the method of Sakurada et al.[7] Brain sections at eight stereotactic planes were selected and tissue with a CBF level below 25 mls/100g/min was measured using an image analysing computer. From this the volume of ischaemic

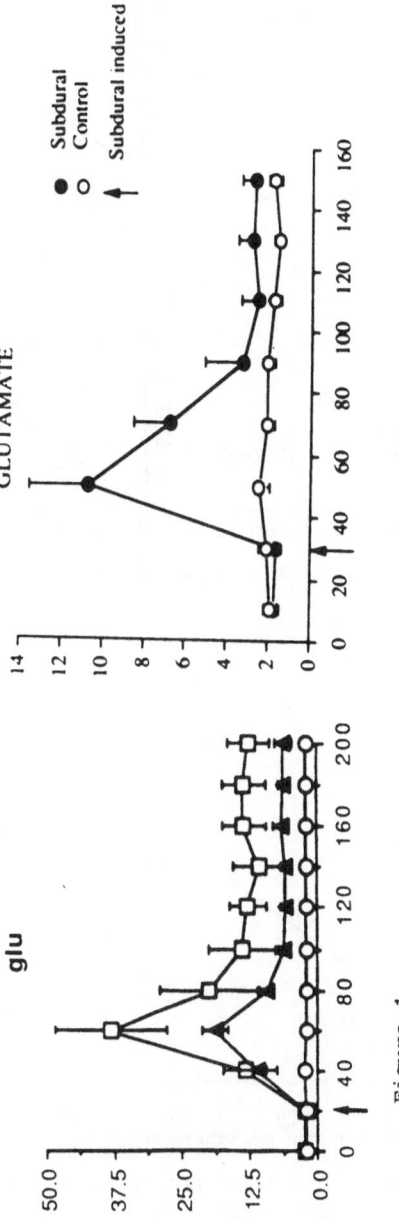

Figure 1

Time course of EAA release after middle cerebral
artery occlusion (cortical probe, left) and subdural
haematoma. (Right) □ -proximal occlusion Δ - distal
occlusion O - sham controls. Dialysate content is
expressed in p mol/min.

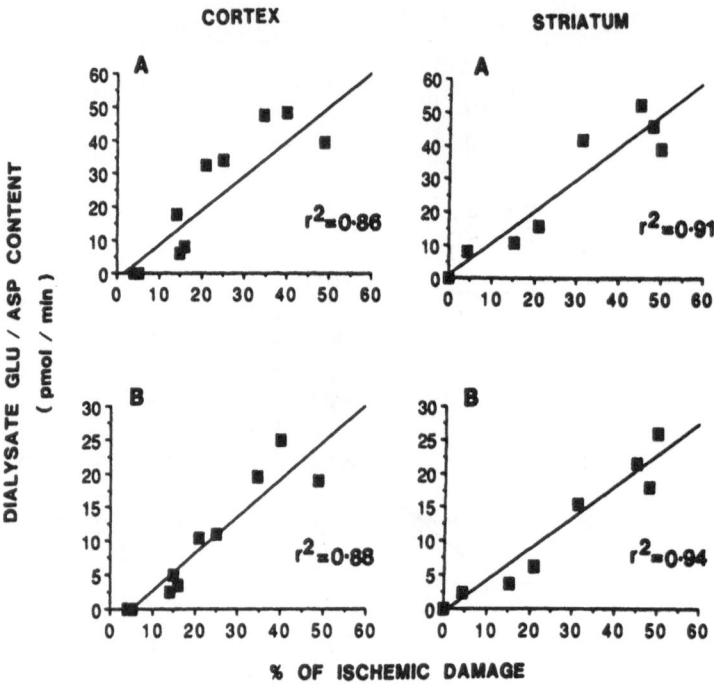

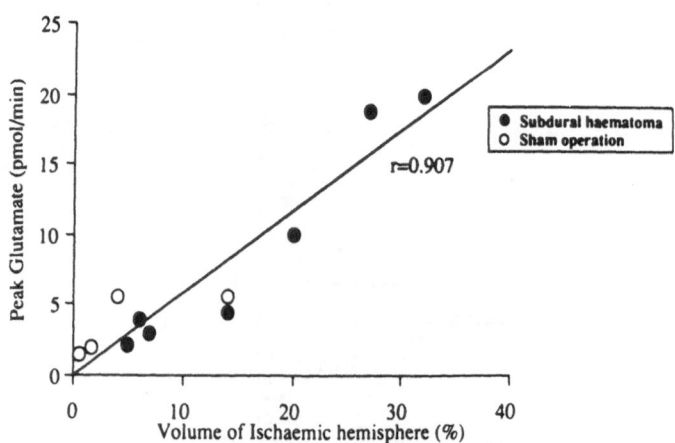

Figure 2   Relationship between EAA release and the volume of ischaemic tissue after MCAO. (Cortex and striatum) A - peak release, B - sustained release (4 hours). The lower plot shows the relationship between peak glutamate release (p mol/min) and volume of ischaemic tissue after subdural haematoma.

tissue for the hemisphere was calculated as above.[4] The relationship between peak glutamate release and the extent of hemispheric ischaemia was tested by linear regression analysis.

RESULTS

EAA release

The pattern but not the magnitude of EAA release was similar for both models, as seen by the probes implanted in the ischaemic cortex and striatum in the animals with proximal MCA occlusion (Fig. 1). The magnitude of glutamate release was up to tenfold higher after proximal MCAO than after SDH.

Volumetric histopathology - MCA occlusion

In control animals a small triangular area of ischaemic damage was seen in cortex at the site of probe insertion. The volume of cortical ischaemic damage after proximal occlusion ranged from 25 to 47% (mean 34.2% $\pm$ 6% SEM) and for striatum from 31 to 48% of total hemisphere volume (mean 42.9% $\pm$ 4% SEM). Distal occlusion produced smaller lesions - cortical range 14 to 21% (mean 16.5 $\pm$ 1%) and in striatum, lesions were minimal.

Volume of ischaemic tissue after SDH

The volume of ischaemic tissue (CBF less than 25 mls/100g/min) ranged from 0.5 to 7% of mean hemisphere volume in the sham. In animals with subdural haematoma, the range was from 2% to 22% of hemisphere volume.

Relationship between peak levels of EAA release and volume of ischaemic tissue

Figure 2 shows the relationship between dialysate glutamate and aspartate content (MCA occlusion series) and the percentage of hemispheric ischaemic damage, for cortex and striatum separately. The relationship between peak cortical glutamate content in the dialysate and the volume of ischaemic tissue after subdural haematoma is also shown. Significant correlation was seen between the volume of ischaemic tissue and peak glutamate content in both these series. The R values for these relationships ranged from 0.907 to 0.969.

DISCUSSION

In the MCA occlusion study reported here both the peak and the sustained efflux of excitatory amino acids correlated with the volume of ischaemic damage seen at four hours.(Fig. 2)[2] Unlike studies in reperfusion global injury models, EAAs remained at levels around 13 times (striatum) and 6 times (cortex) above basal (for glutamate) for four hours, in the proximal MCAO series. Sustained efflux was much less for the distal MCA group.

EAA release in ischaemic cortex was originally sixfold increased, but then fell much closer to basal levels after SDH; (Fig. 1) and the mean ischaemia volume (14%) was similar to that seen in the distal MCAO series (16%)[3]. It is thus tempting to speculate that a volume threshold may exist in the rat, above which EAA release cannot be normalised by either diffusion, and/or residual CBF. (Energy dependent re-uptake systems are unlikely to be functioning in the densely ischaemic core of focal lesions such as these).

Further studies are needed to study factors affecting re-uptake systems and to test whether EAAs can diffuse from the irreparably damaged 'core' of an ischaemic lesion, to damage, or activate surrounding viable or "penumbral" tissue.

REFERENCES

1. R Bullock, SP Butcher, M-H Chen, L Kendall, J McCulloch. Correlation of the extracellular glutamate concentration with extent of blood flow reduction after subdural haematoma in the rat. J. Neurosurg. 74:794-802, (1991)

2. SP Butcher, R Bullock, DI Graham, J McCulloch. Correlation between amino acid release and neuropathological outcome in rat striatum and cortex following middle cerebral artery occlusion. Stroke. 21:I. 1727-1733, (1990)

3. JD Miller, R Bullock, DI Graham, et al. Ischaemic brain damage in a model of acute subdural haematoma. Neurosurgery. 27: 433-439, (1990)

4. KA Osborne, T Shigeno, AM Balowski, et al. Quantitative assessment of early brain damage in a rat model of focal cerebral ischaemia. Journal of Neurology, Neurosurgery and Psychiatry. 50: 402-410, 1987

5. L Persson, L Hillered, U Ponten U, et al. Intracerebral microdialysis for continuous monitoring of neurosurgical patients: preliminary methodological considerations. J. CBF. Metabol. 9 (Suppl.1) S584, (1989)

6. S Rothman. Synaptic release of excitatory amino acid neurotransmitters mediates anoxic neuronal death. J. Neuro. Sci. 4: 1884-1891, (1984)

7. O Sakurada, C Kennedy, JW Jehle, et al. Measurements of local cerebral blood flow with IODO ^{14}C antipyrene. Am. J. Physiol. 234: H59-H66, (1978)

8. N Shimada, R Graf, G Rosner, et al. Ischaemic flow threshold for extracellular glutamate increase in cat cortex. J CBF Metabol. 9: 603-606, (1989)

9. A Tamura, DI Graham, J McCulloch, GM Teasdale. Focal cerebral ischaemia in the rat: 1. Description of technique and early neuropathological consequences following middle cerebral artery occlusion. J. Cereb. Blood Flow Metab. 1: 53-60, (1981)

A PYRIMIDINE DERIVATIVE, BW 1003C87, DECREASES GLUTAMATE RELEASE AND PROTECTS AGAINST ISCHEMIC DAMAGE

*B.S. Meldrum, +J.H. Swan, *M.H. Millan, +M.J. Leach, #R. Gwinn, #K. Kadota, #S.H. Graham and #R.P. Simon

*Institute of Psychiatry, London SE5 8AF, UK, +Wellcome Research Labs, Beckenham, Kent, and #U.C.S.F. Department of Neurology, San Francisco

INTRODUCTION

BW 1003C87, 5-(2,3,5-trichlorophenyl)-2,4-diamino-pyrimidine ethane sulphonate has certain structural features in common with the triazine derivative, lamotrigine which has previously been shown to decrease glutamate release from brain slices and to be anticonvulsant in animals and man (Miller et al., 1986). In *in vitro* tests of blockade of veratrine-induced glutamate release BW 1003C87 is substantially more potent than lamotrigine (IC_{50} 1.6 μM vs 21.0 μM).

An enhanced release of glutamate during cerebral ischemia, and continuing activation of glutamate receptors during the post-ischemic period are thought to contribute importantly to nerve cell death following focal or transient global cerebral ischemia (Benveniste et al., 1984; Choi & Rothman, 1990; Meldrum, 1990). We have, therefore, tested the ability of BW 1003C87 to block glutamate release *in vivo*, and to prevent brain damage in rodent models of global and focal ischemia. The results suggest that drugs decreasing release of glutamate may have therapeutic potential in syndromes involving cerebral ischemia comparable to that earlier proposed for antagonists acting at the post-synaptic glutamate receptor.

METHODS

In Vivo Glutamate Release

The effect of BW 1003C87 on *in vivo* glutamate release was measured in two situations. The effect on veratrine-induced release was studied in male Sprague-Dawley rats under chloral hydrate anaesthesia with body temperature maintained at 37 °C, with microdialysis probes (1 mm membranes) inserted into the dorsal hippocampus and perfused with Ringer's solution pH 7.2 at 2 μl min. The effect of veratrine (500 μM) was compared with that of veratrine (500 μM) + BW 1003C87 100 μM added to the dialysis fluid on the glutamate content of 10 min dialysate samples.

The effect of BW 1003C87 on glutamate release induced by focal cerebral ischemia was determined in anaesthetised rats. A microdialysis probe (with a 4 mm membrane) was inserted into one caudate nucleus 90 min prior to ipsilateral MCA occlusion. Probes were perfused at 2 µl/min and dialysate collected for 120 min after MCA occlusion.

Transient Incomplete Global Ischemia

In male Wistar rats under halothane and N_2O/O_2 anaesthesia with body temperature maintained at 37 ^{o}C animals were subjected to 10 min of bilateral common carotid artery occlusion as described by Smith et al., (1984). After a 7 day survival period rats were perfusion-fixed with formalin/acetic acid/methanol. Stained (cresyl violet) coronal sections were examined by light microscopy and a grid used to count pyramidal neurons in hippocampus CA1. Cell counts were compared in ventilated control rats, rats with 10 min ischemia, rats with 10 min ischemia receiving BW 1003C87, (5 mg/kg i.v.) 0 and 4 h after ischemia, and rats with 10 min ischemia receiving BW 1003C87 (20 mg/kg i.v.) 0 and 4 h after ischemia.

Middle Cerebral Artery Occlusion

In male Sprague-Dawley rats under chloral hydrate anaesthesia the middle cerebral artery was occluded by cauterization according to Tamura et al., 1981 as modified by Shiraishi & Simon 1989. Intravenous infusion of saline or BW 1003C87, 5, 10 or 20 mg/kg began 5 min after MCA occlusion. After 72 h the animals were perfusion-fixed and coronal sections were cut from 2.7 mm anterior to bregma to 4.8 mm posterior to bregma. Stained sections (toluidine blue) at six levels were used to compute infarct volumes in basal ganglia and cortex.

RESULTS

Glutamate Release

In anaesthetised rats with microdialysis probes in the dorsal hippocampus application of veratrine hydrochloride, 0.5 mM, via the probe increased the glutamate concentration in the dialysate by 482±108%. With co-infusion of veratrine and BW 1003C87, 0.1 mM, the increase was only 57%±14%, n=5; p=0.008 (Mann-Whitney 'U' test).

In anaesthetised rats with microdialysis probes in the caudate nucleus the glutamate content of the dialysate was greatly increased in the 2 h after MCA occlusion . This increase was markedly attenuated by the administration of BW 1003C87, 20 mg/kg, i.v.

Transient Incomplete Global Ischemia

The loss of hippocampal pyramidal cells induced by ischemia was not modified by the lower dose of BW 1003C87.

Significant protection was, however, produced by administering BW 1003C87, 20 mg/kg i.v. twice (at times 0 and 4h after ischemia). The mean number of pyramidal cells/3 grid lengths was 377±17 (n=9) in ventilated controls, 140±35 (n=9) in ischemic rats and 279±41 (n=9) in ischemic rats receiving BW 1003C87, 20 mg/kg at 0 and 4 h (p<0.050).

Middle Cerebral Artery Occlusion

The volume of cortex infarcted appeared to be less after BW 1003C87 5 or 10 mg/kg. Statistically significant reduction in the volume of cortex infarcted was, however, found only after the i.v. administration (at + 5 min) of BW 1003C87, 20 mg/kg.

DISCUSSION

The potent blockade of veratrine-induced glutamate release shown by BW 1003C87 *in vitro* is reflected *in vivo* by similar blockade of veratrine-induced glutamate release and by a reduced glutamate efflux during ischemia. The mechanism of this effect is not definitively known but an action on voltage-sensitive sodium channels has been proposed in the case of lamotrigine (Leach et al., 1986). The precise importance of the intra-ischemic increase in extracellular glutamate concentration in determining neuronal death is not known. Concentrations of glutamate that are toxic to neurons in culture have been reported during prolonged global ischemia and during focal ischemia (Benveniste et al., 1984; Hagberg et al., 1985; Globus et al., 1988; Graham et al., 1988; Butcher et al., 1990; Shimada et al., 1990). The temperature dependence of brain damage following global ischemia correlates with the temperature dependence of ischemia-induced increases in extracellular glutamate concentration. Removing a glutamatergic input to CA1 pyramidal neurons by lesioning the CA3 neurons protects CA1 neurons against delayed cell loss following global ischemia, an effect that can be reversed by the focal injection of glutamate (Benveniste et al., 1989). However regionally selective ischemic brain damage does not correlate with any local differences in the ischemia induced elevations in extracellular glutamate concentration (Globus et al., 1991). The contribution that glutamate makes to pathological outcome undoubtedly differs in transient global ischemia, and in the core and the penumbra of focal ischemia. The marked effect that BW 1003C87 had on ischemia-induced increases in extracellular glutamate levels in the caudate nucleus was not associated with any protection of the striatum when BW 1003C87 was given directly after MCA occlusion.

In transient global ischemia it is clear that certain triggering or predisposing events involving glutamate and its receptors happen at the time of the ischemia, but that activation of glutamate receptors in the post-ischemic period contributes importantly to delayed neuronal death in the hippocampus. Both NMDA and non-NMDA receptors are involved in these processes. NMDA receptor antagonists have a weak protective action against hippocampal and striatal damage if given prior to or directly after brief incomplete global ischemia (Rod & Auer, 1989; Warner et al., 1991; Wieloch et al., 1989; Swan & Meldrum 1990). Non-NMDA antagonists can protect against delayed cell death in the hippocampus if effective plasma levels are maintained for a prolonged period post-ischemically; the effect is present but reduced when therapy begins 1 or 2h after transient global ischemia (Sheardown et al., 1990; Diemer et al., 1990). Cortex and striatum can also be protected by non-NMDA antagonists acting during the first 3 hours after global ischemia (Le Peillet et al., 1991). The effect we describe here with BW1003C87 is comparable to the effect of glutamate receptor blockade. Thus the effect of BW 1003C87 on post-ischemic glutamate release requires study.

The volume of cortex showing infarction after MCA occlusion can be reduced by administration of either NMDA or non-NMDA antagonists post-occlusion (McCulloch et al., 1991; Smith & Meldrum, 1991; Gill & Lodge 1991). The cerebroprotective effect we observe here with BW 1003C87 is comparable to that found with post-synaptic glutamate antagonists. The enhanced extracellular glutamate levels in the ischemic penumbra are the result of several different processes, possibly including a spreading depression-like effect. Here also the action of BW 1003C87 requires further study.

Previous attempts to protect against ischemic damage by reducing the release of glutamate have mainly concerned drugs acting as adenosine agonists (Evans et al., 1987; Von Lubitz et al., 1988). Such drugs however show marked cardiovascular and other side effects. Lamotrigine is finding increasing use in the treatment of epilepsy and in this context appears to be relatively free from side effects.. Lamotrigine protects against the excitotoxic action of kainate in the striatum (McGeer & Zhu, 1990). BW 1003C87 is significantly more potent than lamotrigine at blocking glutamate release, and in these preliminary studies shows clear protective effects in rodent models of global and focal cerebral ischemia. BW 1003C87 has significant antifolate activity and may not itself be appropriate for clinical trial. Nevertheless it provides evidence that an approach based on reduction of glutamate release may be useful in the therapy of cerebral ischemic damage.

ACKNOWLEDGEMENT

We thank the Chest, Heart & Stroke Association and the Leverhulme Trust for financial support. We thank Dr M. Nobbs, Wellcome Foundation laboratories, for the synthesis and supply of BW 1003C87.

REFERENCES

Benveniste, H., Jorgensen, M., Sandberg, M., Christensen, T., Hagberg, H. and Diemer, N.H., 1989, Ischemic damage in hippocampal CA1 is dependent on glutamate release and intact innervation from CA3. J. Cereb. Blood Flow Metab. 9:629-639, 1989.

Butcher, S.P., Bullock, R., Graham, D.I. and McCulloch, J., 1990, Correlation between amino acid release and neuropathologic outcome in rat brain following middle cerebral artery occlusion. Stroke 21:1727-1733.

Choi, D.W., Rothman, S.M., 1990, The role of glutamate neurotoxicity in hypoxic-ischemic neuronal death. Ann. Rev. Neurosci. 13:171-182.

Evans, M.C., Swan, J.H., Meldrum, B.S. (1987), An adenosine analogue, 2-chloroadenosine, protects against long term development of ischaemic cell loss in the rat. Neurosci. Lett. 83:287-292.

Gill, R., Lodge, D., 1991, The neuroprotective action of 2,3-dihydroxy-6-nitro-7-sulfamoyl-benzo(F)quinoxaline (NBQX) in a rat focal ischaemia model. Brit. J. Pharmacol. 102:61P.

Globus, M.Y-T., Busto, R., Dietrich, W.D., Martinez, E., Valdes, I., Ginsberg, M.D., 1988, Effect of ischemia on the in vivo release of striatal dopamine, glutamate, and γ-aminobutyric acid studied by intracerebral microdialysis. J. Neurochem. 51:1455-1464.

Globus, M.Y-T., Busto, R., Martinez, E., Valdes, I., Dietrich, D., Ginsberg, M.D., 1991, Comparative effect of transient global ischemia on extracellular levels of glutamate, glycine, and γ-aminobutyric acid in vulnerable and non-vulnerable brain regions in the rat. J. Neurochem. 57, 470-478.

Graham, S.H., Shiraishi, K., Panter, S.S., Simon, R.P., Faden, A.I., 1990, Changes in extracellular amino acid neurotransmitters produced By focal cerebral ischemia. Neurosci. Lett. 110:124.

Hagberg, H., Lehmann, A., Sandberg, M., Nystroem, B. Jacobson, I., Hamberger, A. 1985, Ischemia-induced shift of inhibitory and excitatory amino acids from intra- to extracellular compartments. J Cerebr. Blood Flow Metab., 5:413-419.

Leach, M.J., Marden, C.M., Miller, A.A., 1986, Pharmacological studies on Lamotrigine, a novel potential antiepileptic drug: 2. Neurochemical studies on the mechanism of action. Epilepsia 27(5):490-497.

Le Peillet, E., Arvin, B., Moncada, C., Meldrum, B.S. The non-NMDA antagonists, NBQX and GYKI 52466 protect against cortical and striatal cell loss following transient global ischaemia in the rat. Brain Res. (in press).

McCulloch, J., Bullock, R., Teasdale G.M., 1991, Excitatory amino acid antagonists : opportunities for the treatment of ischaemic brain damage in man. In: "Excitatory Amino Acid Antagonists", B.S. Meldrum, eds., Blackwell Scientific Publications, Oxford, pp287-326.

McGeer, E.G., Zhu, S.G., 1990, Lamotrigine protects against kainate but not ibotenate lesions in rat striatum. Neurosci. Lett. 112:348-351.

Meldrum, B.S. 1990, Protection against ischaemic neuronal damage by drugs acting on excitatory neurotransmission. Cerebrovasc. Brain Metab. Rev. 2:27-57.

Meldrum, B., Garthwaite, J., (1990). Excitatory amino acid neurotoxicity and neurodegenerative disease. Trends Pharmacol. Sci. 11:379-387.

Miller, A.A., Sawyer, D.A., Roth, B., Peck, A.W., Leach, M.J., Wheatley, P.L., Parsons, D.N., Morgan, R.J.I., 1986, Lamotrigine. In "New Anticonvulsant Drugs", B.S. Meldrum and R.J. Porter, eds., pp.165-177, John Libbey, London.

Paxinos, G., Watson, C., 1982, The rat brain in sterotaxic coordinates, 2nd Ed, Sidney, Academic Press.

Rod, M.R., Auer, R.N., 1989, Pre- and postischemic administration of dizocilpine (MK-801) reduces cerebral necrosis in the rat. Can. J. Neurol. Sci. 16:340-344.

Sheardown, M.J., Nielsen, E.O., Hansen, A.J., Jacobsen, P., Honoré, T., 1990, 2,3-dihydroxy-6-nitro-7-sulfamoyl-benzo(F)quinoxaline : a neuroprotectant for cerebral ischemia. Science 247:571-574.

Shimada, N., Graf, R., Rosner, G. and Heiss, W-D., 1990, Differences in ischemia-induced accumulation of amino acids in the cat cortex. Stroke 21:1445-1451.

Shiraishi, K., Simon, R.P., 1989, A model of proximal middle cerebral artery occlusion in rat. J. Neurosci. Methods 30: 169-174.

Smith, A.L., Bendek, G., Dahlgren, N., Rosen, I., Wieloch, T., Siesjö, B.K., 1984. Models for studying long-term recovery following fore-brain ischemia in the rat. 2. A 2-vessel occlusion model. Acta Neurol. Scand. 69:385-401.

Smith, S., Meldrum, B.S. 1991. The effect of the non-N-methyl-D-aspartate antagonist, GYKI 52466, on lesion volume and neurological deficit following focal cerebral ischemia in the rat. Stroke (submitted).

Swan, J.H., Meldrum, B.S., 1990, Protection of NMDA antagonists against selective cell loss following transient ischaemia. J. Cereb Blood Flow Metab. 10:343-351.

Tamura, A., Graham, D.I., McCulloch, J., Teasdale, G.M., 1981, Focal cerebral ischaemia in the rat: 1. Description of technique and early neuropathological consequences following middle cerebral artery occlusion. J. Cereb. Blood Flow Metabol. 1:53-60.

Warner, M.A., Neill, K.H., Nadler, J.V., Crain, B.J., 1991, Regionally selective effects of NMDA receptor antagonists against ischemic brain damage in gerbil. J. Cerebr. Blood Flow Metab. 11:600-611.

Wieloch, T., Gustafson, I., Westerberg, E., 1989, The NMDA antagonist, MK-801, is cerebroprotective in situations where some energy production prevails but noit under conditions of complete energy deprivation. J. Cerebr. Blood Flow Metab. 9(1):S6.

Von Lubitz D.K.J.E., Dambrosia, J.M., Kempski, O., Redmond, D.J., 1988, Cyclohexyl adenosine protects against neuronal death following ischemia in the CA1 region of gerbil hippocampus. Stroke 19:1133-1139.

PILOCARPINE-INDUCED ACTIVATION OF IN VIVO ASPARTATE AND

GLUTAMATE RELEASE IN DORSAL HIPPOCAMPUS

Maria H. Millan, Astrid G. Chapman and
Brian S. Meldrum

Department of Neurology, Institute of Psychiatry
De Crespigny Park, London SE5 8AF, U.K.

INTRODUCTION

An increased release of aspartate and glutamate, observed during ischemia in numerous studies, is believed to be a causal factor in the subsequent 'excitotoxic' neuropathology (Benveniste et al., 1984; see Meldrum, 1990). A similar mechanism is expected to be involved in seizure-induced brain injury (see Meldrum, 1991). Increased *in vivo* release of aspartate and glutamate has been reported to be associated with some forms of experimental seizures (Dodd & Bradford, 1976; Lehmann, 1987) and with ictal events in patients undergoing assessment for temporal lobectomy (Mattson et al., 1991; Do et al., 1991; Ronne-Engström et al., 1991). However, there have also been several studies that have failed to show an increased release of aspartate and glutamate in different brain regions following electrical stimulation or administration of convulsant drugs (Korf & Venema, 1985; Lehmann et al., 1985; Millan et al., 1991). A concomitant enhancement of excitatory amino acid (EAA) reuptake under these conditions, shown by Perschak and Cuénod (1990), would contribute to the lack of increased extracellular levels of aspartate and glutamate.

Cholinergic agents (e.g. pilocarpine) injected systemically or focally into the brain induce limbic seizures in rats (Turski et al., 1983; Pirreda & Gale, 1985), which are inhibited by EAA antagonists injected focally into some limbic structures or the basal ganglia output stations (De Sarro et al., 1986; Millan et al., 1986; Patel et al., 1987). These findings implicate that stimulation of muscarinic receptors may result in hyperactivity of the EAA transmitter system.

The hippocampus plays a key role in the development and sustaining of limbic seizures in animal models of epilepsy (Collins et al., 1983) and in humans (complex partial seizures; Wieser, 1988). In this study the influence of pilocarpine delivered via microdialysis probe on the extracellular levels of aspartate and glutamate in the dorsal hippocampus was investigated.

The Role of Neurotransmitters in Brain Injury, Edited by
M. Globus and W.D. Dietrich, Plenum Press, New York, 1992

METHODS

Drugs were dissolved in Ringer solution and the pH was adjusted to 7.2. Pilocarpine hydrochloride crystalline atropine sulfate, tetrodotoxin (TTX), and ethylene glycol-bis (ß-aminoethyl ether) N,N,N',N'-tetraacetic acid (EGTA) were purchased from Sigma.

Male Sprague-Dawley rats of (250-260g) were anaesthetized with 600 mg/kg, i.p. of chloral hydrate and placed in a stereotaxic frame. A concentric dialysis probe (3-3.5 mm long membrane tip; Cuprophane 0.65 mm diameter) was inserted into the dorsal hippocampus (Paxinos & Watson, 1982) atlas coordinates: A+4.0, V+4.5 (from intraaural line) L+3.4), where it crossed vertically the CA_1, dendate gyrus and CA_4 cell layers. The probe was perfused with Ringer solution (1.57 mM $CaCl_2$, 4.0 mM KCl and 147 mM NaCl, pH 7.2) at 2 µl/min rate. Following a 90 min equilibration period, 3 baseline 20 µl (= 3x10 min) samples were collected, after which drugs were applied via the probe according to the following protocol (n = 4-6 per group):
Group 1: 10 mM pilocarpine (20 min application).
Group 2. 10 mM pilocarpine + 20 mM atropine (20 min).
Group 3. 10 mM pilocarpine + 1 µM TTX (20 min).
Group 4. 10 mM pilocarpine (20 min) in animals perfused with Ca^{++} free medium containing 2 mM EGTA throughout the experiment.
Following drug application, perfusion with Ringer solution was resumed, and 3 more samples (3 x 20 µl) were collected. The probes were perfused with 5% toluidine blue at the end of experiments to mark the hippocampal placement using frozen cryostat sections.

Amino acid levels in the 20 µl dialysate samples were determined following pre-column OPA derivatization and separation on a Spherisorb ODS2 5µ column (25 x 0.46 cm), using a 20-55% methanol gradient in 0.1 M sodium acetate buffer, pH 5.8 (2.5% in tetrahydrofuran), and a Spectra-Physics HPLC system consisting of a ternary SP8800 pump, a Kratos FS950 fluorimeter and a ChromJet integrator. Levels of aspartate, glutamate, serine, glutamine, glycine, taurine and alanine were assessed by this method, while GABA levels could not be reliably determined with the HPLC conditions used. The mean values of extracellular aspartate or glutamate in the 3 pre-drug baseline samples were statistically compared (Student's t-test) with those in the post-drug samples within each group.

In vitro amino acid recovery through the probes was estimated by placing the dialysis probes in a 10 µM solution of a standard amino acid mixture prior to the hippocampal placement, and collecting 3 x 20 µl samples (pumping rate 2 µl/min) for amino acid analysis. The mean recovery value for aspartate was 14.0±1.6% and for glutamate 15.8±1.6% (n=12).

RESULTS

Concentration of aspartate in the hippocampal dialysate was 0.44±0.06 µM and of glutamate 1.2±0.26 µM (n=10) under resting conditions. Perfusion of 10 mM pilocarpine resulted in rapid and marked increase in the levels of glutamate and aspartate (to 279.9% and 243.7%, respectively, of baseline values) during the first 10 min of pilocarpine application

(Table 1). The extracellular aspartate and glutamate values remained significantly elevated during the 20 min of pilocarpine application, and returned gradually to baseline control levels following the restoration of perfusion by Ringer solution (Table 1). The extracellular levels of the other amino acids analyzed (glycine, serine, glutamine, taurine, alanine) did not show any consistent response to pilocarpine application.

Perfusion with calcium free medium containing 20 mM EGTA did not significantly influence baseline levels of the amino

Table 1. Effect of pilocarpine administration on levels of aspartate and glutamate in hippocampal perfusate.

	Aspartate	Glutamate
	(% of baseline level)	
Baseline level		
(µM in perfusate)	0.44±0.06	1.20±0.26
	=100.0± 13.6%	=100.0± 21.7%
During 10 mM pilocarpine administration:		
0-10 min	243.7± 48.3***	279.9±155.6**
10-20 min	258.2± 48.0***	199.2± 58.1***
During post-pilo recovery:		
0-10 min	488.6±271.7**	160.2± 39.9**
10-20 min	181.1± 76.5	112.6± 13.2
20-30 min	112.1± 23.5	103.2± 22.1
During 10 mM pilocarpine co-administered with:		
20 mM atropine		
0-10 min	95.1± 37.5	91.4± 8.0
10-20 min	99.2± 23.5	101.1± 13.8
1 µM TTX		
0-10 min	89.3± 14.8	89.0± 7.7
10-20 min	130.4± 23.4	89.2± 5.4
2 mM EGTA (calcium free)		
0-10 min	61.2± 17.2	67.9± 2.0***
10-20 min	112.7± 36.6	67.2± 5.7***

Values are expressed as percent of mean baseline values for aspartate and glutamate (baseline dialysate concentrations, µM, are given in top line of table). n = 4-6 per group. Asterisks denote values significantly different from corresponding baseline values, **$p<0.01$, ***$p<0.001$.

acids, but completely prevented the pilocarpine-evoked enhancement in the extracellular levels of glutamate and aspartate (Table 1). The actual decrease in extracellular glutamate during calcium-free pilocarpine stimulation can be accounted for by an enhanced (calcium-independent) glutamate reuptake under these conditions (Perschak & Cuénod, 1990; Nicholls & Atwell, 1990). Co-administration of 10 mM pilocarpine along with 1 µM TTX or with 20 mM atropine also

prevented the pilocarpine-induced increases of in
extracellular aspartate and glutamate levels (Table 1).

DISCUSSION

Activation of muscarinic receptors within the
hippocampus facilitates release of the excitatory
transmitters, aspartate and glutamate, in a calcium dependent
and tetrodotoxin sensitive manner indicating that neuronal
release is taking place. The reversal of the pilocarpine-
induced activation by the antagonist, atropine, confirms the
cholinergic involvement. There have been no previous reports
of *in vivo* release of EAAs due to the activation of the
cholinergic system.

The anatomical organisation of glutamatergic and
cholinergic projections within the hippocampus facilitates
the interaction between these two systems in many hippocampal
regions. Cholinergic innervation of the hippocampus
originates in the medial septum, and the main EAA innervation
originates in the entorhinal cortex via the perforant path.
Both projections synaps mainly on dentate gyrus granule
cells. Within the hippocampus the mossy fibers and Schäffer
collaterals also use EAAs as transmitters, while CA_3
pyramidal neurons, and to a lesser extent the CA_1 region
receive cholinergic inputs (see Ottersen and Storm-Mathisen,
1986 and Woolf et al., 1984).

In vivo electrophysiological studies point to an
excitant role of the cholinergic system in the hippocampus.
Electrical stimulation of the septo-hippocampal cholinergic
pathway releases acetylcholine in the hippocampus (Dudar,
1975) and produces excitation of pyramidal and granule
hippocampal neurons, which is inhibited by muscarinic
antagonists (Bilkey & Goddard, 1985, Krnjevic et al., 1988;
Wheal & Miller, 1980).

There are several studies on ischaemia and epilepsy
indicating that cholinergic system has a facilitatory effect
on EAA system in the hippocampus. Pilocarpine induces limbic
motor seizures in rats, which are suppressed by focal
injections of EAA antagonists (Turski et al., 1983, Patel
1987), while lower doses of pilocarpine are proconvulsant
(Stringer & Lothman 1991). Electric kindling is delayed by
systemic injection of the muscarinic antagonist atropine
(Lupica & Berman 1988) or by ibotenate lesion of the basal
forebrain cholinergic system (Cain & Steward 1990).

In a forebrain ischemia model in rat the lesion of
septo-hippocampal pathway (fimbria/fornix) protects against
CA_1 ischaemic neuronal damage (Buchan & Pulsinelli 1990).

SUMMARY

Perfusion of the dorsal hippocampus with 10 mM
pilocarpine via a dialysis probe resulted in a transient
2.5-3 fold increase in extracellular aspartate and glutamate
levels. The evoked response was calcium dependent, TTX
sensitive, and blocked by atropine. As cholinergic agonists
are known to produce epileptic seizures in experimental
animals, the cholinergic stimulation of excitatory amino acid
release in the hippocampus may be involved in the
epileptogenic activity of pilocarpine.

REFERENCES

Benveniste, H., Drejer, J., Schousboe, A., and Diemer, N.H. (1984) Elevation of the extracellular concentrations of glutamate and aspartate in rat hippocampus during transient cerebral ischemia monitored by intracerebral microdialysis. J. Neurochem. 43, 1369-1374.

Bilkey, D.K. and Goddard, G.V. (1985) Medial septal facilitation of hippocampal granular cell activity is mediated by inhibition of inhibitory neurons. Brain Res. 361, 89-106.

Buchan, A.M. and Pulsinelli, W.A. (1990) Septo-hippocampal deafferentation protects CA1 neurons against ischemic injury. Brain Res. 512, 7-14.

Cain, D.P. and Stewart, D.J. (1990) Ibotenic acid lesions of the basal forebrain cholinergic system retard amygdala kindling. Pharmacol. Biochem. Behav. 36, 207-210.

Collins, R.C., Tearse, R.G., and Lothman, E.W. (1983) Functional anatomy of limbic seizures: focal discharges from medial entorhinal cortex in rat. Brain Res. 280, 25-40.

De Sarro, G., Patel, S., and Meldrum, B.S. (1986) Anticonvulsant action of a kainate antagonist γ-D-glutamylaminomethylsulphonic acid injected focally into the substantia nigra and entopeduncular nucleus. Eur. J. Pharmacol. 132, 229-236.

Do, K.Q., Klancnik, J., Gähwiler, B., Perschak, H., Wieser, H.G., and Cuénod, M. (1991) Release of EAA: animal studies and epileptic foci studies in human. In: Excitatory Amino Acids , 677-685. Edited by Meldrum, B.S., Moroni, F., Simon, R.P., and Woods, J.H., New York, Raven Press.

Dodd, P.R. and Bradford, H.F. (1976) Release of amino acids from the maturing cobalt-induced epileptic focus. Brain Res. 111, 377-388.

Dudar, J.D. (1975) The effect of septal nuclei stimulation on the release of acetylcholine from the rabbit hippocampus. Brain Res. 83, 123-133.

Korf, J. and Venema, K. (1985) Amino acids in rat striatal dialysates: methodological aspects and changes after electroconvulsive shock. J. Neurochem. 45, 1341-1348.

Krnjevic, K., Ropert, N., and Casullo, J. (1988) Septohippocampal disinhibition. Brain Res. 438, 182-192.

Lehmann, A. (1987) Alterations in hippocampal extracellular amino acids and purine catabolites during limbic seizures induced by folate injections into the rabbit amygdala. Neuroscience 22, 573-578.

Lehmann, A., Hagberg, H., Jacobson, I., and Hamberger, A. (1985) Effect of status epilepticus on extracellular amino acids in the hippocampus. Brain Res. 359, 147-151.

Lupica, C.R. and Berman, R.F. (1988) Atropine slows olfactory bulb kindling while diminished cholinergic innervation does not. Brain Res. Bull. 20, 203-209.

Mattson, R.H., During, M.J., Scheyer, R., Katz, A., Cramer, J.A., Spencer, D.D., and Toftness, B.R. (1991) Studies of antiepileptic drug pharmacology using human intracerebral microdialysis. Neurology 41, Suppl.1, 385.

Meldrum, B.S. (1990) Protection against ischaemic neuronal damage by drugs acting on excitatory neurotransmission. Cerebrovascular and Brain Metabolism Reviews 2, 27-57.

Meldrum, B.S. (1991) Excitatory amino acid neurotransmission

in epilepsy and anticonvulsant therapy. In: Excitatory
Amino Acids, Edited by Meldrum, B.S., Moroni, F., Simon,
R.P., and Woods, J.H., New York, Raven Press.

Millan, M.H., Patel, S., Mello, L.M., and Meldrum, B.S. (1986)
Focal injection of 2-amino-7-phosphonoheptanoic acid into
prepiriform cortex protects against pilocarpine-induced
limbic seizures in rats. Neurosci. Lett. 70, 69-74.

Millan, M.H., Obrenovitch, T.P., Sarna, G.S., Lok, S.Y.,
Symon, L., and Meldrum, B.S. (1991) Changes in rat brain
extracellular glutamate concentration during seizures
induced by systemic picrotoxin or focal bicuculline
injection: an in vivo dialysis study with on-line
enzymatic detection. Epilepsy Res. (in press).

Nicholls, D. and Attwell, D. (1990) The release and uptake of
excitatory amino acids. TIPS 11, 462-468.

Ottersen, O.P. and Storm-Mathisen, J. (1986) Excitatory amino
acid pathways in the brain. In: Advance in Experimental
Medicine and Biology, Vol. 203, 263-284. Edited by
Schwarz, R. and Ben-Ari, Y., New York, Plenum Press.

Patel, S., Millan, M.H., and Meldrum, B.S. (1987)
Neurotransmission in the pedunculopontine nucleus and
pilocarpine-induced motor limbic seizures in rats.
Neurosci. Lett. 74, 243-249.

Paxinos, G. and Watson, C. (1982) The rat brain in stereotaxic
coordinates, New York, Academic Press.

Perschak, H. and Cuénod, M. (1990) In vivo release of
endogenous glutamate and aspartate in the rat striatum
during stimulation of the cortex. Neuroscience 35,
283-287.

Piredda, S. and Gale, K. (1985) A crucial epileptogenic site
in the deep prepiriform cortex. Nature 317, 623-625.

Ronne-Engström, E., Carlsson, H., Flink, R., Hillered, L.,
Spännare, B., and Ungerstedt, U. (1991) In vivo studies
of the human epileptic focus using intracerebral
microdialysis. J. Cereb. Blood Flow Metab. 11, Suppl.2,
s656.

Stringer, J.L. and Lothman, E.W. (1991) Cholinergic and
adrenergic agents modify the initiation and termination
of epileptic discharges in the dentate gyrus.
Neuropharmacology 30, 59-65.

Turski, W.A., Cavalheiro, E.A., Schwarz, M., Czuczwar, S.J.,
Kleinrok, Z., and Turski, L. (1983) Limbic seizures
produced by pilocarpine in rats: behavioural,
electroencephalographic and neuropatholoical study.
Behav. Br. Res. 9, 315-335.

Wheal, H.V. and Miller, J.J. (1980) Pharmacological
identification of acetylcholine and glutamate excitatory
systems in the dentate gyrus of the rat. Brain Res. 182,
145-155.

Wieser, H.G. (1988) Human limbic seizures, origin and pattern
of spread. In: Anatomy of Epileptogenesis. Current
Problems in Epilepsy. Vol.6, 127-138. Edited by Meldrum,
B.S., Ferrendelli, J.A., and Wieser, H.G., London, John
Libbey.

Woolf, N.J., Eckenstein, F., and Butcher, L.L. (1984)
Cholinergic systems in the rat brain: I. Projections to
the limbic telencephalon. Brain Res. Bull. 13, 751-784.

TEMPERATURE DEPENDENCE OF GLUTAMATE RELEASE DURING TRANSIENT ISCHEMIA IN THE GERBIL AND THE EFFECT OF REPEATED OCCLUSIONS

Nobuhito Saito, Thaddeus S. Nowak, Jr., and Igor Klatzo

Stroke Branch
National Institute of Neurological Disorders and Stroke
NIH, Bldg. 36, Rm. 4D04, Bethesda, MD 20892

INTRODUCTION

Release of glutamate and other neurotransmitters into the extracellular space has been documented in various ischemia models, and has been suggested to play a role in proposed 'excitotoxic' mechanisms of ischemic neuronal injury (Benveniste et al., 1984; Globus et al., 1988). In recent years a striking cumulative effect of brief repeated ischemic insults has been demonstrated that depends on the interval between ischemic insults (Tomida et al., 1987; Nowak et al., 1990; Vass et al., 1990).

The present studies characterize glutamate release during repeated ischemic insults using intracerebral microdialysis. In view of increasing evidence for the importance of brain temperature in determining the magnitude of ischemic injury (Busto et al., 1987; Busto et al., 1989; Buchan et al., 1990; Kuroiwa et al., 1990), these experiments include an evaluation of this variable during repeated ischemia, and of its effect on glutamate release.

MATERIALS AND METHODS

Experimental Ischemia

Gerbils (female, 50-70 g) were maintained under anesthesia with 1.25 % halothane in 30 % O_2 / 70 % N_2O and the left femoral artery was catheterized with PE-10 and PE-50 tubing to monitor blood pressure. Animals were placed in a stereotaxic holder and the microdialysis probe was inserted into the striatum (2.3 mm lateral, 0.8 mm anterior of bregma, 4 mm depth). After stabilization for 90 min the animals were subjected to single or repeated ischemic insults of 5 or 15 min duration, as indicated for individual experiments, by applying tension to a suture loop previously placed around both carotid arteries. Rectal and temporalis muscle temperatures were monitored with thermistor probes, and maintained with a heating lamp. Changes in cerebral blood flow (CBF) were continuously monitored by laser-Doppler flowmetry. Halothane levels were reduced to 0.5 % during ischemia and recirculation to minimize the fall in blood pressure.

Microdialysis

The probes used in these studies were type CMA/10 with a membrane length of 2 mm. They were perfused with a solution consisting of 145 mM Na, 1.2 mM Ca, 2.7 mM K, 1.0 mM Mg and 152.1 mM Cl, buffered at pH 7.4 with 2 mM phosphate, using a CMA/100 syringe pump (BAS Inc.) at a flow rate of 2 ml/min. Samples were collected at 5 min intervals during ischemia and the first 15 min recirculation, and at 15 min intervals during

The Role of Neurotransmitters in Brain Injury, Edited by
M. Globus and W.D. Dietrich, Plenum Press, New York, 1992

other periods. Glutamate recoveries were evaluated for each experiment and ranged from 6.6 to 13.5 percent, obtained at room temperature with a glutamate standard prepared in dialysis buffer stabilized against convection with 0.1 % agarose.

Glutamate Measurement

Glutamate was determined by an enzymatic cycling method utilizing glutamate-oxaloacetate transaminase (GOT) and glutamate dehydrogenase (GDH) (Lowry and Passonneau, 1972). Briefly, 5 µl aliquots of dialysate or glutamate standard were added to 50 µl of a reagent consisting of 50 mM imidazole acetate (pH 6.7), 10 mM ammonium acetate, 0.2 mM ADP, 0.2 mM oxaloacatate, 0.015 mM DPNH, 5 µg/ml each of GOT and GDH and 0.02% albumin. After 1 h at 37°C the reactions were quenched on ice and remaining NADH was destroyed by the addition of 10 ml 1 N HCl. To each tube was added 1 ml 6 N NaOH to convert the NAD+ formed in the reaction to fluorescent products, which were then quantitated. Glutamate values obtained for a given animal were normalized by a factor corresponding to the recovery of the probe, and expressed as the equivalent extracellular concentration (mM).

RESULTS

Repeated Occlusions

With regard to the general features of the model, CBF showed virtually complete ischemia during the time of each occlusion and a brief return toward normal values upon release, followed by the characteristic postischemic hypoperfusion (not shown). These laser-Doppler results are in good agreement with the pattern previously obtained by standard radioactive tracer methods (Tomida et al., 1987). With maintained rectal temperatures during the period of microdialysis, there were the expected decreases in head temperature of approximately 1°C during each 5 min ischemic insult. These were superimposed on a progressive fall of approximately 0.8°C in temporalis muscle temperature during the course of the 5 h experiment, under conditions of continuous anesthesia and maintained rectal temperature at 37°C, perhaps resulting in part from the prolonged postischemic hypoperfusion that occurred during this period.

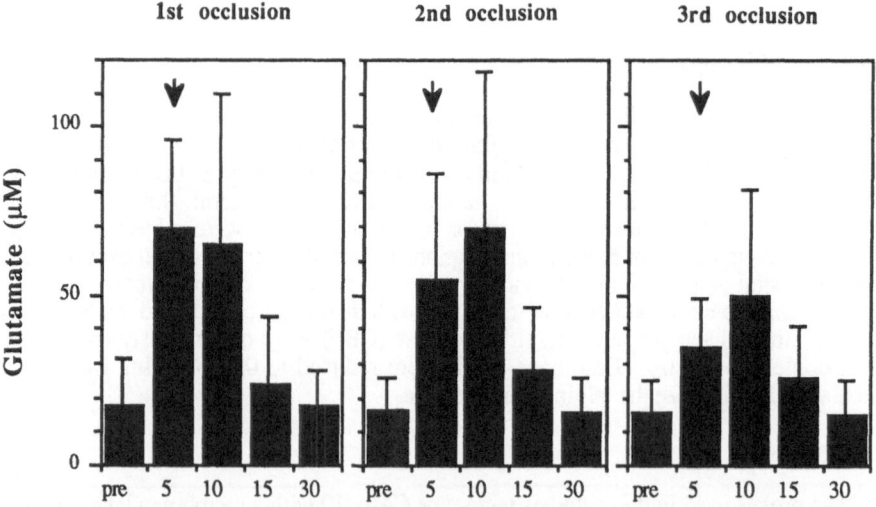

Fig. 1. Glutamate levels following repeated ischemic insults. Glutamate was determined as described in the text on dialysates collected before each occlusion (pre, 15 min sample), during occlusion (arrow) and early recirculation (three 5 min samples) and between 15 and 30 min recirculation. Bars indicate means ± SD.

Table 1. Effect of Temperature on Glutamate Release during 15 min Ischemia.

	Temperature (°C)			
Rectal	30.0 ± 0.5	33.2 ± 0.7	37.2 ± 0.2	39.8 ±0.2
Head	29.3 ± 1.8	33.3 ± 0.7	35.9 ± 0.3	38.9 ± 0.3
	Glutamate in dialysate (μM, mean ± SD)			
Preischemic	14 ±12	10 ±19	13 ± 17	15 ±12
5 min	51 ± 20	36 ± 42	38 ± 22	59 ± 20
10 min	82 ±16	75 ± 52	120 ± 39	187 ± 42
15 min	116 ± 31	140 ± 70	170 ± 79	280 ± 57*

(*, $p \leq 0.05$, Dunnett's test)

Repeated 5 min occlusions resulted in repeated bursts of glutamate release (Fig. 1). Peak glutamate levels were comparable after a second ischemic insult, but tended to be lower after the third insult in these experiments. Comparable results have been obtained in cortex and thalamus. Glutamate levels returned to baseline after each occlusion and remained so through 2 h of perfusion after the final insult. Other animals subjected to repeated insults with only brief anesthesia prior to occlusion showed a characteristic rise in body temperature to 39-40°C during the period of reperfusion after each ischemic insult. This transient hyperthermia had not been mimicked in the animals maintained under continuous anesthesia during microdialysis. While temperature fell toward control levels during reanesthesia for subsequent occlusions, the gradual decrease in head temperature seen during the microdialysis study would not be expected in the repeated ischemia model as originally characterized.

Effect of Varied Temperature

In view of other studies demonstrating effects of temperature on glutamate release, the importance of this variable was evaluated in the gerbil model. Ischemia of 15 min duration was employed in these studies. Groups of 6 animals were heated or cooled so that temporalis muscle temperature at the time of occlusion averaged 30, 33, 36 and 39°C. Head temperature fell in all groups during occlusion, with the greatest decrease in the high temperature group in which the thermal gradient with room air was most extreme. High mortality was observed in gerbils maintained at 39°C, and temperature was therefore allowed to drop during recirculation in this group to minimize losses.

Basal glutamate levels were at the limits of sensitivity of the assay employed in these studies, and no effect of temperature on this parameter could be detected. Glutamate levels in the dialysate increased throughout the 15 min occlusion in all groups (Table 1). Peak values tended to be higher with increasing temperature. Peak levels of 116 and 140 μM were attained in groups of animals maintained at 30 and 33°C, increasing to 170 μM and 280 μM at 36 and 39°C, respectively. Notably, there appeared to be a temperature-independent component of glutamate release in these experiments that remained evident at temperatures below 33°C, and effects on the modest glutamate release that occurred during the first 5 min of ischemia did not achieve statistical significance. Effects of increasing temperature on glutamate levels were observed at 10 min and 15 min time points. Only the group maintained at 39°C showed values that differed significantly from those of the other ischemic groups. The relationship between temperature and glutamate release is more apparent when data are evaluated for individual animals. Peak glutamate levels plotted as a function of head temperature (Fig. 2) show a striking dependence on temperature in the physiological range (35-39°C). The two sets of points represent temperatures obtained before and at the end of the 15 min ischemic interval, since head temperature fell during the occlusion in all groups. Final head temperature appear to provide the best correlation with glutamate release.

DISCUSSION

Brief 5 min ischemic insults resulted in transient, moderate increases in extracellular glutamate in striatum and other brain regions, as determined by in vivo microdialysis. Increases of comparable magnitude occur following each of three occlusions repeated at 1 hour intervals, but with a tendency toward a delayed and somewhat reduced peak dialysate glutamate concentration after the third ischemic period (Fig. 1). These observations differ from our preliminary studies of cortical glutamate release in the same model (Saito et al., 1991) in which cumulative increases in dialysate gultamate were frequently seen with repeated occlusions. We attribute the differences in these results to improved experimental technique resulting in reduced local tissue injury during the course of study. A delayed increase in striatal glutamate release following repeated ischemia in the rat has been reported (Lin et al., 1991) and was considered to reflect the onset of cumulative tissue injury characteristic of repeated ischemia models (Tomida et al., 1987). While secondary glutamate release may occur under such conditions we have also observed similar glutamate release in control animals maintained for long periods (unpublished observation) and have therefore restricted the present studies to relatively short intervals during which stability of the preparation can be demonstrated. The present experimental conditions fail to reproduce those under which the cumulative effect of repeated occlusions was first characterized. Notably, persistent hyperthermia is observed following release of occlusions in transiently anesthetized animals (Kuroiwa et al., 1990) that does not occur with the conditions of anesthesia employed for microdialysis, during which there is even a progressive fall in head temperature (data not shown). To evaluate the importance of this variance, the effect of temperature on glutamate release was evaluated in the gerbil ischemia model.

Glutamate release during a 15 min ischemic insult was directly related to head temperature in the physiological range (Fig. 2), in general agreement with studies in the rat (Busto et al., 1989; Sternau et al., this volume). No significant dependence on temperature was evident at 5 min of ischemia, although the method used in the present study may not be sensitive enough to resolve the expected differences. These differ from those in the rat in that

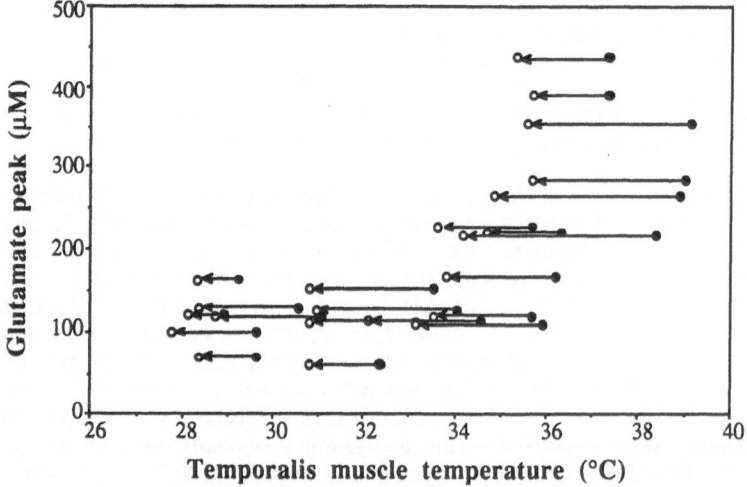

Fig. 2. Peak glutamate levels plotted as a function of head temperature before (solid circles) or at the end of (open circles) 15 min ischemia.

they appear to indicate a basal glutamate release that is insensitive to temperature reduction in the range studied, with striking dependence only at pre-ischemic head temperatures above 36°C. It should be noted that rectal temperature was maintained in the present study, while head (temporalis muscle) temperature was monitored but not specifically controlled during the ischemic period, and glutamate release was better correlated with the final head temperature at the end of the occlusion. In addition, the gradient between head and deep brain temperature may vary at different rectal temperatures, so that the relationship defined in Fig. 2 cannot be taken to indicate the true temperature dependence of striatal glutamate release. Future studies in animals with chronically implanted microdialysis probes are required to clarify the relationship between glutamate release and cumulative damage after repeated ischemic insults, under conditions that adequately reproduce the temperature changes known to occur in this model.

REFERENCES

Benveniste, H., Drejer, J., Schousboe, A., and Diemer, N. H., 1984, Elevation of the extracellular concentrations of glutamate and aspartate in rat hippocampas during transient cerebral ischemia monitored by intracerebral microdialysis, J. Neurochem., 43: 1369.

Buchan, A., and Pulsinelli, W. A., 1990, Hypothermia but not N-methyl-D-aspartate Antagonist, MK-801, attenuates neuronal damage in gerbils subjected to transient global ischemia, J. Neurosci., 10: 311.

Busto, R., Dietrich, W. D., Globus, M. Y-T., Valdes, I., Scheinberg, P., and Ginsberg, M. D., 1987, Small differences in intraischemic brain temperature critically determine the extent of ischemic neuronal injury, J. Cereb. Blood Flow Metab., 7: 729.

Busto, R., Globus, M. Y-T., Dietrich, W. D., Martinez, E., Valdes, I., and Ginsberg, M. D., 1989, Effect of mild hypothermia on ischemia-induced release of neurotransmitters and free fatty acids in rat brain, Stroke, 20: 904.

Globus, M. Y-T., Busto, R., Dietrich, W. D., Martinez, E., Valdes, I., and Ginsberg, M. D., 1988, Effect of ischemia on in vivo release of striatal dopamine, glutamate, and γ-aminobutyric acid studied by intracerebral microdialysis, J. Neurochem., 51: 1455.

Lin, B., Globus, M. Y-T., Dietrich, W. D., Busto, R., Martinez, E., Kraydieh, S., and Ginsberg, M. D., 1991, Different morphological and neurochemical sequelae of global ischemia: comparison of single- and multiple-insult paradigms, J. Cereb. Blood Flow Metab., 11, suppl 2: S854.

Kuroiwa, T., Bonnekoh, P., and Hossman, K-A., 1990, Prevention of postischemic hyperthermia prevents ischemic injury of CA1 neurons in gerbils, J. Cereb. Blood Flow Metab., 10: 550.

Lowry, O. H., and Passonneau, J. V., 1972, "A Flexible System of Enzymatic Analysis," Academic Press, New York.

Nowak, T. S. Jr, Tomida, S., Pluta, R., Xu, S., Kozuka, M., Vass, K., Wagner, H. G., and Klatzo, I., 1990, Cumulative effect of repeated ischemia on brain edema in the gerbil. Biochemical and physiological correlates of repeated ischemic insults, Adv. Neurol., 52: 1.

Saito, N., Chang, C., Kawai, K., Joó, F., Nowak, T. S. Jr, Mies, G., Ikeda, J., Nagashima, G., Ruetzler, C., Lohr, J. M., Spatz, M., and Klatzo, I., 1990, Role of neuroexcitation in development of blood-brain barrier and edematous changes following cerebral ischemia and traumatic brain injury, Acta Neurochir., suppl 51: 186.

Sternau, L. L., Globus, M. Y-T., Martinez, E., Dietrich, W. D., Busto, R., and Ginsberg, M. D., 1991, Ischemia-induced neurotransmitter release: effect of mild intraischemic hyperthermia, in: The Role of Neurotransmitters in Brain Injury, M. Y-T. Globus, W. D. Dietrich, eds., Plenum Publishing Corp., New York.

Tomida, S., Nowak, T. S. Jr, Vass, K., Lohr, J. M., and Klatzo, I., 1987, Experimental model for repetitive ischemic attacks in the gerbil: the cumulative effect of repeted ischemic insults, J. Cereb. Blood Flow Metab., 7: 773.

Vass, K., Tomida, S., Hossman, K-A., Nowak, T. S. Jr, Klatzo I., 1990, Microvascular disturbances and edema formation after repetitive ischemia of gerbil brain, Acta Neuropathol. (Berl.) 75: 288.

ISCHEMIA-INDUCED NEUROTRANSMITTER RELEASE:
EFFECTS OF MILD INTRAISCHEMIC HYPERTHERMIA

Linda L. Sternau, Mordecai Y.-T. Globus, W. Dalton Dietrich,
Elena Martinez, Raul Busto and Myron D. Ginsberg

Cerebral Vascular Disease Research Center
Department of Neurology
University of Miami School of Medicine
Miami, FL, 33101.

INTRODUCTION

Small differences in intraischemic brain temperature can significantly affect the neuropathological and functional outcome following transient global forebrain ischemia. Lowering intraischemic brain temperature by 2-3°C in the rat prevents the delayed neuronal damage that typically occurs in zones of selective vulnerability following 20 minutes of ischemia (Busto et al, 1987), and improves the behavioral outcome following a 10 min insult (vanDijk et al, 1991). Conversely, raising the intraischemic brain temperature to 39°C significantly worsen ischemic outcome and increased the vulnerability of resistant brain areas (Dietrich et al, 1990a). The protective effect of intraischemic hypothermia was not associated with the reduction in the degree of ischemia as reflected by the magnitude of energy metabolites depletion or cerebral blood flow reduction (Busto et al, 1987; Busto et al, 1989), suggesting that some other mechanism prevails.

Excessive neurotransmitter release, such as glutamate, has been documented to be involved in the development of ischemic neuronal injury (Meldrum, 1989). Decreases in temperature are known to inhibit the biosynthesis, release and uptake of neurotransmitters (Boels et al, 1985; Haikala et al, 1986; Okuda et al, 1986; Vanhoutte et al, 1981). In a recent a study we have demonstrated that mild reduction of brain temperature during ischemia was associated with inhibition of ischemia-induced glutamate release (Busto et al, 1989; Globus et al, 1988a). These findings suggest that the possible mechanism underlying the beneficial effect of hypothermia may be reduction of excitotoxic induced ischemic damage. Therefore, the detrimental consequences of hyperthermia may be associated with increased levels of extracellular glutamate during ischemia and recirculation. We have recently determined that ischemia-induced increases in glutamate occur in the extracellular fluid of several sites that are not destined for ischemic damage (Globus et al, 1990). This suggest that not only the release of glutamate but the release of other neurotransmitters leading to the imbalance between excitation and inhibition determines

The Role of Neurotransmitters in Brain Injury, Edited by
M. Globus and W.D. Dietrich, Plenum Press, New York, 1992

the histopathological outcome of transient ischemia. Among these neurotransmitters GABA and glycine may play a major role (Johnson and Ascher, 1987; Kleckner and Dingledine, 1988; Roberts et al, 1976). Since the effects of temperature may extend to affect the ischemia-induced changes in these neurotransmitters, we included them in this study.

MATERIAL AND METHODS

Fasted male Wistar rats (250-300 g) were subjected to 20 min of global forebrain ischemia induced by bilateral temporary occlusion of the carotid arteries combined with systemic hypotension (45 mm Hg) (Smith et al, 1984). Femoral vessels were cannulated under halothane anesthesia and ligatures placed around each common carotid artery. Rats were then intubated and mechanically ventilated to maintain blood gases. A 4 mm microdialysis probe was then stereotaxically implanted in the striatum and subsequently perfused with Ringer's solution at a flow rate of 2 ul/min. Two hours after microdialysis probe implantation, collection of dialysate was begun for baseline levels for 30 minutes. Ischemia was then induced for 20 minutes and recirculation established for the subsequent 4 hours. Ten minute samples were obtain prior, during and after ischemia. The rats were then sacrificed by pentothal overdose and probe position verified. The dialysate samples were frozen at -80°C and later analyzed for glutamate, GABA and glycine by HPLC-EC technique (Allison et al, 1984).

Two ischemic groups were investigated – rats in which the intraischemic brain temperature was maintained at 36.5-37°C (normothermia, n = 5) and at 38.5-39°C (hyperthermia, n = 6). Cerebral hyperthermia was obtained by adjusting the thermostatically regulated heat lamp placed above the head. Brain temperature was measured with a thermocouple probe placed stereotactically into the opposite striatum. Brain temperature in all rats was 37°C before and following the ischemic insult. The rectal temperature was monitored and maintained at 36.5-37°C throughout the experiment by a second heating lamp.

RESULTS

The effects of ischemia on the levels of glutamate, GABA and glycine in both temperature groups are illustrated in the Figure 1A-C. In animals whose intraischemic brain temperature was maintained at 37°C there was a 21-fold increase in glutamate levels which return to normal by 20-30 min of recirculation and remained unchanged through the whole reperfusion period (4h). In animals whose intraischemic brain temperature was maintained at 39°C a significant higher magnitude of surge in extracellular glutamate levels was observed (37-fold, $p = 0.0152$). Glutamate levels during the recirculation period tended to stay higher in the hyperthermic group, however, no statistical significant difference was demonstrated. Ischemia induced an increase in extracellular GABA levels in both the normothermic and hyperthermic animal groups (41 and 50-fold increase, respectively). The magnitude of increase in the hyperthermic group was significantly higher ($p = 0.039$). Normothermic ischemia induced a 3-fold increase in glycine levels and the levels remained elevated (2-fold increase) during the first 2h of recirculation. A higher magnitude of increase in extracellular glycine levels was observed in the hyperthermic group

(7-fold, p = 0.0414), and the levels remained high for the whole recirculation time period (3-4 fold). In an effort to derive a quantitative descriptor which might reflect the composite magnitude of aminoacid neurotransmitter changes with ischemia, we previously defined an "excitotoxic index" as [extracellular glutamate] X [extracellular glycine] / [extracellular GABA] (Globus et al, 1991a; Globus et al, 1991b). The excitotoxic data are presented in Figure 1D. The changes in the excitotoxic index were even more dramatic than the changes in the individual neurotransmitters. At 37°C, there was a 2-fold rise, whereas at 39°C, there was a 20-fold increase over baseline during the recirculation phase (p = 0.0476).

DISCUSSION

This study demonstrates that the magnitude of ischemia-induced increase of glutamate is augmented with mild intra-ischemic cerebral hyperthermia. The source of the increased extracellular glutamate levels during ischemia is not clear. Extracellular levels of glutamate are determined by the balance between release and uptake. The abnormal surge in extracellular levels of glutamate during ischemia is the result of depolarization-induced presynaptic release and a perturbed uptake

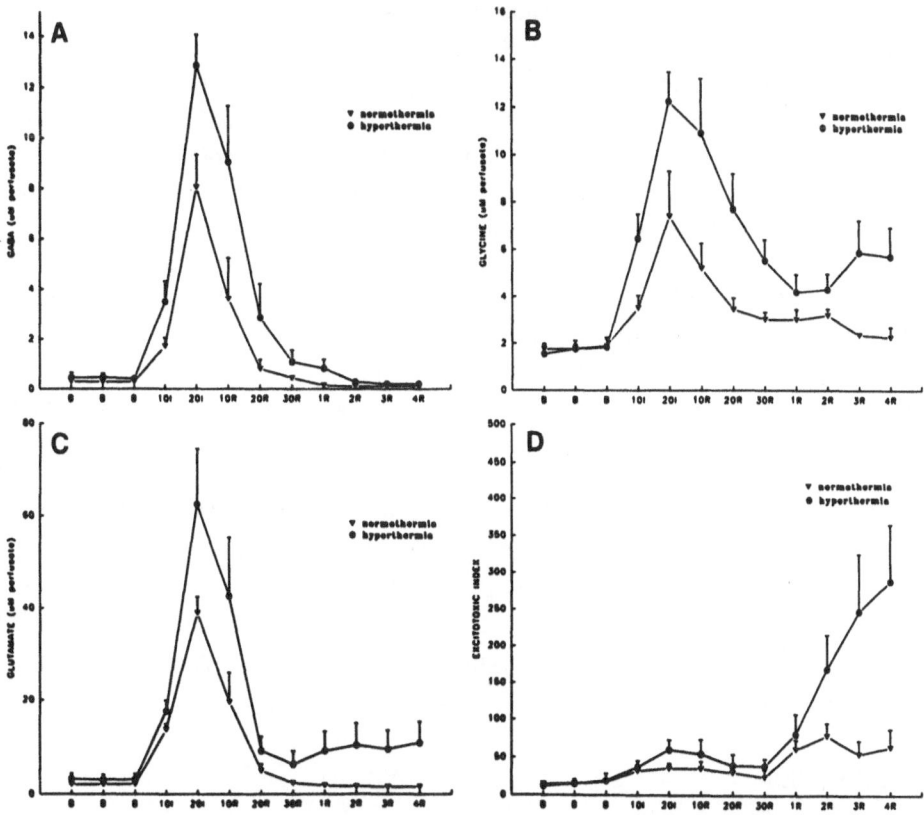

Figure 1: Time course changes in the perfusate levels (uM) of (A) GABA, (B) glycine, and (C) glutamate sampled from the dorsolateral striatum of animals subjected to 20 min of ischemic under normothermic and hyperthermic condition. (D) Time course changes in the excitotoxic index computed from the extracellular levels of glutamate, glycine, and GABA in the same groups of animals.

process. Hypothermia has been demonstrated to affects the release and uptake of several neurotransmitters (Boels et al, 1985; Haikala et al, 1986; Okuda et al, 1986; Vanhoutte et al, 1981), and to inhibit the ischemia-induce surge in extracellular neurotransmitters (Busto et al, 1989; Globus et al, 1988b). Conversely, hyperthermia may adversely augment this surge by increasing the magnitude of release through a damaged presynaptic membrane and/or by further disturbing the uptake process. An additional potential source of glutamate is the metabolic pool generated mainly from the tricarboxylic acid. Hyperthermia can alter the enzymatic reactions leading to greater amounts of intracellular glutamate released during ischemia. Finally, elevated brain temperature during ischemia alters the integrity of the blood brain barrier (BBB) (Dietrich et al, 1990b). Consequently, systemic glutamate may leak through a compromise BBB and contribute to the elevated extracellular glutamate.

Interest in the possible role of GABA and glycine in ischemic brain injury has arisen because of the possible scenario that an imbalance between excitation and inhibition of vulnerable neurons prevails in the postischemic phase. GABA has been shown to increase the influx of chloride into the cell causing hyperpolarization of the membrane, thus inhibiting the cells to fire (Roberts et al, 1976). Thus, GABA serves as an endogenous neuromodulator which can counteract postischemic neuronal hyperactivity induced by glutamate. Glycine has been demonstrated to facilitate glutamate activation of the NMDA receptor (Johnson and Ascher, 1987). Furthermore, studies have demonstrated that glycine may even be critical for the NMDA activation by glutamate (Kleckner and Dingledine, 1988). Alteration of glycine activity may therefore lead to profound changes in neuronal activity mediated by NMDA. Our results demonstrate that in addition to glutamate, both GABA and glycine release were significantly augmented by mild intraischemic hyperthermia. The excitotoxic index serves as an quantitative descriptor which reflects the composite magnitude of changes in several neurotransmitter with ischemia (Globus et al, 1991a; Globus et al, 1991b). Our results demonstrate that the excitotoxic index is significantly raised during the recirculation phase in the hyperthermic group, implying that temperature may augment the imbalance between excitation and inhibition during the late recirculation period. Since these neurochemical changes occur when early signs of irreversible neuronal damage occurs (Dietrich et al, 1991), they may play a significant role in the acceleration and the increased magnitude of ischemic damage observed in animals with moderate intraischemic hyperthermia.

ACKNOWLEDGMENTS

This grant is supported by AHA grant. Dr. Sternau is a recipient of a Clinician Scientist Award by the AHA.

REFERENCES

Allison, L.A., Mayer, G.S., and Shoup, R.E., 1984, O-Phthalaldehyde derivatives of amines for high-speed liquid chromatography/electrochemistry, *Anal Chem* 56:1089-1096.

Boels, P.J., Verbeuren, T.J., and Vanhoutte., 1985, Moderate cooling depresses the accumulation and the release of newly synthesized catecholamines in isolated canine saphenous veins, *Experientia* 41:1374-1377.

Busto, R., Dietrich, W.D., Globus, M.Y.-.T., Valdes, I., Scheinberg, P., and Ginsberg, M.D., 1987, Small differences in intra-ischemic brain temperature critically determine the extent of ischemic neuronal injury, *J Cereb Blood Flow Metab* 7:729-738.

Busto, R., Globus, M.Y.-.T., Dietrich, W.D., Martinez, E., Valdes, I., and Ginsberg, M.D., 1989, Effect of mild hypothermia on ischemia-induced release of neurotransmitters and free fatty acids in rat brain, *Stroke* 20:904-910.

Dietrich, W.D., Busto, R., Valdes, I., and Loor, Y., 1990a, Effects of normothermic versus mild hyperthermic forebrain ischemia in rats, *Stroke* 21: 1318-1325.

Dietrich, W.D., Busto, R., Halley, M., and Valdes, I., 1990b, The importance of brain temperature in alterations of the blood-brain barrier following cerebral ischemia, *J Neuropathol Exp Neurol* 49:486-497.

Dietrich, W.D., Halley, M., Valdes, I., and Busto, R., 1991, Interrelationship between increased vascular permeability and acute neuronal damage following temperature-controlled brain ischemia in rats, *Acta Neuropathol* 81:615-625.

Globus, M.Y.-.T., Busto, R., Dietrich, W.D., Martinez, E., Valdes, I., and Ginsberg, M.D., 1988a, Effect on ischemia on the in vivo release of striatal dopamine, glutamate, and gama-aminobutyric acid studied by intracerebral microdialysis, *J Neurochem* 51:1455-1464.

Globus, M.Y.-.T., Busto, R., Dietrich, W.D., Martinez, E., Valdes, I., and Ginsberg, M.D., 1988b, Intra-ischemic extracellular release of dopamine and glutamate is associated with striatal vulnerability to ischemia, *Neurosci Lett* 91:36-40.

Globus, M.Y.-.T., Busto, R., Martinez, E., Valdes, I., and Dietrich, W.D., 1990, Ischemia induces release of glutamate in regions spared from histopathological damage in the rat, *stroke* 21[supp:III43-III46.

Globus, M.Y.-.T., Busto, R., Martinez, E., Valdes, I., Dietrich, W.D., and Ginsberg, M.D., 1991a, Comparative effect of transient global ischemia on extracellular levels of glutamate, glycine, and GABA-aminobutyric acid in vulnerable and nonvulnerable brain regions in the rat, *J Neurochem* 57:470-478.

Globus, M.Y.-.T., Ginsberg, M.D., and Busto, R., 1991b, Excitotoxic index – a biochemical marker of selective vulnerability, *Neurosci Lett* 127:39-42.

Haikala, H., Karmalahti, T., and Ahtee, L., 1986, The nicotine-induced changes instriatal dopamine metabolism of mice depend on body temperature, *Brain Res* 375:313-319.

Johnson, J.W., and Ascher, P., 1987, Glycine potentiates the NMDA response in cultured mouse brain neurons, *Nature* 325:529-531.

Kleckner, N.W., and Dingledine, R., 1988, Requirement for glycine in activation of NMDA-receptors expressed in Xenopus oocytes, *Science* 241:835-837.

Meldrum, B., 1989, Excitotoxicity in ischemia: An overview, *in* "Cerebrovascular Diseases – Sixteenth Research (Princeton) Conference" (M.D. Ginsberg and W.D. Dietrich, eds.), 47-60, Raven Press, New York.

Okuda, C., Saito, A., Miyazaki, M., and Kuriyama, K., 1986, alterations of the turnover of dopamine and 5-hydroxytryptamine in rat brain associated with hypothermia, *Pharmacol Biochem Behav* 25:79-83.

Roberts, E., Chase, T.N., and Tower, D.B. , 1976, "GABA in Nervous System Function," Raven Press, New York.

Smith, M.-.L., Auer, R.N., and Siesjo, B.K., 1984, The density and distribution of ischemic brain injury in the rat following 2-10 min of forebrain ischemia, *Acta Neuropathol* 64:319-332.

vanDijk, F., Green, E.J., Dietrich, W.D., Busto, R., McCabe, P.M., Markgraf, C. G., Globus, M.Y.-.T., and Alonso, O., 1991, Protective effects of brain hypothermia on behavior following global cerebral ischemia in rats, *Neurosci Abs* 17:1080.

Vanhoutte, P.M., Verbeuren, T.J., and Webb, R.C., 1981, Local cerebral modulation of the adrenergic neuroeffector interaction in the blood vessel wall, *Physiol Rev* 61:151-247.

GLUTAMATE LEVELS IN CEREBROSPINAL FLUID: DO THEY REFLECT *IN VIVO* MODIFICATIONS IN NEUROLOGICAL DISORDERS?

Carlo Ferrarese, Nicoletta Pecora, Ildebrando Appollonio, Maura Frigo, Angelo Mamoli[1], Massimo Camerlingo[1], Salvatore Pittalis and Lodovico Frattola

Department of Neurology, University of Milan, San Gerardo Hospital
Via Donizetti 106, 20052 Monza, Italy
[1]Department of Neurology, Ospedali Riuniti, Bergamo, Italy

INTRODUCTION

Current evidence strongly suggests that glutamic acid is not only an intermediate of energy metabolism, but also the most important excitatory neurotransmitter in mammalian brain (Fonnum, 1984; Roberts et al., 1981). Moreover, high concentrations of glutamate can exert specific neurotoxic effects both in vivo and in vitro, which ultimately lead to neuronal degeneration and death due to continuous and uncontrolled excitatory activity (Olney and Sharpe, 1969).
It is not surprising, therefore, that glutamate and other excitatory amino-acids are involved in many physiological functions as well as in the pathogenesis of different neuropsychiatric disorders (Table 1). Dysfunctions of these neurotransmitter systems have been demonstrated in various neurological disorders such as epilepsy (McDonald et al., 1991), diseases of the basal ganglia (Carlsson and Carlsson, 1990), senile dementia of Alzheimer's type (Greenamyre et al., 1985 and 1988; Ellison et al., 1986; Chalmers et al., 1990), amyotrophic lateral sclerosis (Perry et al., 1987; Plaitakis and Caroscio, 1987; Plaitakis et al., 1988), progressive supranuclear palsy (Perry et al., 1988) and cerebral ischemia (Rothman and Olney, 1986; Choi, 1988 and 1990). However, most of these studies were performed in autopsied brains and spinal cords, which reveal changes characteristic only of end-stage disease (Blin et al., 1990), or in animal models which are sometimes not very close to human disorders.

Table 1. EAA NEUROTRANSMISSION IN BRAIN

AREAS	PHYSIOLOGY	PATHOLOGY
Cortex	cognition	dementia, epilepsy
Basal ganglia	motor control	Huntington, Parkinson
Cerebellum	coordination	olivopontocerebellar atrophy
Hippocampus	memory	amnesias, epilepsy
Limbic system	emotional responses	anxiety, depression, schizophrenia
Spinal cord	motor and sensory functions	spasticity, amyotrophic lateral sclerosis

The Role of Neurotransmitters in Brain Injury, Edited by
M. Globus and W.D. Dietrich, Plenum Press, New York, 1992

Measurements of CSF levels of glutamate and other amino acids have been performed in patients, but conflicting results have been obtained (Young, 1990). In fact, the method of determining amino acid levels in CSF is still controversial; many of them are also variably related to the plasma concentration as well as to age and sex (McGale et al., 1977). Moreover, whereas catecholamine metabolites are end-products which accumulate in CSF and may be indexes of turnover of their precursors, amino-acids are bioactive compounds which can be transformed in different ways. Thus, the physiological significance of modifications of CSF levels of amino-acids, particularly glutamate, remains to be established.

Different methodological approaches have been utilized in the various studies to clarify this issue, and may account for the discrepancies that have been detected (Young, 1990; Lau et al., 1990). For these reasons, we studied CSF levels of amino-acids in control subjects and in patients with various neurological disorders using an improved HPLC technique which allows amino-acids to be determined at picomolar levels (Patrizio et al., 1989). We analyzed in particular any changes of glutamate levels under various conditions of storage and different procedures of enzyme inactivation, to assess the reliability of CSF glutamate measurements as in vivo indexes of EAA neurotransmission.

GLUTAMATE MEASUREMENTS IN CSF: ASSESSMENT OF RELIABILITY

In our study 300 μl CSF were derivatized with the same volume of derivatizing solution (10 ml 0.4 M borate buffer pH 9.5 containing 50 μl of 0.5 mg/ml o-phtaldialdehyde -OPA- dissolved in methanol and 5 μl 2-mercaptoethanol). 15 μl of 5 μM α-aminoadipic acid was employed as internal standard both in CSF samples and in the amino-acid standard solution (200 μl = 100 pmol of each amino-acid). CSF and amino-acid standard solution were injected after 2 minutes of derivatization. The elution of amino-acids from a C_{18} reverse-phase column (Waters 30 cm x 4.9 mm, flow rate 1.5 ml/min) was obtained by a multistep gradient of two solvents (solvent A: 0.1 M Na-acetate buffer pH 7.2, solvent B: methanol:tetrahydrofuran 97:3 vol:vol). Fluorimetric detection was with excitation and emission wavelengths of respectively 254 and 418 nm (Shimadzu RF 535) and analysis of chromatographic peaks was performed with a Shimadzu C-R3A integrator.

Different studies have demonstrated that HPLC after precolumn derivatization is the most sensitive and reliable method to measure free amino-acids in CSF. However, it is still uncertain which is the best method of collection and storage of CSF samples, since modifications of these procedures have yielded up to 100 fold differences of CSF glutamate levels in the different studies (Table 2). Such differences may be largely explained by two factors: a) metabolic instability of glutamate in CSF, with the possibility of intrathecal or in vitro glutamate formation and/or degradation, according to activation of different enzymes; b) artifactual in vitro increase of glutamate after the addition to CSF of acids to inactivate enzymes.

Table 2. CSF GLUTAMATE LEVELS IN THE LITERATURE

AUTHORS	REFERENCE	GLUTAMATE LEVELS (pmol/ml)
Lakke et al.	Neurology,26:489,1976	15,800
McGale et al.	J.Neurochem.29:291.1977	26,100
Engelsen et al.	Neurosci.Letters,62:97,1985	2,900
Spink et al.	Anal.Biochem.158:79,1986	480
Manyam et al.	Arch.Neurol.45:48,1988	3,460
Pitkaen et al.	J.Neural Transm.76:221,1989	183
Perry et al.	Ann.Neurol.28:12,1990	200
Rothstein et al.	Ann.Neurol.28:18,1990	2,900

Since these phenomena can bias the results in any control or patient population, we employed different strategies to clarify the functional meaning of modifications of CSF glutamate levels. We first observed that glutamate levels in native (untreated) CSF changed with time. When CSF was left at room temperature, a 50% decrease of glutamate levels occurred within 15 minutes and was maintained for one hour, followed by a late and progressive increase of the amino-acid levels (+ 250 % after 24 hours). This pattern of time-related glutamate changes may indicate that two types of enzymatic processes occurr in native CSF: degradation of free glutamate and formation of new glutamate from glutamine or proteins. A slow increase of glutamate levels was also observed when CSF was stored at -80^{0}C; for this reason results obtained on stored untreated CSF may be compared only if the storage time is similar for the different groups of patients.

Different procedures have been employed to inactivate the enzymes in CSF: sulfosalicylic acid (Lakke and Teelken, 1976), perchloric acid (Lundquist et al., 1989) or freeze-thawing cycles (Spink et al., 1986). Results obtained with these procedures may have confused the issue. In fact, we observed a time-dependent increase of glutamate levels in acid-treated CSF (+50% after 10 minutes; +130% after 30 minutes; +370% after one hour). A sulfosalicylic acid induced increase of glutamate has previously been observed and has been linked to acid hydrolysis of glutamine (Spink et al., 1986).

We believe that the only possibility of obtaining glutamate levels stable over time is to immediately inactivate the enzymes with acid and neutralize the acid thereafter. CSF samples treated in this way present glutamate levels stable in different storage conditions.

MODIFICATIONS OF CSF GLUTAMATE LEVELS IN NEUROLOGICAL DISORDERS

Despite the methodological pitfalls, measurements of CSF levels of amino-acids have been performed in various neurological disorders in the past 15 years. Not surprisingly, conflicting results have been observed by different authors. Increased, decreased and unchanged levels of glutamate have been observed in CSF of epileptic patients (Mutani et al., 1974; Plum, 1974; Crawford and Chadwick, 1987; Pitkaen et al., 1989). A defect of the amino-acid transport system from CSF to plasma has been hypothesized in Parkinson's disease and in other extrapyramidal disorders (Lakke and Teelken, 1976). Reduced CSF levels of glutamate have been observed in Alzheimer's dementia (Smith et al., 1985; Tosca et al., 1990), in which a degeneration of glutamatergic pathways has been described (Greenamyre et al., 1988). However, amyotrophic lateral sclerosis is the neurodegenerative disorder that has been most extensively studied and has generated most controversies in recent years. A systemic defect in glutamate metabolism was originally hypothesized (Plaitakis et al., 1982, 1987); subsequently various authors have analyzed CSF levels of glutamate in this disease, with conflicting results (Perry et al., 1990; Rothstein et al., 1990; Lau et al., 1990). Such discrepancies could be avoided with a standardized methodology of CSF processing, storage and analysis.

We analyzed the levels of glutamate in CSF of controls (patients hospitalized for lumbar disk herniation) and patients with different neurological disorders (multiple sclerosis, acute stroke, ALS).

Preliminary results from CSF stored untreated at -80^{0}C for similar time periods showed a significant decrease of glutamate levels in acute stroke patients compared to controls (-60%). No significant change was observed in multiple sclerosis, and decreased values were recorded in the few ALS patients. In acute stroke patients CSF glutamate levels were lower later after stroke (from 2 to 10 days), then earlier (between 6 and 24 hours after stroke). There was an inverse relation between glutamate levels and time after stroke. A correlation study between CSF glutamate levels and size and location of the infarcted area is in progress, to investigate the hypothesis that the fall of CSF glutamate levels may be due to the destruction of glutamatergic pathways.

However, these data are still open to different interpretations. It may be difficult to study the pathologic release of glutamate from ischemic neurons through analysis of CSF levels. In vivo studies of animals with intracerebral microdialysis have shown, in the case of intermittent four-vessels occlusion, an immediate release from ischemic neurons and a rapid return to normal interstitial glutamate levels after termination of ischemia (Benveniste et al., 984; Busto et al., 1989); in the model of permanent middle cerebral artery occlusion a return to basal glutamate interstitial levels is observed within three hours (Bullock et al., 1991). Processes such as glial or neuronal uptake in surrounding regions may be activated to counteract excessive glutamate exposure in ischemic areas. Thus the late decrease of glutamate in CSF could be due to these compensatory mechanisms. However, it is possible that a glutamate peak in CSF immediately after stroke is not detected by CSF measurements, which cannot be performed earlier than 4-5 hours after stroke; at this time a transient increase of glutamate could already be blunted by enzymatic degradation of the amino-acid.

An alternative hypothesis to explain the decrease of glutamate in CSF of stroke patients is that ischemic or necrotic neurons and astrocytes may release their enzymes in the interstitial fluid and in CSF; increased enzymatic activity in CSF can lead to faster glutamate metabolism. In fact, very early studies documented modifications of CSF levels of different enzymes after stroke (Fleisher et al., 1957; Green et al., 1957, 1958; Lieberman et al., 1957; Royds et al., 1983; Hay et al., 1984). At the time of most of these studies the neurotransmitter role of glutamate was still unknown, and these data can now be analyzed with a different perspective. A change in the enzyme levels occurring during stroke or in other neurological disorders (Florez et al., 1976) may counteract or enhance the potential neurotoxic effect of glutamate (Plaitakis et al., 1988).

In our stroke patients we observed raised glutamic-oxalacetic transaminase levels in CSF, which paralleled the decrease of glutamate levels. Further studies to characterize the enzymatic system involved in the metabolism of glutamate in CSF are in progress. However, it is clear from this preliminary study that, unlike catecholamine metabolites, levels of amino-acids, particularly glutamate, in CSF do not directly reflect changes in their brain levels, but that different secondary processes are involved. Only extensive studies of such processes will reveal the physiological significance of changes of CSF glutamate levels which, until now, are the only indexes of glutamatergic functions clinically available in patients.

The development of new ligands of glutamate receptors suitable for positron emission tomography (PET) studies may lead to new ways of studying excitatory amino-acid function in vivo (Ferrarese et al., 1990; McCullogh et al., 1991). Possible correlations between glutamate CSF levels and glutamate receptor activation, evaluated by PET, will provide the clinician with knowledge of pre- and post-synaptic glutamatergic involvement in different neurological disorders.

REFERENCES

Benveniste, H., Drejer, J., Schousboe, A., and Diemer, N. H., 1984, Elevation of the extracellular concentrations of glutamate and aspartate in rat hippocampus during transient cerebral ischemia monitored by intracerebral microdialysis, J. Neurochem., 43:1369.

Blin, O., Samuel, D., Nieoullon, A., and Serratrice, G., 1990, Amino acid levels in CSF of ALS patients, Neurology, 40 (S1):317.

Bullock, R., Butcher, S. P., Graham, D. I., and Graham M. T., 1991, Excitatory amino acid release after local ischemia: infarct volume determines EAA release, Abstract book: The role of neurotransmitters in brain injury, Key West, Florida : 9.

Busto, R., Globus, M. Y., Dietrich, W. D., Martinez, E., Valdes, I., and Ginsberg, M. D., 1989, Effect of mild hypothermia on ischemia- induced release of neurotransmitters and free fatty acids in rat brain, Stroke, 20:904.

Carlsson, M., and Carlsson, A., 1990, Interaction between glutamatergic and monoaminergic system within the basal ganglia - implications for schizophrenia and Parkinson's disease, Trends in Neurosci., 13:272.

Chalmers, D. T., Dewar, D., Graham, D. I., Brooks, D. N., and McCulloch, J., 1990, Differential alterations of cortical glutamatergic binding sites in senile dementia of the Alzheimer type, Proc. Natl. Acad. Sci. (USA), 87:1352.

Choi, D. W., 1988, Glutamate neurotoxicity and diseases of the nervous system, Neuron, 1:623.

Choi, D. W., 1990, Cerebral hypoxia: some new approaches and unanswered questions, J. Neurosci., 10:2493.

Crawford, P. M., and Chadwick, D. W., 1987, GABA and amino acid concentrations in lumbar CSF in patients with treated and untreated epilepsy, Epilepsy Res.,1:328.

Ellison, D. W., Beal, M. F., Mazurek, M. F., Bird, E. D., and Martin, J. B., 1986, A postmortem study of amino acids neurotransmitters in Alzheimer's disease, Ann. Neurol., 20:616.

Ferrarese, C., Guidotti, A., Costa, E., Rice, K. C., de Costa, B. R., Miletich, R. S., and Di Chiro, G., 1991, In vivo study of excitatory aminoacid neurotransmission fluorothienylcycloexylpiperidine, a ligand suitable for positron emission tomography, J. Cer. Blood Flow Metab., 11 (S2):873.

Fleisher, G. A., Wakim, K. G., and Goldstein, N. G., 1957, Glutamic oxalacetic transaminase and latic dehydrogenase in serum and cerebrospinal fluid of patients with neurologic disorders, Proc. Staff. Meet. Mayo Clin., 32:188.

Florez, G., Cabeza, A., Gonzales, J. G., and Ucar, S., 1976, Changes in serum and cerebrospinal fluid enzyme activity after head injury, Acta Neurochir., 35:3.

Fonnum, F., 1984, Glutamate: a transmitter in mammalian brain, J. Neurochem., 42:1.

Green, J. B., O'Doherty, D. S., Oldewurtel, H. A., and Foster, F. M., 1957, Cerebrospinal fluid transaminase comcentrations in clinical cerebral infarctions, New Eng. J. Med., 256:220.

Green, J. B., Oldewurtel, H. A., and O'Doherty, D. S., 1958, Cerebrospinal fluid transaminase and latic dehydrogenase activities in neurologic disease, Arch. Neurol. Psychiatr., 80:148.

Greenamyre, J. T., Penney, J. B., Young, A. B., D'Amato, C. J., Hicks, S. P., and Shoulson, I., 1985, Alterations in L-glutamate binding in Alzheimer's and Huntington's diseases, Science, 227:1496.

Greenamyre, J. T., Maragos, W. F., Albin, R. L., Penney, J. B., and Young, A. B., 1988, Glutamate transmission and toxicity in Alzheimer's disease, Prog. Neuro-Psychopharmacol., 12:421.

Hay, E., Royds, J. A., Davis-Jones, G. A., Lewtas, N. A., Timperley, W. R., and Taylor C. B., 1984, Cerebrospinal fluid enolase in stroke, J. Neurol. Neurosurg. Psychiatry, 47:724.

Lakke, J. P. W. F., and Teelken, A. W., 1976, Amino acid abnormalities in cerebrospinal fluid of patients with parkinsonism and extrapyramidal disorders, Neurology, 26:489.

Lau, B. H. S., Hubbard, R. W., Sanchez, A., Will, A. D., and Peterson, G. W., 1990, Effect of sample processing delay on plasma glutamate levels in ALS, Neurology, 40 (S1):316.

Lieberman, J., Daiber, O., Dulkin, S. I., Lobstein, O. E., and Kaplan, M. R., 1957, Glutamic oxalacetic transaminase in serum and cerebrospinal fluid of patients with cerebrovascular acidents, New. Eng. J. Med., 257:1201.

Mc Culloch, J., Wallace, M. C., Laurie, D., Angerson W. J., Burns, H. D., and Gibson, R. E., 1991, The enanced binding of (+)-3-[125]Iodio MK-801 maps excessive glutamate release: potential as a ligand for SPECT, J. Cer. Blood Flow Metab., 11 (S2):6.

Lundquist, C., Blohstrand, C., Hamberger, A., and Wikkelso, C., 1989, Liquid chromatographic separation of cerebrospinal fluid amino acids after precolumn fluorescence derivatization, Acta Neurol. Scand., 79:273.

Mc Donald, J. W., Garofalo, E. A., Hood, T., Sackellares, J. C., Gilman, S., McKeever, P. E., Troncoso, J. C., and Johnston, M. V., 1991, Altered excitatory and inhibitory amino acid receptor binding in hippocampus of patients with temporal lobe epilepsy, Ann. Neurol., 29:529.

McGale, E. H. F., Pye, I. F., Stonier, C., Hutchinson, E. C., and Aber, G. M., 1977, Studies of the inter-relationship between cerebrospinal fluid and plasma amino acid concentrations in normal individuals, J. Neurochem., 29:291.

Mutani, R., Monaco, F., Durelli, L., and Delsedime, M., 1974, Free amino acids in the cerebrospinal fluid of epileptic subjects, Epilepsya, 15:593.

Olney, J. W., and Sharpe, L. G., 1969, Brain lesions in an infant Rhesus monkey treated monosodium glutamate, Science, 166:386.

Patrizio, M., Gallo, V., and Levi, G., 1989, Measurement of amino acid release from cultured cerebellar granule cells by an improved high performance liquid chromatography procedure, Neurochem. Res., 14:627.

Perry, T. L., Hansen, S., and Jones, K., 1987, Brain glutamate deficiency in amyotrophic lateral sclerosis, Neurology, 37:1845.

Perry, T. L., Hansen, S., and Jones, K., 1988, Brain amino acids and glutathione in progressive sopranuclear palsy, Neurology, 38:943.

Perry, T. L., Krieger, C., Hansen, S., and Eisen, A., 1990, Amyotrophic lateral sclerosis: amino acid levels in plasma and cerebrospinal fluid, Ann. Neurol., 28:12.

Pitkanen, A., Matilainen, R., Halonen, T., Kutvonen, R., Hartikainen, P., and Riekkinen, P., 1989, Inhibitory and excitatory amino acids in cerebrospinal fluid of chronic epileptic patients, J. Neural. Transm., 76:221.

Plaitakis, A., Berl, S., and Yahr, M. D., 1982, Abnormal glutamate metabolism in adult-onset degenerative neurological disorders, Science, 216:193.

Plaitakis, A., and Caroscio, J. T., 1987, Abnormal glutamate metabolism in amyotrophic lateral sclerosis, Ann. Neurol., 22:575.

Plaitakis, A., Constantakakis, E., and Smith, J., 1988, The neuroexcitotoxic amino acids glutamate and aspartate are altered in the spinal cord and brain in amyotrophic lateral sclerosis, Ann. Neurol., 24:446.

Plaitakis, A., 1990, Glutamate dysfunction and selective motor neuron degeneration in amyotrophic lateral sclerosis: a hypothesis, Ann. Neurol., 28:3.

Plum, C. M., 1974, Free amino acid levels in the cerebrospinal fluid of normal humans and their variation in cases of epilepsy and Spielmayer-Vogt-Batten disease, J. Neurochem., 23:595.

Roberts, P. J., and Storm-Mathisen, J., Johnston GAR (Eds), 1981, Glutamate: Transmitter in the Central Nervous System, John Wiley & Sons, Chichester.

Rothstein, J. D., Tsai, G., Kuncl, R. W., Clawson, L., Cornblath, D. R., Drachman, D. B., Pestronk, A., Stauch, B. L., and Coyle, J. T., 1990, Abnormal excitatory amino acid metabolism in amyotrophic lateral sclerosis, Ann. Neurol., 28:18.

Rothman, S. M., and Olney, J. W., 1986, Glutamate and the patophysiology of hypoxic-ischemic brain damage, Ann. Neurol., 19:105.

Royds, J. A., Davies-Jones, G. A. B., Lewtas, N. A., Timperley, W., and Taylor C. B., 1983, Enolase isoenzymes in the cerebrospinal fluid of patients with deseases of the nervous system, J. Neurol. Neurosurg. Psychiatry, 46:1031.

Smith, C. C. T., Bowen, D. M., Francis, P. T., Snowden, J. S., and Neary, D., 1985, Putative amino acid transmitters in lumbar cerebrospinal fluid of patients with histologically verified Alzheimer's dementia, J. Neurol. Neurosurg. Psychiatry, 48:469.

Spink, D. C., Swann, J. W., Snead, O. C., Waniewski, R. A., and Martin, D. L., 1986, Analysis of aspartate and glutamate in human cerebrospinal fluid by high-performance liquid chromatography with automated precolumn derivatization, Anal. Biochem., 158:79.

Tosca, P., Verze', S., Dagani, F., Canevari, L., and Zerbi, F., 1990, Determination of glutamate in CSF from patients affected by different types of dementia, J. Neurol.,:S34.

Young, A. B., 1990, What's the excitement about excitatory amino acids in amyotrophic lateral sclerosis?, Ann. Neurol., 28:9.

GLUTAMATE, ASPARTATE AND GABA RELEASE FROM HIPPOCAMPAL CA1 SLICES DURING IN VITRO ISCHEMIA IS CALCIUM-INDEPENDENT

Stephan P. Burke and Charles P. Taylor

Department of Pharmacology
Parke-Davis Pharmaceutical Research Div.
Warner-Lambert Company
Ann Arbor, MI

INTRODUCTION

The excitatory neurotransmitter glutamate is the most abundant amino acid in the brain. Intracellular glutamate is distributed throughout the cytosol at millimolar concentrations whereas extracellular levels must be kept in low micromolar amounts to prevent excitotoxic effects. High affinity transporter proteins on the plasma membranes of both neurons and glia remove glutamate left from synaptic release and also glutamate from cerebrospinal fluid. The glutamate transporter uses stored energy from the transmembrane sodium and potassium ion gradients and membrane voltage that are maintained at considerable energy cost to the cell. Increasing evidence suggests that neuronal damage from energy depletion or ischemia is in large part caused by prolonged activation of glutamate receptors (Choi, 1990). The mechanism(s) that lead to heightened levels of extracellular glutamate have not been clearly defined but may involve a reversal of uptake carriers due to a loss of the normal ionic gradients and membrane depolarization (Ericinska, 1987, Nicholls and Attwell, 1990). Unlike the release of glutamate from synaptic vesicles, this release may be independent of extracellular calcium (Kauppinen et al., 1988).

Conditions of hypoxia and hypoglycemia greatly increase the overflow of glutamate from brain tissue, both in vivo (Benveniste et al., 1984) and in vitro (Ikeda et al., 1989). We studied amino acid release in vitro from a subregion of hippocampal slices. Previously, minislices of the hippocampal CA1 area have been used to study amino acid release from depolarization (Burke and Nadler, 1988, Nadler et al., 1990). The anatomical homogeneity within this preparation may simplify the interpretation of changes in superfusate amino acid levels.

METHODS

Hippocampi were isolated from adult male rats and 400 μm transverse slices were cut on a McIlwain chopper. The CA1 area minus the stratum lacunosum-moleculare was isolated with microscalpal cuts made under a dissecting microscope. The resulting minislices consist primarily of CA1 pyramidal cell bodies and dendrites and glutamatergic synaptic terminals from CA3 pyramidal cells as well as GABAergic interneurons. Minislices were transferred to small-volume chambers and superfused (1 ml/min) with warmed Elliott's medium gassed with 95%/5% O_2/CO_2. After a 60-min recovery period, superfusate was collected in 5-min fractions. After the collection of basal fractions, medium gassed with 95% N_2/5% CO_2 and lacking glucose was introduced for 20 min. Two more 5-min fractions were collected following the reintroduction of normal medium. Superfusate fractions were derivatized with ortho-phthalaldehyde and analyzed by high performance liquid chromatography with fluorescence detection. Tissue protein content was determined with bicinchoninic acid (Pierce; Smith et al., 1985) and all amino acid release is expressed per mg of tissue protein.

RESULTS

Due to the small amount of tissue (0.2 mg protein) present in each chamber, basal levels of glutamate (GLU), aspartate (ASP) and GABA were not usually detectable, whereas low levels of taurine (TAU) and glutamine (GLN) were present in basal release fractions. Fig. 1 shows a typical time profile for amino acids in the superfusate over the course of an experiment. A large release of GLU is observed beginning in the second or third 5-min collection period of "in vitro ischemia" (IVI). ASP and GABA also peak quickly, and all three transmitter amino acids return to near basal levels rapidly upon switching back to normal medium. GLN levels decrease during IVI and show a slight recovery after IVI, whereas TAU shows a modest increase and consistently remains slightly elevated after IVI.

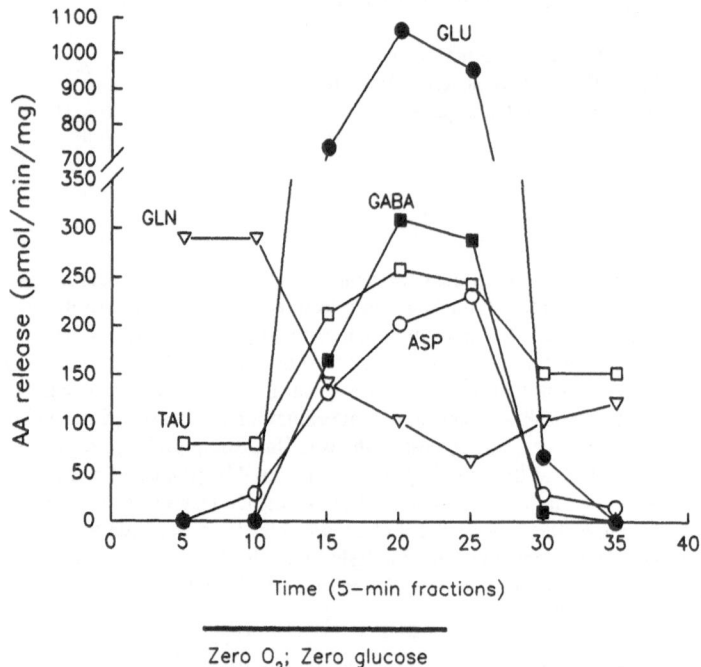

Fig. 1. Amino Acid Release During "In Vitro Ischemia."
The individual points represent the amount of the indicated amino acid in a 5 ml fraction collected over the preceeding 5 min period.

In order to determine the calcium dependence of amino acid release the CaCl$_2$ in the medium was replaced by 2.6 mM MgCl$_2$ for a period of 15 min preceeding and during IVI. Fig. 2A shows that omission of Ca did not alter the peak release of GLU evoked by IVI whereas release in response to a one-min pulse of 50 mM KCl was diminished by more than 90%. In both cases results were similar with measurement of ASP and GABA (Table 1).

Fig. 2B shows that reducing temperature of the superfusate from 37° to 31° reduced maximal GLU overflow by half in the maximal five-min fraction of IVI, whereas the same hypothermia did not effect overflow from 50mM KCl. Results were similar for ASP and GABA (Table 1).

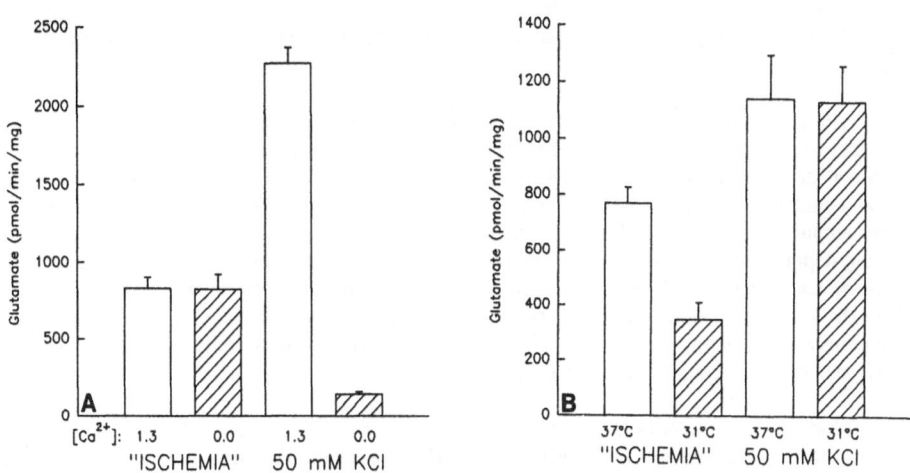

Fig. 2. A: Calcium-dependence of GLU Release. Release is expressed "per minute" therefore absolute amount of GLU in 5 min IVI fraction appears less than GLU released from 1 min pulse of 50 mM KCl. B: Temperature-dependence of GLU Release.

Table 1. ASP and GABA Release under different conditions.

| | | ASP | GABA |
		(pmol / min / mg protein)	
"In vitro ischemia"	$+Ca^{2+}$	250±54	237±19
	$-Ca^{2+}$	242±21	230±25
50 mM KCl	$+Ca^{2+}$	342±35	277±23
	$-Ca^{2+}$	62±13	31± 2
"In vitro ischemia"	37°C	259±44	171±18
	31°C	110±16	77±14
50 mM KCl	37°C	177±22	189±20
	31°C	205±24	153±16

DISCUSSION

The large release of GLU from IVI in the current study agrees with another study employing whole hippocampal slices and microdialysis of the hippocampus in vivo (Ikeda et al., 1989). IVI caused glutamate release that was gradual in onset but then sustained for at least 20 min. In comparison, the release of transmitter amino acids caused by 50 mM potassium was large and almost immediate but short-lasting. The rapid return of amino levels to baseline upon return of energy substrates suggests a return of some degree of homeostatic mechanisms, although this remains to be determined with independent methods. The decline of GLN levels during IVI may reflect impairment of ongoing energy-dependent synthesis and release of GLN by glia, the principal source of extracellular GLN. It is noteworthy that TAU remains elevated since it is presumed to modulate osmotic balance (Wade et al., 1988).

The lack of any calcium-dependence for IVI release may indicate that exocytosis does not contribute significantly to the observed GLU release. A recent report with electron microscopic immunohistochemistry for glutamate (Torp et al., 1991) indicates that ischemia in vivo preferentially depletes glutamate from cytosol of cell bodies and dendrites and not from presynaptic vesicles of the hippocampus, supporting the idea that ischemic release is not vesicular and thus is not expected to be calcium-dependent. It is likely that calcium-independent release of amino acids during ischemia comes from reversal of high-affinity transporter processes (Ericinska, 1987, Sanchez-Prieto and Gonzalez, 1988, Nicholls and Attwell, 1990). Alternatively it may be that synaptic vesicles, which sequester GLU by an energy dependent mechanism (Naito and Ueda, 1985), no longer contain an observable pool of GLU once cytosolic energy stores are depleted.

Moderate hypothermia in vivo eliminates the release of GLU detected by microdialysis in response to ischemia (Busto et al., 1989). Since a slight reduction in temperature in vitro also reduced GLU overflow from IVI, the slice preparation may be suitable to study the marked temperature dependency of ischemic glutamate release. Temperature dependency may reflect a characteristic of the underlying release mechanism or of accumulation of intracellular sodium but could also be related to the indirect effects of slowed depletion of cellular energy stores.

Pyramidal cells of the hippocampal CA1 area are extremely sensitive to ischemic insult. In vitro studies of this area in comparison to other hippocampal or brain areas may reveal the contribution of GLU release to pathology following ischemia.

The slice model is more amenable to manipulation of oxygen tension, glucose concentration, ions and application of drugs than in vivo preparations monitored with microdialysis. In addition, slices preserve many features of the microenvironment between cells and can be used to model the impressive changes during ischemia (Hansen, 1985) that may be quite different in tissue culture preparations.

It is not known whether neurons and glia possess the same or structurally different GLU uptake carrier or whether such differences might be exploited experimentally.

Our results agree with studies of synaptosomal preparations showing calcium-independent release of glutamate following energy depletion (Sanchez-Prieto and Gonzalez, 1988, Kauppinen et al., 1988). Our current results with simulated ischemia in vitro are consistent with massive GLU release caused by reversal of the sodium-dependent uptake carrier (Nicholls and Attwell, 1990) from loss of both transmembrane voltage and sodium gradients (Hansen, 1985, Taylor et al., 1991). Similar sodium-dependent transport mechanisms also exist for GABA, glycine, and other amino acids, and aspartate is known to be a substrate for the glutamate transporter (Ericinska, 1987, Nicholls and Attwell, 1990). Further experiments looking at the ionic dependence of IVI release and employing known inhibitors of uptake carriers are needed to confirm their involvement in the pathologic release of amino acids.

REFERENCES

H. Benveniste, J. Drejer, A. Schousboe, and N. Diemer, 1984, Elevation of the extracellular concentrations of glutamate and aspartate in rat hippocampus during transient cerebral ischemia monitored by intracerebral microdialysis, J. Neurochem, 43:1369-1374.

S.P. Burke and J.V. Nadler, 1988, Regulation of glutamate and aspartate release from slices of the hippocampal CA1 area: effects of adenosine and baclofen, J. Neurochem, 51:1541-1551.

R. Busto, M. Y-T. Globus, W. D. Dietrich, E. Martinez, I. Valdes and M. D. Ginsberg, 1989, Effects of mild hypothermia on ischemia-induced release of neurotransmitters and free fatty acids in rat brain, Stroke, 20:904-910.

D.W. Choi, 1990, Methods for antagonizing glutamate neurotoxicity, Cerebrovasc. Brain Metab. Rev., 2:105-147.

M. Ericinska, 1987, The neurotransmitter amino acid transport systems. A fresh outlook on an old problem, Biochem. Pharmacol., 36:3547-3555.

A.J. Hansen, 1985, Effect of anoxia on ion distribution in the brain, Physiol. Rev., 65:101-148.

M. Ikeda, T. Nakazawa, K. Abe, T. Kaneko, and K. Yamatsu, 1989, Extracellular accumulation of glutamate in the hippocampus induced by ischemia is not calcium dependent-in vitro and in vivo evidence, Neurosci. Letters, 96:202-206.

R.A. Kauppinen, H.T. McMahon and D.G. Nicholls, 1988, Ca^{2+}-dependent and Ca^{2+}-independent glutamate release, energy status and cytosolic free Ca^{2+} concentration in isolated nerve terminals following metabolic inhibition: Possible relevance to hypoglycaemia and anoxia, Neuroscience, 27:175-182.

J.V. Nadler, D. Martin, G.A. Bustos, S.P. Burke, and M.A. Bowe, 1990, Regulation of glutamate and aspartate release from the Schaffer collaterals and other projections of CA3 hippocampal pyramidal cells, in Progress in Brain Research, Vol. 83: Understanding the Brain Through the Hippocampus, (Storm-Mathisen, J., Zimmer, J., and Ottersen, O.P., eds.) pp.115-130, Elsevier, Amsterdam.

S. Naito and T. Ueda, 1985, Characterization of glutamate uptake into synaptic vesicles, J. Neurochem., 44:99-109.

D. Nicholls and D. Attwell, 1990, The release and uptake of excitatory amino acids, Trends Pharmacol. Sci., 11:462-468.

G. Pines and B.I. Kanner, 1990, Counterflow of L-glutamate in plasma membrane vesicles and reconstituted preparations from rat brain, Biochem., 29:11209-11214.

J. Sanchez-Prieto and P. Gonzalez, 1988, Occurrence of a large Ca^{2+}- independent release of glutamate during anoxia in isolated nerve terminals (synaptosomes), J. Neurochem, 50:1322-1324.

P.K. Smith, R.I. Krohn, G.T. Hermanson, A.K. Mallia, F.H. Gartner, M.D. Provenzano, E.K. Fujimoto, N.M. Goeke, B.J. Olson and D.C. Klenk, 1985, Measurement of protein using bicinchoninic acid, Anal. Biochem., 150:76-85.

C. P. Taylor, J. J. Geer and S. P. Burke, 1991, Glutamate release and calcium influx induced in rat neocortical cultures by reversal of the transmembrane sodium gradient, (see this volume).

R. Torp, P. Andine, H. Hagberg, T. Karagulle, T.W. Blackstad, and O.P. Ottersen, 1991, Cellular and subcellular redistribution of glutamate-, glutamine- and taurine-like immunoreactivities during forebrain ischemia: a semiquantitative electron microscopic study in rat hippocampus, Neuroscience, 41:433-447.

J.V. Wade, J.P. Olson, F.E. Samson, S.R.V. Nelson and T.L. Pazdernik, 1988, A possible role for taurine in osmoregulation within the brain. J. Neurochem, 51:740-745.

GLUTAMATE RELEASE AND CALCIUM INFLUX INDUCED IN RAT NEOCORTICAL

CULTURES BY REVERSAL OF THE TRANSMEMBRANE SODIUM GRADIENT

Charles P. Taylor, Joann J. Geer and Stephan P. Burke

Department of Pharmacology
Parke-Davis Pharmaceutical Research Div.
Warner-Lambert Co.
Ann Arbor, MI

INTRODUCTION

Glutamate is widely recognized as an excitotoxic neurotransmitter in the brain. Abundant evidence indicates that glutamate is released in large concentrations during ischemia, abnormally activating glutamate receptors and causing calcium influx that is a major mediator of cell death during ischemia (review: Choi, 1990). Under normal physiological conditions, glutamate is released in a calcium-dependent manner from synaptic vesicles and excites postsynaptic neurons. However, ischemia changes cytoplasmic energy stores and transmembrane ionic gradients (Hansen, 1985), altering many processes including the calcium dependence of glutamate release (Sanchez-Prieto and Gonzalez, 1988, Nicholls and Attwell, 1990). The glutamate transporter is bidirectional and electrogenic, relying on concentration gradients of sodium and potassium across the cell membrane and also on membrane potential to determine the direction of glutamate movement. Other authors have suggested that ionic changes during ischemia may cause the glutamate transporter to operate in reverse (Ericinska, 1987, Nicholls and Attwell, 1990). The present experiments examine artificial reversal of the high-affinity glutamate transporter in cultured neurons. Our aim was to assess the possibility that this process underlies glutamate release and activation of NMDA receptors during ischemia.

METHODS

Neocortical cultures were prepared from cerebral cortices of Sprague Dawley rat fetuses on the 18th day of gestation. Each cortex was bisected and stirred for 15 min. at 21°C. in 0.1% trypsin (Sigma, type VI) in Hanks Balanced Salt Solution (HBSS) lacking Ca^{2+} and Mg^{2+} (Sigma H2387). The tissue was rinsed 3X with plating medium consisting of Minimal Essential Media (MEM) with Earle's salts and L-glutamine (Gibco 410-1100EB), supplemented with D-glucose (30mM), sodium bicarbonate (26mM), 10% horse serum (Gibco) and 5% fetal bovine serum (FBS) (Gibco). Cells were dissociated by trituration. The resulting suspension was diluted with plating media to a concentration of 10 million cells/ml and 0.5ml was dispensed into each well of a 12 well plate containing 0.5 ml of plating media/well. Plates had previously been coated with poly-L-lysine. Cells were maintained for 14-22 days at 37°C in a humidified 5% CO_2 atmosphere.

One day after plating, the culture media was replaced with plating media lacking FBS. On the third day after plating, non-neuronal cell division was reduced with 20 μM 5-fluoro-2'-deoxyuridine and 50 μM uridine. Occasional feedings were performed by removing 0.5 ml media and replacing it with fresh media.

The Role of Neurotransmitters in Brain Injury, Edited by
M. Globus and W.D. Dietrich, Plenum Press, New York, 1992

For experiments, 14-22 day old cultures were washed 2X and then equilibrated for 30 min. at 37°C with Hepes/Tris buffered HBSS (composition in mM: 137 NaCl, 5.4 KCl, 0.8 MgCl$_2$, 1.8 CaCl$_2$, 25.0 Hepes, 0.8 KH$_2$PO$_4$, 6.0 glucose). Radioactivity (1 μCi/ml ^{45}Ca^{2+}) was added with fresh HBSS with or without drug and incubated for 10 min. at 37°C. The reaction was terminated by washing 3X with ice-cold saline and drying. Cells were solublized in 1% SDS and radioactivity measured by liquid scintillation counting. Protein content was determined with the biuret/bicinchoninic acid method (Smith et al., 1985, Pierce BCA protein assay). Zero sodium, or decreased sodium solutions were made by replacing NaCl isoosmotically. Drugs were solublized in water or DMSO.

RESULTS

Substitution of various agents for NaCl in the bathing medium caused increased calcium influx over control levels. Substitution with choline chloride caused calcium influx that was more than 10-

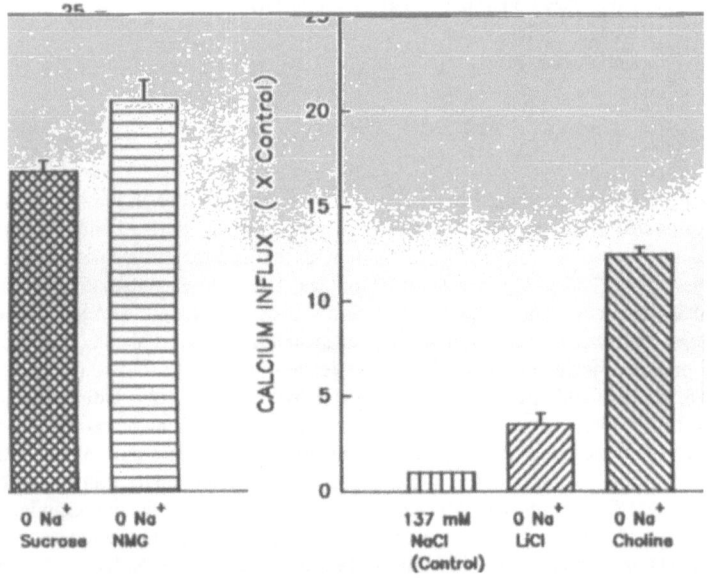

Fig. 1. Sodium chloride was replaced iso-osmotically with lithium chloride, choline chloride, sucrose or N-methyl-D-glucamine (NMG) and cultures were exposed for 10 min at 37°C. ^{45}Ca^{2+} influx was measured over 10 min and expressed as increase over control (137 mM NaCl). Bars denote mean and standard error (N = 6 or greater).

fold increased above control levels (Fig. 1). With LiCl substitution for NaCl, calcium influx increased significantly, but only to about one-third of that seen with choline chloride. Substitution with other agents (sucrose, N-methyl-D-glucamine) caused slightly larger calcium influx than choline chloride.

The time course of calcium influx from choline substitution is shown in Fig. 2. Entry was rapid over the first 10 minutes and was followed by a plateau, suggesting a saturable process. The dependence of calcium entry on extracellular sodium concentration is shown in Fig. 3, with significant calcium flux seen only with external sodium concentrations below 20 mM.

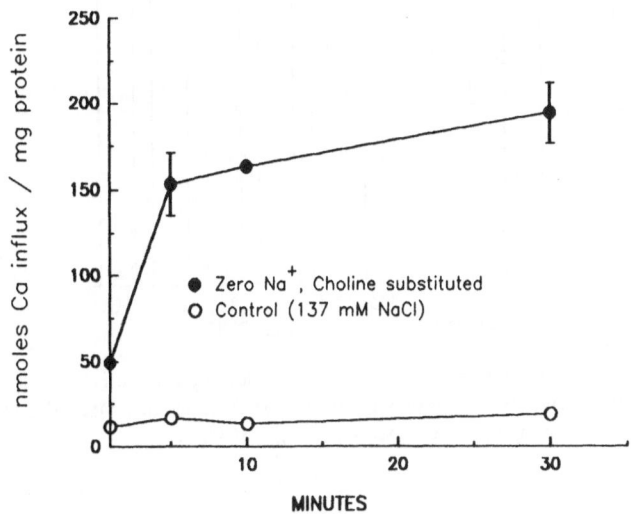

Fig. 2. Time course of calcium entry following substitution of sodium with choline; mean
and standard errors shown, $N \geq 6$. Standard error smaller than symbol
if not shown.

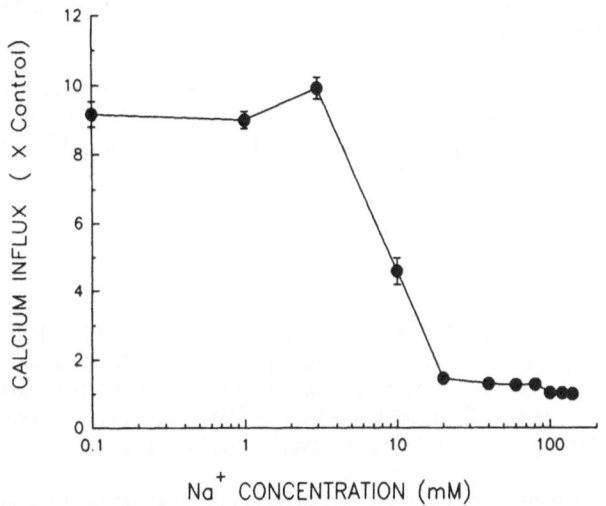

Fig. 3. Dependence of calcium entry on sodium concentration; note sharp change at
approximately 10 mM sodium.

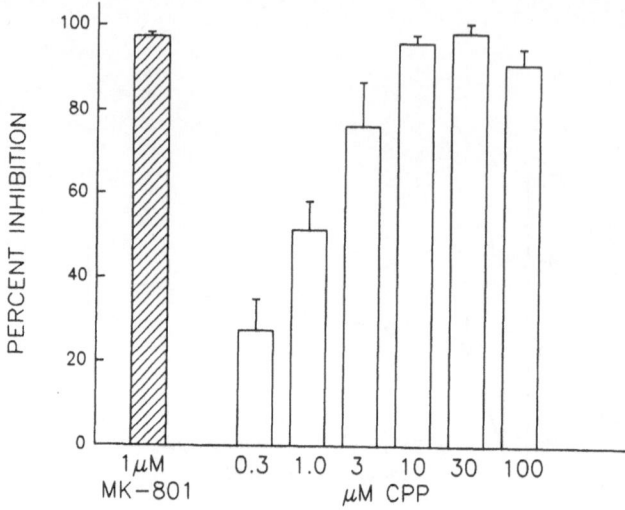

Fig. 4. Calcium influx from substitutution of sodium by choline is blocked
completely by the non-competitive NMDA antagonist MK-801 and dose-
dependently by the competitive antagonist CPP.

Fig. 4 demonstrates that calcium influx from sodium substitution is blocked by both competitive and non-competitive NMDA receptor antagonists. The data with 3-[(+/-)-2-carboxypiperazin-4-yl]-propyl-1-phosphonic acid (CPP) were fit to the logistic equation resulting in an IC_{50} of 0.88 µM with a Hill coefficient of 1.01.

Calcium influx from lowered sodium was unaffected by 1 µM tetrodotoxin or by 100 µM amiloride (data not shown). A 10 min exposure to zero sodium media did not release the cytosolic enzyme lactate dehydrogenase (LDH) into culture supernatant when measured 20 hr later, indicating that this treatment was not lethal to cultured cells (data not shown).

Under conditions identical to these experiments, glutamate concentrations were measured in culture supernatant using HPLC techniques (Burke and Taylor, 1991). These experiments will be described in detail elsewhere, but glutamate increased from approximately 1.0 to 6.0 µM when sodium was reduced below 20 mM and this glutamate efflux was not significantly altered by the addition of 100 µM CPP.

DISCUSSION

We found that lowering sodium concentrations causes massive calcium influx in neocortical cultures. Other groups observed calcium influx through voltage sensitive calcium channels in response to a potassium challenge and report only a 2-3 fold increase in calcium influx above control levels (Martin-Moutot et al., 1990). The magnitude of calcium influx with sodium substitution is similar to that seen with application of 100 µM glutamate (unpublished results).

Several observations suggest that calcium influx in our experiments was secondary to glutamate release and subsequent activation of NMDA receptors. First, direct measurement with HPLC methods show glutamate concentrations in culture supernatant to be elevated about six-fold above that in cultures with normal sodium. Second, the magnitude of calcium influx in our experiments was similar to that seen following addition of high concentrations of exogenous glutamate. Third, selective

NMDA-type glutamate antagonists (MK-801, CPP) blocked calcium influx without altering glutamate release and at concentrations that selectively inhibit cell death from addition of glutamate or NMDA (Choi, 1988). The IC_{50} for CPP (0.88 µM) in our experiments was also similar to those inhibiting calcium influx from addition of 100 µM glutamate (F. W. Marcoux, personal communication), suggesting that glutamate efflux in our experiments activates NMDA receptors to a similar degree as exogenous glutamate. Fourth, calcium influx with lithium substituted for sodium was less than that from choline substitution. Previous studies show that lithium partially substitutes for sodium at the high-affinity glutamate transporter (Pines and Kanner, 1990) and suggest that calcium flux in our studies is secondary to reversal of the glutamate transporter. Finally, high micromolar concentrations of amiloride (a known blocker of the sodium/calcium exchanger) did not alter calcium influx.

The significance of a slightly greater calcium flux with sucrose or N-methyl-D-glucamine (NMG) substitutions in comparison to choline chloride requires further investigation, but suggests that chloride may be involved.

The concentration dependence of choline substitution for sodium indicates a steep increase in calcium influx with 50% of maximal response near 10 mM sodium and maximal responses with 3 mM sodium or less. This suggests that significant glutamate release is only seen when the transmembrane gradient for sodium ions is reversed (intracellular sodium concentrations are approximately 10 mM).

Removing external sodium in our experiments was not intended to mimic actual physiological or pathological conditions. However, our results demonstrate that reversal of the high-affinity sodium-dependent glutamate transporter causes rapid and massive glutamate release from neocortical cultures and subsequently activates calcium influx via NMDA-type glutamate receptors. Extracellular sodium concentrations do not approach zero during ischemia (Hansen, 1985) but the transmembrane sodium gradient nearly comes to equilibrium because intracellular sodium concentration increases simultaneously with an extracellular decrease. In addition, extracellular potassium rises to very high concentrations and membranes depolarize, all conditions that favor reversal of glutamate transport (Nicholls and Attwell, 1990). In vitro it is difficult to mimic all of these ionic changes without causing many other types of biochemical changes, therefore omission of extracellular sodium is a reasonable model.

Reversing the sodium gradient for a 10-min period induced a large efflux of glutamate and influx of calcium, but did not kill cells as assessed by LDH release 20 hr later. This is not surprising since glutamate concentrations measured in the culture supernatant were below 10 µM and since normal energy substrate was present throughout the experiment. Thus it is presumed that glutamate and calcium homeostasis were rapidly reestablished.

A previous study with synaptosomes (Sanchez-Prieto and Gonzalez, 1989) shows that glutamate efflux caused by depletion of cytoplasmic ATP stores was calcium insensitive, in contrast to normal vesicular glutamate release. Likewise, a study with hippocampal slices in vitro (Burke and Taylor, 1991) shows that "ischemia"-induced glutamate overflow is calcium insensitive, in contrast to glutamate overflow from 50 mM potassium. Our current results suggest that during ischemia, calcium-independent glutamate release and subsequent calcium influx and cell death may all result from reversal of the normal glutamate uptake process.

REFERENCES

S. P. Burke and C. P. Taylor, 1991, Glutamate, aspartate and GABA release from hippocampal CA1 slices during in vitro ischemia is calcium-independent. (see this volume)

D. W. Choi, 1990, Methods for antagonizing glutamate neurotoxicity, Cerebrovasc. Brain Metab. Rev. 2:105-147.

M. Ericinska, 1987, The neurotransmitter amino acid transport systems; a fresh outlook on an old problem, Biochem. Pharmacol. 36:3547-3555.

A. J. Hansen, 1985, Effect of anoxia on ion distribution in the brain, <u>Physiol. Rev.</u> 65:101-148.

M. Ikeda, T. Nakazawa, K. Abe, T. Kaneko and K. Yamatsu, 1989, Extracellular accumulation of glutamate in the hippocampus induced by ischemia is not calcium dependent -- in vitro and in vivo evidence, <u>Neurosci. Lett.</u> 96:202-206.

N. Martin-Moutot, M. Seagar and F. Couraud, 1990, Subtypes of voltage-sensitive calcium channels in cultured rat brain neurons, <u>Neurosci. Lett.</u> 115:300-306.

D. Nicholls and D. Attwell, 1990, The release and uptake of excitatory amino acids, <u>Trends Pharmacol. Sci.</u> 11:462-468.

G. Pines and B. I. Kanner, 1990, Counterflow of L-glutamate in plasma membrane vesicles and reconstituted preparations from rat brain, <u>Biochem.</u> 29:11209-11214.

J. Sanchez-Prieto and P. Gonzalez, 1988, Occurrence of a large Ca^{2+}-independent release of glutamate during anoxia in isolated nerve terminals (synaptosomes), <u>J. Neurochem.</u> 50:1322-1324.

P. K. Smith, R. I. Krohn, G. T. Hermanson, A. K. Mallia, F.H. Gartner, M. D. Provenzano, E. K. Fujimoto, N. M. Goeke, B. J. Olson and D. C. Klenk, 1985, Measurement of protein using bicinchoninic acid, <u>Anal. Biochem.</u> 150:76-85.

TOPOGRAPHICAL DISSOCIATION OF CALCIUM ACCUMULATION FOLLOWING HYPOXIC-HYPOGLYCEMIC GLUTAMATE RELEASE, NMDA RECEPTOR CONCENTRATION AND DELAYED NEURONAL DEATH IN GERBIL HIPPOCAMPUS

Kiyoshi Kataoka[1], Akira Mitani[1], Fumito Kadoya[1]
Yukio Yoneda[2], Kiyokazu Ogita[2] and Riyo Enomoto[2]

Department of Physiology, Ehime University
School of Medicine, Shigenobu, 791-02 Ehime[1] and
Department of Pharmacology, Setsunan University
Hirakata, 573-01 Osaka, Japan[2]

INTRODUCTION

Understanding of early mechanism of ischemic neuronal death seems crucial for therapeutic clue in acute stage of stroke. Although the mechanism is not clarified, glutamate neurotoxicity and intraneuronal calcium accumulation seem important. In this communication, we describe results of our topographical and temporal profile studies on these events as well as distribution of NMDA receptor concentration using gerbil hippocampal neurons.

DELAYED NEURONAL DEATH

For an ischemic insult, bilateral reversible occlusions of the carotid artery for 5min were employed.The function of pyramidal neurons can be followed with their spontaneous discharges. By single unit recording method, it was demonstrated that the discharges totally disappeared approx. 10 sec after the insult, which lasted even after reperfusion until they gradually restored the preischemic range 15 - 20 min later[1]. A chronic implantation of multiple units recording electrode revealed further change of this recovered function[2]; namely this lasted 60 - 70 hrs followed by gradual decrement of the discharge rate which eventually disappeared approx. 80 hrs after the ischemic insults (Fig. 1) Histochemical examination at this stage shows a marked region-selective neuronal damage to hippocampal CA1 sector. This is the delayed neuronal death[3] and has so far been widely employed as an experimental model.

GLUTAMATE RELEASE

It has been reported by microdialysis studies that

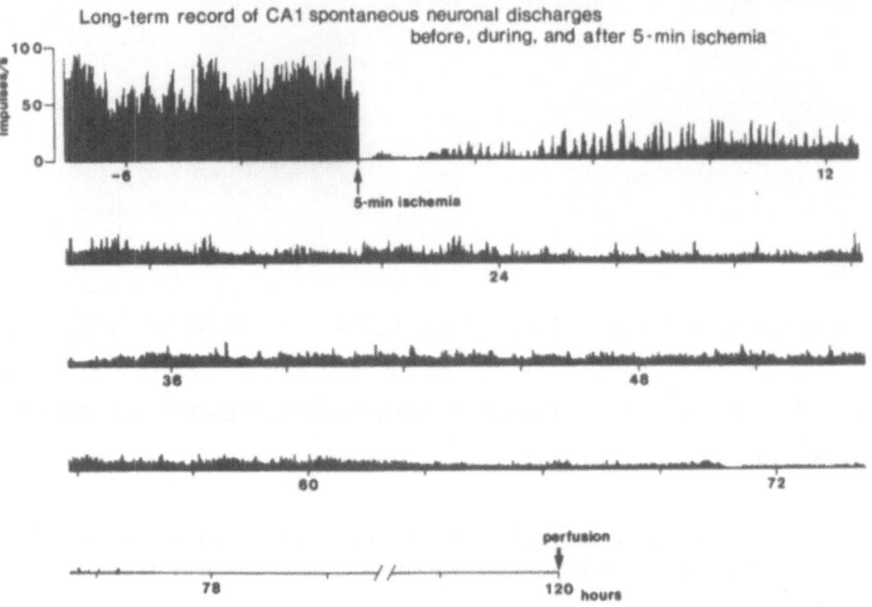

Fig. 1 Histogram of the frequency of discharges of multiple-unit activities recorded from CA1 neurons of gerbil hippocampus (40s bins). Recordings were carried out in non-anesthetised, freely moving state throughout. Ordinate, number of spike discharges per second; abscissa, time (hour) before and after the onset of 5-min ischemia.

ischemia induced a dramatic increase of extracellular concentration of glutamate[4]. An electron microscopical examination coupled with oxalate pyroantimonate technique demonstrated that ischemic CA1 neurons specifically retain calcium ions[5]. Based on these findings we then investigated the temporal profile and regionality of glutamate release and calcium accumulation on hypoxic - hypoglycemic hippocampal slices of the gerbil with the aid of an image system consisting of an epifluorescence microscope - a super sensitive video camera - an image processor. Hypoxia - hypoglycemia was introduced by switching the superfusion medium from normoxic Ringer solution into O_2 - and glucose - omitted Ringer solution. For glutamate release studies, NAD[+] and glutamate dehydrogenase enzyme were added to the medium. When glutamate was released, dehydrogenation underwent in conjunction with NADH production whose strong fluorescence (excitation wave length; 360nm, emission wave length; >450nm) was analyzed[6]. When hypoxia - hypoglycemia was introduced, a universal region-non-selective glutamate release started within 20sec, which consistently increased at least for several min(Fig. 2). This region-non-selectivity was verified by in vivo microdialysis studies. When microdialysis probes were implanted both CA1 and CA3 separately and dialysates were analyzed for glutamate before, during and after the ischemic insult, dramatic rises of this amino acid were noted within 60sec after ischemia, reaching 15-20 folds preischemic level at 5min insult which then returned to preischemic range promptly after reperfusion. It is worth noting here that there is practically no difference

58

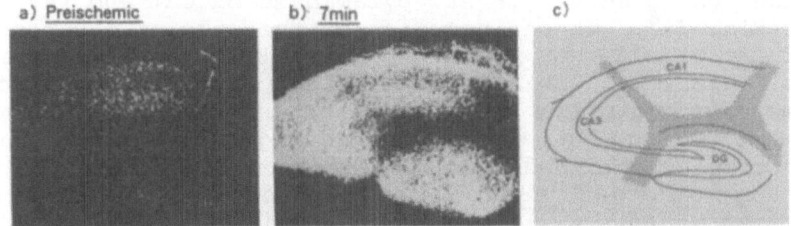

Fig. 2 Hypoxia-induced release of glutamate in a gerbil hippocampal slice. Ratio images were obtained 0s(a) and 420s(b) after the beginning of hypoxia. c: a schematic drawing of the gerbil hippocampal slice shown in a-b. Note: hypoxia-induced increase in fluorescence is observed throughout the slice including CA1 region, CA3 region and the dentate gyrus.

of the pattern and the maximum level of extracellular glutamate rise between CA1 and CA3 (unpublished data). These results indicate that such dramatic glutamate release singly may not explain the delayed neuronal death.

CALCIUM ACCUMULATION

For calcium accumulation studies, slices were preloaded with rhod-2, a calcium chelating fluorescence dye. An excitation wave length at 550 nm and an emission wave length at > 580 nm was used. When hypoxia - hypoglycemia was introduced, increases of calcium fluorescence was observed approx. 150sec which appeared region-selective to st. radiatum, st. oriens, st. lacunosum and st. moleculare of the

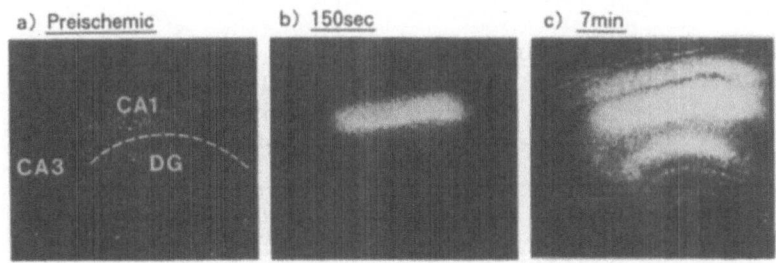

Fig. 3 Hypoxia-induced calcium accumulations in the hippocampal slice. Ratio images were obtained from fluorescence images prior to hypoxia(a), 150s(b) and 7min(c) after the beginning of hypoxia. Hippocampal subregions, CA1, CA3, and dentate gyrus(DG) are indicated in a), and also on the same position for b) and c).

CA1 sector followed by dorsal dentate gyrus. Moderate increase was observed in st. pyramidale of CA1 and far less in CA3(Fig. 3)[8]. These increases of the calcium fluorescence were dramatically depressed when MK-801, a NMDA receptor channel blocker, was added in the hypoxic-hypoglycemic medium. On the other hand, NMDA itself induced a similar calcium accumulation during the normoxic superfusion(unpublished data). These results indicated that when hippocampal slices were exposed to hypoxia - hypoglycemia, an in vitro ischemic condition, extracellular calcium was incorporated mainly through NMDA receptor channels. This calcium incorporation was probably triggered by NMDA receptor binding of released glutamate as well as membrane depolarization, which also may be relevant to delayed neuronal death.

REGIONAL DISTRIBUTION OF NMDA RECEPTOR CHANNEL

From the MK-801 sensitive, region-selective calcium accumulation in ischemic hippocampus, the issue of regionality of NMDA receptor concentration became important which may explain the region-selectivity of the delayed neuronal death. Using autoradiographical imaging method for MK-801 binding, it has already been reported that the hippocampal NMDA receptor is concentrated rather selectively in CA1 sector[9]. In the present investigation, we attempted to re-evaluate this finding by a biochemical technique. From hippocampi of the gerbil, subregions of CA1, CA3 and dentate gyrus were quickly dissected, pooled and subjected to membrane preparation for previously reported [³H] MK-801 binding studies[10]. The binding activities of MK-801 were largely different from region to region; namely the hippocampus as well as the cerebral cortex were categorized in a group of the highest activity (unpublished data). As shown in Table 1, hippocampal subregional binding activity was found not to be highly selective. The highest activity was found in CA1 followed by dentate gyrus and then CA3. However even the CA3 region contained an activity reaching to approx. 70% of that of CA1, a finding which suggests that

Table 1. Subregional Binding Activity of MK-801 in Gerbil Hippocampus: A Biochemical Study

Region	[³H]MK-801
CA1	1525 ± 112
CA3	1149 ± 79
Dentate Gyrus	1425 ± 81

fmol/mg Protein. Mean ± S.E.(n=6-8)

Membrane preparations were made from gerbil hippocampal subregions by homogenization and repeated washings. After treated with Triton X-100, samples were incubated with ligand at 2 C.

NMDA receptor distribute rather ubiquitously in the hippocampus.

CONCLUSION

In the gerbil hippocampus, ischemic insults induce region-non-selective glutamate release followed by region-selective calcium incorporation to CA1 and dorsal dentate gyrus, mainly through NMDA receptor channels. However NMDA receptor concentrations were rather uniform. The delayed neuronal death was produced only in CA1. Thus, the topographical dissociation of these issues became obvious which indicates that another yet unknown but important mechanism is involved in early stage of delayed neuronal death.

REFERENCES

1. A. Mitani, H. Imon and K. Kataoka, High frequency discharges of gerbil hippocampal CA1 neurons shortly after ischemia. Brain Res. Bull. 23: 569 (1989).
2. H. Imon, A. Mitani, Y. Andou, T. Arai and K. Kataoka, Delayed neuronal death is induced without postischemic hyperexcitability: Continuous multiple-unit recording from ischemic CA1 neurons. J. Cereb. Blood Flow Met. in press (1991).
3. T. Kirino, Delayed neuronal death in the gerbil hippocampus following ischemia. Brain Res. 239: 87 (1982).
4. H. Benveniste, J. Dreger, A. Schousboe and N.H. Diemer, Elevation of the extracellular concentrations of glutamate and aspartate in rat hippocampus during transient cerebral ischemia monitored by intracerebral microdialysis. J. Neurochem. 43: 1369 (1984).
5. R.P. Simon, T. Griffiths, M.C. Evans, J.H. Swan and B.S. Meldrum, Calcium overload in selectively vulnerable neurons of the hippocampus during and after ischemia: an electron microscopy study in the rat. J. Cereb. Blood Flow Met. 4: 350 (1984).
6. A. Mitani, F. Kadoya, Y. Nakamura and K. Kataoka, Visualization of hypoxia-induced glutamate release in gerbil hippocampal slice. Neurosci. Lett. 122: 167 (1991).
7. A. Mitani, H. Imon, K. Iga, H. Kubo and K. Kataoka, Gerbil hippocampal extracellular glutamate and neuronal activity after transient ischemia. Brain Res. Bull. 25: 319 (1990).
8. A. Mitani, F. Kadoya and K. Kataoka, Distribution of hypoxia-induced calcium accumulation in gerbil hippocampal slice. Neurosci. Lett. 120: 42 (1990).
9. D.T. Monaghan and C.W. Cotman, Distribution of N-methyl-D-aspartate sensitive L-^{3}H-glutamate-binding sites in rat brain. J. Neurosci. 5: 2909 (1985).
10. Y. Yoneda and K. Ogita, Microbial methodological antifacts in [^{3}H] glutamate receptor binding assays. Annal. Biochem. 177: 250 (1989).

Chapter 2

The Role of Glutamatergic Receptor in Brain Injury

DIZOCILPINE (MK-801) IN CEREBRAL ISCHEMIA

Roland N. Auer

Department of Pathology, The University of Calgary
Health Sciences Building, 3330 Hospital Drive N.W., Rm. 2508
Calgary, Alberta, Canada T2N 4N1

INTRODUCTION

Dizocilpine maleate (MK-801) is a drug with potential use in cerebral ischemia. However, controversy concerns its effectiveness in various ischemic settings, and the acceptability of side effects and potential toxicity. This article will review data from our laboratory and from the literature, amalgamating the existing evidence and pointing to questions in need of resolution before the status of the drug is clear for clinical use.

With the advent of the knowledge that excess glutamate released by ischemia may play a role in hypoxic/ischemic brain damage[1], it became logical in the pursuit of treatments for acute ischemia to search for drugs that block glutamate receptors. The first subtype of glutamate receptor shown to play a role in the pathogenesis of ischemic neuronal death was the N-methyl-d-aspartate (NMDA) receptor[2]. Competitive, polar NMDA antagonists do not easily cross the blood brain barrier, but non-competitive, hydrophobic antagonists such as MK-801 do cross, blocking at the receptor operated ion channel rather than the agonist recognition site (Figure 1). This channel is permeable to several ions, including calcium. MK-801 has a very high affinity for the channel binding site, with a K_d of 37 nM[3]. The fact that this drug binds selectively to open channels[4], where glutamate is released and ischemic damage is occurring, and that it enters the brain after parenteral injection, together make it appealing as a potential therapeutic drug in acute cerebral ischemia.

DIZOCILPINE IN FOCAL ISCHEMIA

When discussing the effects of any drug in ischemia, the various kinds of brain ischemia (Table 1) should be distinguished. Ischemia can be either focal or global. With respect to reflow, ischemia can be either permanent or temporary. Permutation gives four general types of ischemia (Table 1), in which the effects of drugs, and the pathophysiology of damage, may not be identical. For example, during either temporary or even permanent focal ischemia, blood-borne drug can reach the borderzone ischemic territory.

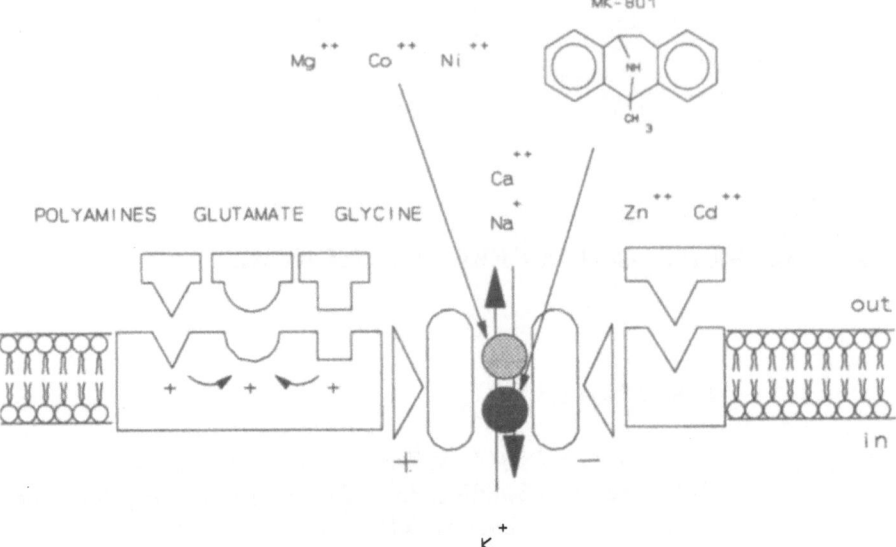

Figure 1. NMDA receptor and allosteric regulatory sites, both positive and negative. MK-801 binds to a separate site from Mg^{++} and other divalent cations within the ion channel.

TABLE 1

BASIC TYPES OF ISCHEMIA

Type of Ischemia	Clinical Examples
Transient Focal	Thrombo-Embolism
Transient Global	Cardiac Arrest
Permanent Focal	Calcific Athero-Embolism
Permanent Global	Brain Death

In focal ischemia, dizocilpine has been shown to be effective in reducing infarct size, whether the drug has been given before[5], or up to 2 hours after[6,7] ischemia. The occlusions of the middle cerebral artery were permanent in these studies. It is unlikely that damage in the central territory of dense, permanent ischemia due to arterial occlusion are treatable. Degradation of tissue in permanent, complete ischemia closely resembles autolysis[8], but at the border of a permanent focal occlusion, some flow still exists. We conclude that MK-801 rescues tissue in this zone of partial ischemia where tissue is salvagable, giving rise to infarcts which are smaller by 50% in experimental animals[6,7].

Results with dizocilpine in global ischemia have been far less promising. Using both four vessel occlusion models[9] and the two vessel occlusion model[10], Buchan and co-workeres have shown no effect of dizocilpine in reducing hippocampal CA1 pyramidal cell necrosis. Neocortical damage was not produced in this model. In a dog model[11] and primate model[12] of 17 minutes cardiac arrest, no beneficial effect of dizocilpine was shown. The former study[11] even produced extra-hippocampal damage and no promising trends were seen, although the number of animals per group was small (n=4).

Using 10 minutes of transient forebrain (2-vessel occlusion, 2-VO) ischemia, a model which spares the structures supplied by the vertebro-basilar circulation, we were able to show a protective effect against neuronal necrosis whether MK-801 was given before ischemia, or 20 minutes after ischemia[13]. No protection was seen at 2 hrs or 24 hrs postischemia, unlike the results of Gill and co-workers[14]. Brain temperature was not monitored in these earlier experiments with MK-801[13,15], and results from other laboratories have suggested that only the hypothermia induced by MK-801, rather than the drug itself, causes the neuroprotective effect[9,10,16]. *Pre-ischemic* hypothermia is potently neuroprotective[17,18]. However, *postischemic* hypothermia has no effect if induced 30 minutes after ischemia[19,20]. It is thus unlikely that delayed hypothermia can explain results of drugs administered late in the recovery period, when hypothermia itself has no effect.

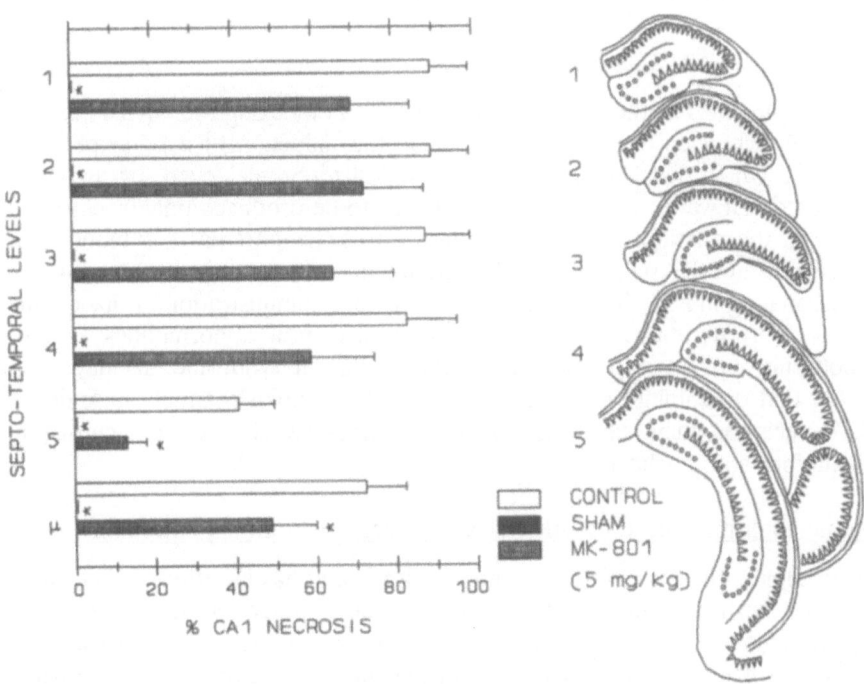

Figure 2. Septo-temporal distribution of intra-hippocampal necrosis with and without MK-801 given 20 minutes after transient forebrain ischemia.

Results in transient forebrain ischemia with MK-801

These considerations prompted us to repeat our initial experiments with delayed (20 minute postischemia) MK-801 administration using monitoring and strict control of brain and body temperature[21]. Using sub-serial sectioning techniques, and examination at multiple standardized coronal levels, neuroprotection was chiefly seen in the temporal portion of the hippocampus (Figure 2), and in the cerebral cortex. Prior to sacrifice, animals received extensive neuropsychologic testing. Overall brain necrosis was reduced enough to give a significant improvement in neuropsychologic tests of learning and memory[21].

The distribution of the protection within the hippocampus was noteworthy, and may shed some light on mechanism of action. Neuroprotection was not seen in septal areas, subjected to dense cerebral ischemia (levels 1-4, Figure 2), but only in the temporal hippocampus, where ischemia is incomplete in this model due to the unoccluded vertebro-basilar circulation. When total number of dead cells (sum of numerators at all coronal levels) was divided by the total CA1 cell count at all levels (sum of the denominators), a significant overall effect was seen $(0.01 < p < 0.05)$[21]. The results suggest that the CA1 neurons of the hippocampus, with their dense complement of NMDA receptors, may only be amenable to rescue by NMDA antagonists if ischemia is not dense or severe.

ADVERSE EFFECTS WITH DIZOCILPINE

While weighing the adverse effects of dizocilpine vs potential benefit, it should be recalled that ischemic infarction is often lethal, due to the development of edema, brain herniation, and death within the first 1-2 days. In judging the risk/benefit ratio of giving the dizocilpine for cerebral ischemia, the devastating and potentially lethal nature of the untreated disorder must be borne in mind.

Psychotomimetic Side Effects

Dizocilpine acts as the same receptor as PCP in blocking the NMDA associated ion channel. The side effects of PCP include the induction of a schizophrenia-like syndrome[22], which has been mistaken for schizophrenia even by experienced psychiatrists[22]. However, such a syndrome seems to be produced only by *chronic* PCP abuse, and, the relationship of this syndrome to schizophrenia is dubious. Current evidence indicates that schizophrenia is accompanied by structural abnormalities of the brain originating in early life[23]. It seems unlikely that administration of dizocilpine for relatively short durations could lead to such structural abnormalities as seen in schizophrenia[23], or to a permanent, enduring clinical syndrome. It may well be, however, that dizocilpine, resembling ketamine in its open channel blockade of the NMDA receptor may cause transient ketamine-like side effects consisting of hallucinations and psychosis.

Hypermetabolism induced by MK-801 and Structural Neuronal Alterations

Dizocilpine induces limbic hypermetabolism[24], and has an effect on the structure of cerebral cortical neurons. Dizocilpine has been described to cause reversible neuronal vacuolation[25]. The effect and has not been shown to lead to neuronal necrosis, although this has been suggested at the highest dose of $\geq$ 5mg/kg[26]. In resolving this question of whether or not dizocilpine gives rise to neuronal death, rigorous scientific criteria for defining necrosis morphologically must be used. These include distinct acidophilia and karyorrhexis or cytorrhexis under the light microscope, and confluent paerikaryal (not dendritic) membrane breaks, and mitochondrial flocculent densities under the electron microscope[27].

If dizocilpine is indeed shown to kill a few neocortical neurons in rodents, these findings would have to be extended to higher animals. Even if a few neurons in the limbic region are damaged, this would have to be weighed against the powerful demonstrated effect of pre- or post-ischemic dizocilpine in reducing infarct size in focal ischemia[5-7]. It may be that the number of neurons killed is so small, that this, and any transient psychosis, might be acceptable in exchange for a significant reduction in infarct size. Lastly, combination of dizocilpine with voltage sensitive calcium channel antagonists such as nimodipine[28] or non-NMDA EAA antagonists such as NBQX[29] may lead to either increased neuroprotection, or a reduction in the dose of MK-801 which needs to be given to achieve the same neuroprotective benefit.

Dedication

During the time period that this article was in preparation, the author's father had a large thromboembolic stroke, in the author's presence, arising from stable atrial fibrillation of 18 yrs duration. Although MK-801 was available from the author's laboratory within 45 minutes, the drug was not yet accepted treatment, and could not be given in spite of a possible reduction in infarct size[6]. CT scan was normal at 2 hrs, but a large infarct developed later, in major portions of the right hemisphere. Herniation did not ensue. This article is dedicated to my father, who is left incapacitated with a complete paralysis of face, arm and leg, visuo-sensory and other deficits. The resolution of the scientific questions concerning therapy for ischemic stroke is urgent, in view of patients who are daily devastated by stroke.

REFERENCES

1. S. M. Rothman, J. W. Olney, Excitotoxicity and the NMDA receptor. Trends Neurosci 10:299-302, (1987).
2. R. P. Simon, J. H. Swan, T. Griffiths, B. S. Meldrum, Blockade of N-methyl-D-aspartate receptors may protect against ischemic damage in the brain. Science 226:850-852, (1984).
3. E. H. F. Wong, J. A. Kemp, T. Priestley, A. R. Knight, G.N. Woodruff, L. L. Iversen, The anticonvulsant MK-801 is a potent N-methyl-D-aspartate antagonist. Proc Natl Acad Sci USA 83:7104-7108, (1986).
4. J. E. Huettner, B. P. Bean, Block of N-methyl-D-aspartate-activated current by the anticonvulsant MK-801: selective binding to open channels. Proc Natl Acad Sci 85:1307-1311, (1988).
5. E. Ozyurt, D. I. Graham, G. N. Woodruff, J. McCulloch, Protective effect of the glutamate antagonist, MK-801 in focal cerebral ischemia in the cat. J Cereb Blood Flow Metabol 8:138-143, (1988).
6. C. K. Park, D. G. Nehls, D. I. Graham, G. M. Teasdale, J. McCulloch, Focal cerebral ischaemia in the cat: treatment with the glutamate antagonist MK-801 after induction of ischaemia. J Cereb Blood Flow Metabol 8:757-762, (1988).
7. C. K. Park, D. G. Nehls, D. I. Graham, J. McCulloch, The glutamate antagonist MK-801 reduces focal ischemic damage in the rat. Ann Neurol 24:543-551, (1988).
8. J. H. Garcia, Y. Kamijyo, Cerebral infarction. Evolution of histopathologic changes after occlusion of a middle cerebral artery in primates. J Neuropathol Exp Neurol 33:408-421, (1974).

9. A. Buchan, W. A. Pulsinelli, Hypothermia but not the N-methyl-d-aspartate antagonist, MK-801, attenuates neuronal damage in gerbils subjected to transient global ischemia. J Neurosci 10:311-316, (1990).

10. A. Buchan, H. Li, W. A. Pulsinelli, The N-methyl-D-aspartate antagonist, MK-801, fails to protect against neuronal damage caused by transient, severe forebrain ischemia in adult rats. J Neurosci 11:1049-1056, (1991).

11. F. Sterz, Y. Leonov, P. Safar, G. T. Shearman, S. W. Stezoski, H. Perch, Effect of the excitatory amino acid receptor blocker MK-801 on overall and neurologic outcome after prolonged cardiac arrest in dogs. Anesthesiology 71:907-918, (1989).

12. W. L. Lanier, W. J. Perkins, B. R. Karlsson, J. H. Milde, B. W. Scheithauer, G. T. Shearman, J. D. Michenfelder, The effects of dizocilpine maleate (MK-801), an antagonist of the N-methyl-D-aspartate receptor, on neurologic recovery and histopathology following complete cerebral ischemia in primates. J Cereb Blood Flow Metabol 10:252-261, (1990).

13. M. R. Rod, R.N. Auer, Pre- and post ischemic administration of the NMDA receptor antagonist dizocilpine maleate (MK-801) reduces ischemic brain necrosis in the rat. Can J Neurol Sci 16:340-344, (1989).

14. R. Gill, A. C. Foster, G. N. Woodruff, MK-801 is neuroprotective in gerbils when administered in the post-ischaemic period. Neuroscience 25:847-855, (1988).

15. R. Gill, A. C. Foster, G. N. Woodruff, Systemic administration of MK-801 protects against ischemia-induced hippocampal neurodegeneration in the gerbil. J Neurosci 7:3343-3349, (1987).

16. D. Corbett, S. Evans, C. Thomas, D. Wang, R. Jonas, MK-801 reduces cerebral ischemic injury by inducing hypothermia. Brain Res 514:300-304, (1990).

17. R. Busto, W. D. Dietrich, M. Y-T. Globus, I. Valdés, P. Scheinberg, M. D. Ginsberg, Small differences in intraischemic brain temperature critically determine the extent of ischemic neuronal injury. J Cereb Blood Flow Metabol 7:729-738, (1987).

18. H. Minamisawa, C-H. Nordström, M-L. Smith, B. K. Siesjö, The influence of mild body and brain hypothermia on ischemic brain damage. J Cereb Blood Flow Metabol 10:365-374, (1990).

19. H. L. Rosomoff, Hypothermia and cerebrovascular lesions. Arch Neurol Psychiatr 78:454-464, (1957).

20. R. Busto, W. D. Dietrich, M. Y-T. Globus, M. D. Ginsberg, Postischemic moderate hypothermia inhibits CA1 hippocampal ischemic neuronal injury. Neurosci Lett 101:299-304, (1989).

21. M. R. Rod, R. M. Auer, I. Q. Whishaw, The relationship of structural ischemic brain damage to neurobehavioural deficit: the effect of postischemic MK-801. Can J Psychol 44:196-209, (1990).

22. S. H. Snyder, Phencyclidine. Nature 285:355-356, (1980).

23. G. W. Roberts, Schizophrenia: the cellular biology of a functional psychosis. Trends Neurosci 13:207-211, (1990).

24. D. G. Nehls, A. Kurumaji, C. K. Park, J. McCulloch, Differential effects of competitive and non-competitive N-methyl-d-aspartate antagonists on glucose use in the limbic system. Neurosci Lett 91:204-210, (1988).

25. J. W. Olney, J. Labruyere, M. T. Price, Pathological changes induced in cerebrocortical neurons by phencyclidine and related drugs. Science 244:1360-1362, (1989).

26. H. L. Allen, L. L. Iversen, Phencyclidine, dizocipine, and cerebrocortical neurons. Science 247:221, (1990).

27. R. N. Auer, H. Kalimo, Y. Olsson, T. Wieloch, The dentate gyrus in hypoglycemia. Pathology implicating excitotoxin-mediated neuronal necrosis. Acta Neuropathol (Berl) 67:279-288, (1985).

28. D. Uematsu, N. Araki, J. H. Greenberg, J. Sladky, M. Reivich, Combined therapy with MK-801 and nimodipine for protection of ischemic brain damage. Neurology 41:88-94, (1991).

29. M. J. Sheardown, E. Ø. Nielsen, A. J. Hansen, P. Jacobsen, T. Honoré, 2,3-dihydroxy-6-nitro-7-sulfamoyl-benzo(F)quinoxaline: a neuroprotectant for cerebral ischemia. Science 247:571-574, (1990).

PROTECTION BY TWO NON-NMDA ANTAGONISTS, NBQX AND GYKI 52466
AGAINST SELECTIVE CELL LOSS FOLLOWING TRANSIENT GLOBAL
ISCHAEMIA (4VO) IN THE RAT

E. Le Peillet, B. Arvin, C. Moncada, and
B.S. Meldrum

Institute of Psychiatry, De Crespigny Park
London, U.K.

INTRODUCTION

Transient global ischaemia in the rodent is characterized biochemically by large increases in extracellular concentrations of excitatory amino acids and several other transmitters and pathologically by an early phase of damage (less than 48h) in cortex and basal ganglia followed by delayed (post 48h) loss of hippocampal CA1 pyramidal neurons. In the striatum the early pathology (less than 10h post reperfusion) involves the small/medium sized neurons. Within the last decade glutamate antagonists selective for NMDA receptor have been shown to be have a weak protective action against hippocampal and striatal damage following transient global ischaemia when given prior to or shortly after ischaemia (Block and Pulsinelli, 1987; Buchan and Pulsinelli, 1991; Rod and Auer, 1989: Swan and Meldrum 1990; Warner et al 1991). This is in contrast to their powerful protective effects in the neocortex following focal ischaemia (for review see Meldrum 1990). More recently two reports have shown that the non-NMDA antagonist 2,3-dihydro-6-nitro-7-sulphamoyl- benzo(F)-quinoxaline (NBQX) protects hippocampal damage against transient global ischaemia in rats and gerbils (Sheardown et al 1990; Diemer et al 1990). Unlike other compounds of the same family (CNQX and DNQX), NBQX is highly selective for non-NMDA receptors as it lacks the interaction with the NMDA receptor via the glycine modulatory site (Honorè 1991). NBQX shows a 20-30 fold greater potency for inhibition of $[^3H]AMPA$ than $[^3H]kainate$ binding to cortical membranes. A muscle relaxant 2,3 benzodiazepine (1-(amino-phenyl)-4-methylendioxy-5H-2,3, benzodiazepine (GYKI 52466) has also recently been shown to have non-NMDA antagonist properties. GYKI 52466 blocks depolarisations induced by iontophoretic application of glutamate, quisqualate and kainate, but not NMDA in rat cortical slices (Tarnawa et al 1990). In vivo GYKI 52466 inhibits monosynaptic spinal reflexes in the cat, and blocks glutamate- but not NMDA- or kainate-induced responses in rat abducens motoneurones (Ouarduoz and Durand 1991).

The Role of Neurotransmitters in Brain Injury, Edited by
M. Globus and W.D. Dietrich, Plenum Press, New York, 1992

We have studied the cerebroprotective action of these two non-NMDA antagonists in the rat 4 vessel occlusion model of Pulsinelli & Brierley 1979, with a 20 minute period of carotid occlusion and histological evaluation after a seven day survival period.

MATERIALS AND METHODS

NBQX was supplied as the Lithium salt (Dr T. Honoré, Novo-Nordisk, Copenhagen). GYKI 52466 was supplied by Dr I. Tarnawa, Budapest. Both compounds were dissolved in distilled water.

Male Wistar rats (250-300g) were anaesthetised with 2% halothane (in 70% Nitrous oxide and 30% oxygen) and both vertebral arteries were permanently occluded by electro-cauterisation within the alar foraminae of the first cervical vertebra. At the same time, both common carotid arteries were isolated and atraumatic clamps were placed around each one. One femoral vein was cannulated to enable the subsequent i.v. administration of fluid. The following day cerebral ischaemia was induced, in the unanaesthetised animal, by tightening the clamps around the carotid arteries for 20 min. Carotid clamping resulted within 1-2 min in a loss of righting reflex. The body and head temperature was maintained at 36.5-37.5 °C during the period of the experiment. Each animal received a bolus injection of vehicle (n=7) or drug, 10 mg/kg, i.v. (via femoral vein) (NBQX n=6 or GYKI 52466 n=6) immediately after the end of the occlusion. This was followed by a slow i.v. infusion of vehicle or NBQX or GYKI 52466, 30 mg/kg, over 3 h. Seven days post-ischaemia animals were perfusion-fixed FAM (Formaldehyde - glacial acetic acid - methanol, 1-1-8). The brains were paraffin embedded and coronal sections (7 µm) were cut and stained with Luxol fast blue and cresyl fast violet for light microscopy. Cortical pathology was assessed at 4 coronal levels 10.7, 9.7, 5.7 and 3.7 mm anterior to the interaural line, I.A.), striatal pathology at 2 levels (10.7 and 9.7 mm anterior to I.A. line), and hippocampal pathology at 2 levels (5.7 and 3.7 mm anterior to I.A. line) in both hemispheres (according to Paxinos and Watson Atlas, 1982). Assessment of ischaemic brain damage was performed by grading cell loss where 0: 0-10%, 1:10-30%, 2:30-50%, 3:50-70%, 4:70-90%, 5:90-100% damage. All values are given as means ±SEM of each score in each hemisphere. Evaluation of significant differences between control (n=14 hemispheres) and treatment groups (n=12 hemispheres) was performed using a non-parametric test (Mann & Whitney).

RESULTS

The highest degree of damage was observed in the hippocampal pyramidal cells in the CA1 subfield at the level A 5.7mm (mean score 4.9±0.1). At the level of A 3.7mm less damage was seen in CA1 region. In CA_2, CA_3 and CA_4 neuronal damage was always less extensive than in CA_1 region (mean score at level A 5.7mm 1.9±0.5, 1.2±0.4 and 1.5±0.5 respectively). The degree of damage in CA_3 and CA_4 in level 3.7mm was similar to level 5.7mm; lesions in the CA_2 area were however less extensive at this level. NBQX produced a significant protection in both levels of CA_3 area examined. The means scores were 0.25±0.1 (p < 0.05 and p < 0.01 compared with controls in level A 5.7mm and A 3.7mm respectively). A significant protective effect of CA_2

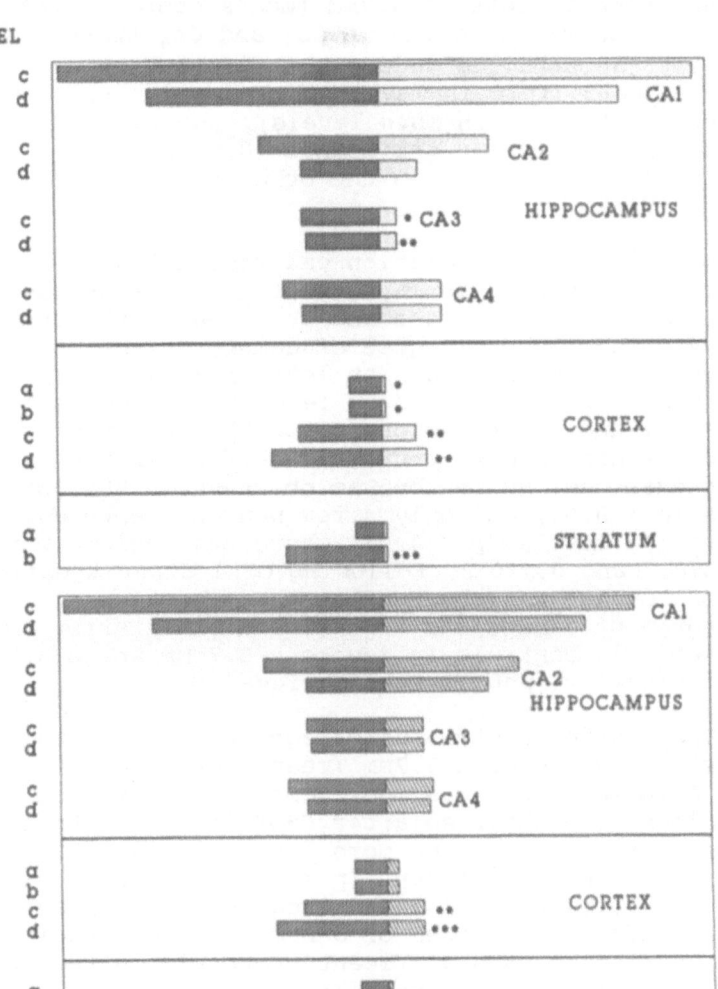

Figure 1. histological evaluation of hippocampus, cortical and striatal pathology at 4 coronal levels (a=10.7 b=9.7 c=5.7 and d=3.7mm anterior to the interaural line) following 20 min of transient global ischaemia in male Wistar rats. Seven days post-ischaemia animals were perfused with FAM. Brains were wax-embedded and coronal sections (7μm) were cut and stained with luxol fast blue and cresyl violet. Assessment of damage was performed by grading cell loss where 0: 0-10%, 1:10-30%, 2:30-50%, 3:50-70%, 4:70-90% and 5:90-100%. Values are given as mean ± SEM of each score in each hemisphere. Evaluation of significant difference between control (n=14 hemispheres) and treatment group (n=12 hemispheres) was performed using a non-parametric test (Mann & Whitney). *** p = < 0.001, ** p = < 0.01, * p = < 0.1

area was seen in level A 3.7mm (means score = 0.6±0.3 p<0.05).
No protection was observed in CA_1 and CA_4 areas of hippocampus.
GYKI 52466 appeared to reduce the severity of damage in CA_1
(3.8±0.5 and 3.1±0.4 in level A 5.7mm and A 3.7mm respectively)
and in CA_3 (0.6±0.2 in both levels); the values failed however
to reach statistical significance (0.05 p < 0.1). No
protection was observed in the other regions of hippocampus
(CA3 and CA4).

Neocortical degeneration was restricted to small and
medium sized neurons in layer 3 of the somatosensory cortex.
Occasionally neuronal loss was observed in the upper border of
layer 4. The highest degree of damage in the cortex was
observed in control groups at level A 3.7mm where the mean
score was 1.7±0.2 followed by level 5.7mm (score 1.2±0.2) and
levels 9.7 and 10.7mm (score 0.5±0.2). Treatment with NBQX
provided a significant protective effect in the four cortical
levels examined. No lesion was observed in the anterior 2
levels (p < 0.05) and only a few neurones were damaged in the
posterior 2 levels (p < 0.01) where the lesions were estimated
at 0.5±0.3 and 0.7±0.2. Following GYKI 52466 treatment some
neuronal damage was observed in the 4 cortical levels examined.
The degree of damage was significantly different from that in
the ischaemic controls in levels A 5.7 mm and A 3.7 mm (mean
score: 0.55±0.15 p<0.01 in both levels).

The degree of striatal damage in control groups was the
highest in the level A 9.7mm (mean score 1.5±0.4). At the other
level A 10.7mm there was less damage (mean score was 0.4±0.2).
No lesions were observed after NBQX in the 2 striatal levels
examined (Fig.3). In the more posterior level (A 9.7 mm) this
was significantly different (p < 0.001) from the ischaemic
control. Following GYKI 52466 treatment only 2 animals (out of
6) exhibited a small area of damage in the dorsolateral portion
of the striatum. A significant protective effect was observed
only in the level A 9.7 mm (p < 0.001).

DISCUSSION

In this study we have shown the protective effects of two
non-NMDA antagonists NBQX and GYKI 52466 in the rat cortex and
striatum following transient global ischaemia. Unlike two other
studies of transient ischaemia in rats (Diemer et al 1990) and
gerbils (Sheardown et al 1990) where the predominant effect of
NBQX was shown in the hippocampus with preservation of CA_1
neurons, in our study little or no protection was observed in
the hippocampus with either NBQX or GYKI 52466. This
discrepancy can probably be explained in terms of the route of
administration (i.v in this study as opposed to i.p in the
other studies) and pharmacokinetics of the drugs. Given i.v
both NBQX and GYKI 52466 penetrate the brain rapidly and have a
short duration of action (Ouarduoz and Durand 1991, Smith et al
1991). Therefore, in our study, where the drugs are infused
over the first 3h after reperfusion, the early phase of
degeneration which occurs principally in the striatum and
cortex is blocked. In contrast in the studies of Sheardown et
al (1990) and Diemer et al (1990) NBQX was given as i.p
injections. In this situation the drug is initially
precipitated and then is slowly released into the circulation
over a longer period (days) giving rise to protection in

hippocampus. As for the mechanism of action for neuroprotection it would appear that a blockade of both AMPA and kainate receptor are possibilities. From binding studies with NBQX and in vivo ionotophoresis studies with GYKI 52466 it is evident that these compounds have a higher affinity for AMPA than kainate receptor. However, the pattern of protection observed in our study (striatum> cortex> CA3> CA1) is possibly consistent with an action on kainate receptors which show a high density in these structures; however lamina 3 of the cortex has a high density of AMPA receptors and a low density of kainate receptors. Recently we have shown (Moncada et al., 1991) that NBQX and GYKI 52466 administered under similar conditions as in this study protect against kainate but not AMPA toxicity in the rat hippocampus. It is notable however that in the latter study the CA3 region (rich in KA receptors) was not protected against KA toxicity.

It is reasonable to conclude that activation of non-NMDA receptors during the first 3 hours after transient global ischaemia contributes to selective neuronal death in the cortex and striatum. A longer period of non-NMDA receptor activation may be involved in delayed cell death in the hippocampus, although activity in the first hour is probably important as protection is greater when treatment begins at 0 hour rather than 1 hour post ischaemia (Sheardown et al 1990). The non-NMDA receptors may be on the vulnerable neurons or they may be on neurons or glia that influence their functions directly or indirectly.

SUMMARY

In Wistar rats subjected to 20 minutes of 4 vessel occlusion the intravenous infusion of either NBQX or GYKI 52466 (antagonists acting selectively on non-NMDA receptors) 40 mg/kg during the first 3 hours post ischaemia protects the cortex and striatum from selective neuronal damage (assessed after seven days).

ACKNOWLEDGEMENTS

We thank the Medical Research Council, the Wellcome Trust and the Bethlem-Maudsley Research fund for financial support.

REFERENCES

Block, G.A., Pulsinelli, W.A., N-Methyl-D-aspartate antagonists: failure to prevent ischemia-induced selective nerve cell death. In: M.E. Raichle, W.J. Powers (eds.), Cerebrovascular diseases. New York, Raven Press, (1987) 37-44.

Buchan, A.M., Li, H, Pulsinelli, W.A. The N-methyl-D-aspartate antagonist, MK-801, fails to protect against neuronal damage caused by transient severe forebrain ischemia in adult rats. J.Neurosci. 11 (1991) 1049-1056

Diemer, N.H., Johansen, F.F., Jorgensen, M.B. N-methyl-D-aspartate and non-N-methyl-D-aspartate antagonists in global cerebral ischemia. Stroke 21 suppl III (1990) III-39-III-42.

Honoré, T. Inhibitors of kainate and AMPA ionophore receptors. In B.S. Meldrum (ed) Excitatory Amino Acid Antagonists Blackwell Scientific, Oxford (1991) 180-194.

Meldrum, B. Protection against ischaemic neuronal damage by drugs acting on excitatory neurotransmission. Cerebrovasc. Brain Metab. Rev. 2 (1990) 27-57.

Moncada, C., Arvin, B., Le Peillet, E., and Meldrum, B.S. The non-NMDA and GYKI 52466 protect aagainst kainate but not (S) AMPA toxicity in the rat hippocampus. Neurosci. Lett. (in press).

Ouarduoz M., Durand, J. GYKI 52466 antagonizes glutamate responses but not NMDA and kainate responses in rat abducensmotoneurones. Neurosci. Lett. 125 (1991) 5-8.

Paxinos G., Watson, C. The Rat Brain in Stereotaxic Coordinates, Academic Press, New York, (1982) 1-13.

Pulsinelli, W.A., Brierley, J.B., A new model of bilateral hemispheric ischemia in the unanesthetized rat. Stroke 10 (1979) 267-272.

Rod, M.R., Auer, R.N., Pre- and postischemic administration of dizocilpine (MK-801) reduces cerebral necrosis in the rat. Can. J. Neurol. Sci. 16 (1989) 340-344.

Sheardown, M.J., Nielsen, E.O., Hansen, A.J., Jacobsen, P., Honoré, T., 2,3-dihydroxy-6-nitro-7-sulfamoyl-benzo(F) quinoxaline: a neuroprotectant for cerebral ischemia. Science 247 (1990) 571-574.

Smith, S.E., Dürmüller, N., Meldrum, B.S., The non-N-methyl-D-aspartate antagonists, GYKI 52466 and NBQX are anticonvulsant in two animal models of reflex epilepsy. Europ. J. Pharmacol. 201 (1991) 179-183.

Swan, J.H., Meldrum, B.S., Protection of NMDA antagonists against selective cell loss following transient ischaemia. J. Cereb Blood Flow Metab. 10 (1990) 343-351.

Tarnawa, I., Engberg, I., Flatman, J.A. GYKI 52466, an inhibitor of spinal reflexes is a potent quisqualate antagonist. In G.Lubec and G.A.Rosenthal (ed.) Amino Acids: Chemistry, Biology and Medicine, ESCOM Science Publishers (1990) 538-546.

Warnar, M.A., Neil, K.H., Nadler, J.V., Crain, B.J. Regionally selective effects of NMDA receptor antagonists against ischaemic brain damage in gerbil. J. Creb. Blood Flow Metab., 11 (1991) 600-610.

GLUTAMATE NEUROTOXICITY AND EFFECT OF ANTAGONISTS

--IN VITRO STUDY USING HIPPOCAMPAL NEURONS

EXPOSED TO HYPOXIA

Eiji Kohmura, Kazuo Yamada, Akira Kinoshita,
Toru Hayakawa

Department of Neurosurgery, Osaka University
Medical School, 1-1-50 Fukushima, Fukushimaku
Osaka 553, Japan

INTRODUCTION

Roles of glutamate in various pathological situations are suggested, though it is an physiological neurotransmitter. Studies using in vivo microdialysis technique showed rapid increase of extracellular glutamate concentration during the ischemic period in models of transient global ischemia, but it declined also rapidly after recirculation (Benveniste et al., 1984; Hagberg et al., 1985). Therefore slowly ongoing neuronal injury such as delayed neuronal loss is difficultly explained by elevated glutamate concentration during the ischemic period. We speculated that neurons might become more vulnerable to glutamate after exposure to ischemia and they might be thereafter injured by lower amount of glutamate. We intended to evaluate possible changes of glutamate neurotoxicity induced by subcritical hypoxia in vitro. We indeed confirmed increased sensitivity to glutamate after subcritical hypoxia(Kohmura et al., 1990). In this paper specific agonists and antagonists for the glutamate receptor subtypes were tested in order to evaluate the underlying mechanism.

METHOD

Culture of hippocampal neurons and exertion of hypoxia in vitro were realized with the same method as reported previously (Kohmura et al., 1990, 1991). In brief, dissociated hippocampal neurons were obtained from 18-day-pregnant rat fetuses (Wistar-strain) and plated at a low density (100 living cells/mm^2) on 12-well culture plates coated with poly-L-lysine. These neurons were maintained for the first 12 hours in Eagle's minimal essential medium with Earl's balanced salts mixture (MEM) supplemented with 10% fetal calf serum. The culture medium was then changed to a serum-free one (MEM supplemented with 1mM pyruvate, 5mg/l insulin, 10mg/l transferrin, 6.3μg/l progesterone, 5.2μg/l selenium, 100mg/l kanamycin sulfate, and 0.25mg/l amphotericin B), which contained no glutamate but 0.8mM Mg^{2+}.

After 3 days in culture, these neurons were extending long neurites but were still growing independently owing to the initial low plating density. The culture plates were placed in a hypoxic chamber, which was originally constructed as a vacuum desiccator (Iuchiseieido, Japan) for two hours to undergo a hypoxic stress. The chamber was suctioned and flushed three times with 5% CO_2/95% N_2 gas in order to reduce the oxygen content within the chamber to less than 1%. The oxygen tension within the culture medium was initially 120 mmHg and it declined exponentially during the hypoxic period to about 40 mmHg in two hours and 25 mmHg in three hours (Kinoshita et al., 1990). Three hours' hypoxic exposure resulted in loss of the majority of neurons within 24 hours, whereas exposure for 2 hours resulted in much less neuronal death. Therefore, we termed this hypoxic stress for two hours as subcritical hypoxia and used for subsequent experiments.

Drugs to be tested were added immediately after the hypoxic period and the culture plates were returned to an incubator (37°C, 5% CO_2/ 95% air with saturated humidity). Surviving neurons were counted 24 hours thereafter under a phase-contrast microscope with the aid of the morphological criteria described previously (Kohmura et al., 1990). Ten microscopic fields (under X100 magnification) were counted for each well and their mean was regarded as the representative value for the well. Six wells were used for statistical analysis (Student t-test). Neuronal survival rate was calculated as a percentage of neurons surviving in corresponding control cultures. All the experiments were repeated twice with similar results. Representative data are presented in this paper (mean±SD, N=6).

We tested as an agonist glutamate, N-methyl-D-aspartate (NMDA; Sigma), α-amino-3-hydroxy-5-methyl-4-isoxazole
propionic acid (AMPA; Cambridge Research Biochemicals), and combination of NMDA and AMPA. As an antagonist CNQX
(Cambridge Research Biochemicals) for the AMPA receptor and MK-801 (Research Biochemicals Incorporated) for the NMDA receptor. Further we tested the effect of nifedipine (Research Biochemicals Incorporated), which is an ionchannel-blocker for the voltage dependent Ca-channel.

RESULTS

Under normal culture conditions, one mM of glutamate resulted in severe neuronal loss, while less concentration of glutamate caused no neuronal toxicity. On the other hand, when glutamate was applied to neurons after two hours' hypoxic stress, less amount of glutamate showed neurotoxicity. In this case, the neuronal survival rate was reduced by, respectively, 10 and 100μM of glutamate to 80.2±7.3 and 56.8±6.7% of control culture exposed to hypoxia but not exposed to glutamate (p<0.01) (Kohmura et al., 1990).

In order to evaluate the mechanism involved, we gave NMDA, AMPA, or both of them instead of glutamate to neuronal culture after the hypoxic stress. We could not observe any augmentation of NMDA toxicity after subcritical hypoxia (Kohmura et al., 1990). One mM of NMDA was necessary to cause significant neuronal loss regardless of hypoxia. Ten μM of AMPA alone, and 10 or 100 μM of NMDA alone was not toxic. However, when we gave them together, they did cause neuronal loss (Fig. 1).

Nifedipine blocks voltage dependent Ca channel. Nifedipine did not reduce neuronal injury caused by glutamate after the subcritical hypoxia in our system. As a next experiment, we gave glutamate and MK-801 after the hypoxic stress. MK-801 (2μM), which could completely prevent neuronal loss caused by 1mM of NMDA, inhibited the neuronal loss caused by 10μM of glutamate sufficiently. And it improved the neuronal survival against 100μM of glutamate to 75.4±16.3% but it was still less than control (Kohmura et al., 1990). These results indicated that receptors other than the NMDA receptor should be involved in the glutamate neurotoxicity.

From the agonist study, importance of the AMPA receptor was suggested. Therefore, we gave CNQX together with glutamate after the hypoxic stress. The neurotoxicity, which would be caused by 10μM of

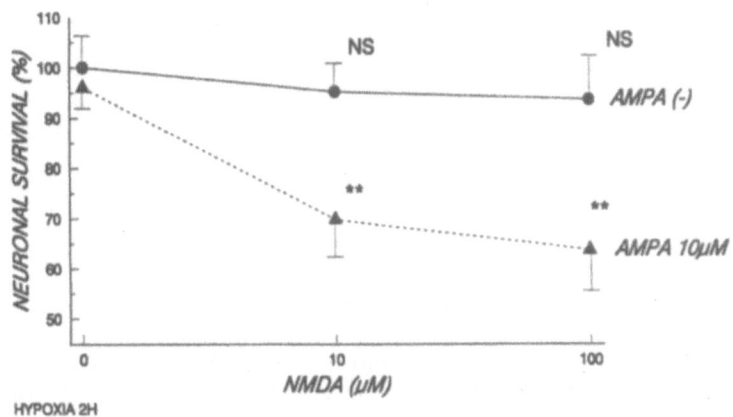

Fig. 1 Effect of NMDA and AMPA. 10 or 100μM of NMDA or 10μM of AMPA caused no neuronal injury by itself. However, when they were given together, they could cause neuronal injury. (**: p<0.01 compared to control)

glutamate, could be completely prevented by giving 0.5μM of CNQX. And higher concentration of CNQX (5μM) could sufficiently suppress the neuronal loss which would be caused by 10 or 100μM of glutamate. At this setting, neuronal survival was 94.2±6.1% (p>0.5, compared to control without glutamate but with hypoxia) (Kohmura et al., 1991).

We gave CNQX with a time-delay from glutamate administration. Till 3 hours of time delay, 5μM of CNQX could reduce neuronal injury which would be caused by 100μM of glutamate. After 5 hours of time delay, however, it was no more effective (Fig. 2). These experiments showed that the AMPA receptor should play an important role in glutamate neurotoxicity after hypoxia presumably in its initiation phase and that CNQX is more effective than MK-801.

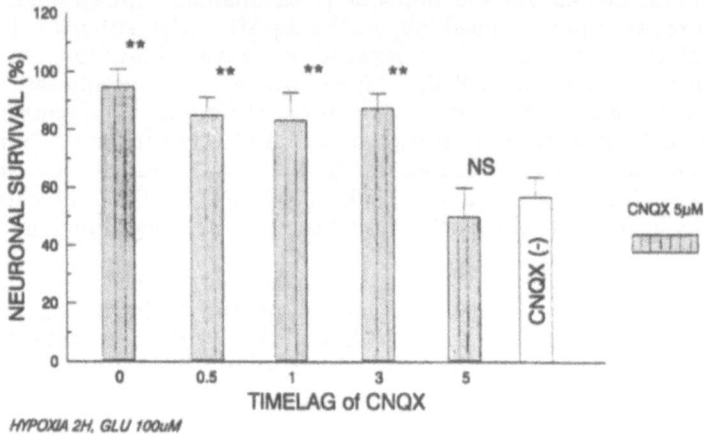

Fig. 2　Effect of delayed administration of
　　　　CNQX (5μM) on glutamate neurotoxicity
　　　　of 100μM after two hours' hypoxia.
　　　　(**: p<0.01 compared to CNQX(-))

DISCUSSION

　　We found that the threshold at which glutamate causes neuronal death is reduced after subcritical hypoxia. This fact implies that neurons, which survived the hypoxic or ischemic period, can be thereafter injured by continuous exposure to low amount of glutamate. Delayed neuronal loss observed in vivo experiments might be caused by such mechanism. Usual synaptic activity after the ischemia and recirculation period could provide enough circumstance to induce neuronal injury.

　　Glutamate is a well-known mixed agonist and activates various subtypes of glutamate receptors. Among these receptors, the role of NMDA receptor has been extensively studied, because it coupled with a Ca^{2+} channel. Our in vitro study showed that activation of the NMDA receptor alone cannot explain the augmentation of the glutamate neurotoxicity after the hypoxic stress. Increase of neurotoxicity could be observed, when both the NMDA and AMPA receptors are stimulated. From the study with specific antagonists we found that the blockade of the AMPA receptor could prevent the increased neuronal injury from glutamate after the hypoxia. Moreover, CNQX was effective even with delayed administration till three hours of time-lag, which suggests that neuronal injury occurs gradually in continuous presence of glutamate and that there exists possible therapeutic window.

　　In physiological condition, transsynaptically released glutamate activates the AMPA receptor and cause membrane depolarization, i.e. fast synaptic transmission, which is followed by repolarization by extruding Na^+ accompanied by ATP consumption. Glutamate can also act on the NMDA receptor, but the Ca^{2+} channel, which is coupled with the NMDA receptor, is normally controlled by Mg^{2+} so as not to allow massive Ca^{2+} influx. Neurons are excited by glutamate but neuronal toxicity cannot be expressed under these circumstances. After subcritical hypoxia, however, cellular energy level is reduced and membrane depolarization might persist with continuous stimulation of the AMPA receptor. Blockage of the NMDA-gated Ca^{2+} channel by Mg^{2+} is known to depend on the membrane potential, and membrane depolarization removes such an effect of Mg^{2+} (Mayer et al., 1984, Nowak et al., 1984). Therefore, Ca^{2+} can enter into the neuron through the NMDA receptor-gated channel and initiate a vicious cycle, probably with activation of protein kinase C, which can be also activated via the metabotropic glutamate receptor, followed by cellular death (Siesjö and Bengtsson 1989). Neurons might die from a higher dose of glutamate due to osmolysis caused by massive influx of Na^+.

Although our results were obtained from in vitro experiments using fetal neurons, in vivo tissue exposed to a brief, dense, transient ischemia might be damaged by low dose of glutamate with the similar mechanism. This neurotoxicity could be closely correlated with both the NMDA and non-NMDA subtypes of glutamate receptors. Sheardown et al. (1990) reported recently that NBQX, another potent inhibitor of the AMPA receptor, can protect CA1 neurons from delayed neuronal loss after 5 minutes of bilateral carotid occlusion in gerbils. The protective effect could be observed even when NBQX was given two hours after ischemia. Their results are agree with ours obtained from in vitro study. Both studies suggested the therapeutic window for the AMPA-antagonist. Prevention of glutamate release or blockage of glutamate receptors might be rational for a therapeutic approach.

REFERENCES

Benveniste H, Drejer J, Schousboe A, Diemer NH (1984) Elevation of the extracellular concentrations of glutamate and aspartate in rat hippocampus during transient cerebral ischemia monitored by intracerebral microdialysis. J Neurochem 43:1369-1374

Hagberg H, Lehmann A, Sandberg M, Nyström B, Jacobson I, Hamberger A (1985) Ischemia-induced shift of inhibitory and excitatory amino acids from intra- to extracellular compartments. J Cereb Blood Flow Metab 5:413-419

Kinoshita A, Yamada K, Hayakawa T (1990) Hypoxic injury of rat cortical neurons in primary cell cultures. Exp Cell Biol 57:310-314

Kohmura E, Yamada K, Hayakawa T, Kinoshita A, Matsumoto K, Mogami H (1990) Hippocampal neurons become more vulnerable tp glutamate after subcritical hypoxia: an in vitro study. J Cereb Blood Flow Metab 10:877-884

Mayer ML, Westbrook GL, Guthrie PB (1984) Voltage-dependent block by Mg^{2+} of NMDA responses in spinal cord neurons. Nature 309:261-263

Nowak L, Bregestovski P, Ascher P, Herbet A, Prochiantz A (1984) Magnesium gates glutamate-activated channels in mouse central neurons. Nature 307:452-465

Sheardown MJ, Nielsen EØ, Hansen AJ, Jacobsen P, Honoré T (1990) 2,3-Dihydroxy-6-nitro-7-sulfamoyl-benzo (F) quinoxaline: a neuroprotectant for cerebral ischemia. Science 247:571-574

Siesjö BK, Bengtsson F (1989) Calcium fluxes, calcium antagonists, and calcium-related pathology in brain ischemia, hypoglycemia, and spreading depression: a unifying hypothesis. <u>J Cereb Blood Flow Metab</u> 9:127–140

PHARMACOLOGICAL EFFECTS OF REMACEMIDE AND MK-801 ON MEMORY AND HIPPOCAMPAL

CA1 DAMAGE IN THE RAT FOUR-VESSEL OCCLUSION (4-VO) MODEL OF GLOBAL ISCHEMIA

J.M. Ordy, B. Volpe, R. Murray, G. Thomas, P. Bialobok, T.M. Wengenack, and W. Dunlap

Fisons Pharm., Rochester, NY 14623, Cornell Med. Coll., NY, NY 10605, Univ. Roch., Rochester, NY 14642, and Tulane Univ., New Orleans, LA 70118

Rationale for Pharmacotherapy of Remacemide in 4-VO Model: Studies have reported memory impairment (25) and a selective and delayed onset of CA1 pyramidal cell necrosis in man after global ischemia (18). Memory tests demonstrated anterograde amnesia in a patient (R.B.) who developed circumscribed memory impairment following an ischemic episode. Prior to death, R.B. showed no signs of cognitive impairments other than memory loss. Histological evaluations revealed selective and bilateral CA1 pyramidal cell loss (28). The 4-VO model is widely used for studying the phenomenon of "selectively vulnerable, delayed onset" pathophysiology of hippocampal CA1 neurons in global ischemia (19,20). Post-ischemic pharmacological treatments may modify the functional and morphological outcome of global ischemia in the hippocampus, involving ischemically compromised but possibly still "viable" CA1 neurons (20). There is credible evidence for a critical role of the hippocampus in a specific type of spatial, representational, or episodic memory (23,24,28). Circumscribed memory impairment with selective CA1 cell necrosis has been reported in the 4-VO model (15,26). Protection of hippocampal CA1 cell damage may represent a valid "neural endpoint", or target for drug intervention (20). However, demonstration of functional drug efficacy, particularly amelioration of memory loss produced by global ischemia involving hippocampal CA1 damage, may represent a clinically more relevant measure of therapeutic outcome (4,15,25). The bilateral organization of memory circuits and pyramidal cell redundancy of hippocampal subfields may complicate detection of significant correlations among degree of memory impairment, CA1 cell damage, and drug effects (15,24). Combined evaluations of drug efficacy may be necessary, encompassing 2 "neural endpoints" of memory loss and CA1 damage, distinct characteristics of global ischemia in man and the 4-VO model.

The role of neurotransmitters in the pathophysiology of ischemic hippocampal damage has become a major focus for pharmacotherapy (1,4,9,20). A significant post-ischemic imbalance between excitation and inhibition, disturbed transmitter release, re-uptake, and changes in post-synaptic receptors and intracellular messengers may all represent molecular mechanisms of selectively vulnerable CA1 neurons, and thus targets for drug intervention (1,4,9). According to the "excitotoxic hypothesis", glutamate overstimulation of NMDA receptors during ischemia may allow a lethal influx of calcium through ion channels linked to the NMDA receptors resulting in

CA1 cell death (12). The CA1 subfield contains high concentrations of NMDA receptors (6). Studies have reported that the non-competitive NMDA recepto antagonist dizocipiline maleate (MK-801) may protect vulnerable neurons against ischemia, and that this represents a new drug therapy for stroke, one not dependent upon modification of cerebral blood flow (4). Hippocampa NMDA receptors have also been implicated in synaptic plasticity, or the induction of long-term potentiation (LTP), a proposed mechanism of learning and memory (5,13).

Specific aims of this drug development program were to evaluate the neuroprotective efficacy of remacemide on: 1) memory of 4-VO memory-impaired rats in a T-maze and 2) post-4-VO CA1 pyramidal neurons by histological evaluation of CA1 cell necrosis, and number of "viable" CA1 neurons, in topographically comparable locations of the CA1 subfield. Remacemide, (±) 2-amino-N-(1-methyl-1,2-diphenylethyl)-acetamide, is an anticonvulsant currently undergoing phase II clinical trials. A preclinica pharmacological profile of remacemide as an anticonvulsant has been published (17). Remacemide is a weak NMDA receptor antagonist. Its primar desglycinate metabolite, FPL 12495, has been found to exhibit potent inhibition of [^{3}H] MK-801 binding. Ki (μM) values of remacemide, FPL 12495 and MK-801 were 70, 0.45, 0.01, respectively. NMDA receptor-mediated neuroprotective effects of MK-801 have been reported in many *in vitro* studies, and in some global and focal models of ischemia (4,12). Non-competitive NMDA antagonists, including MK-801, however, may block LTP and impair memory (13), interact with the phencyclidine (PCP) binding site in the NMDA receptor/channel, and produce disturbing PCP-like behavioral signs (27). They may also produce other, possibly non-NMDA receptor-mediated effects on temperature, plasma glucose, and glucocorticoids (4,8). Comparisons were made of remacemide and MK-801 in normal, non-4-VO rats on: 1) memory, 2) PCP-like signs, 3) stereotypic behavior, and 4) ataxia. Comparisons were also made on: 5) body temperature, 6) plasma glucose, and 7) glucocorticoids, variables known to modulate the severity of ischemic CA neuronal damage (8).

Subjects, Methods, Data Analyses: Sprague-Dawley male rats, weighing 250-300g were used as subjects. Transient, 15 or 30 min of global, reversible, bilateral forebrain ischemia was produced by the 4-VO model (19), modified for producing bilaterally symmetrical hippocampal pyramidal cell damage, restricted to CA1 (15 min), or to CA1-CA3 (30 min) subfields (16). Body temperature of 4-VO rats was monitored and maintained at 37.5°C during ischemia and reperfusion. In the 4-VO model, advantages for drug studies also include absence of confounding effects of anesthesia, low incidence of seizures, rapid post-ischemic recovery, and awake, freely moving rats, available for neurological testing. Spatial memory testing was conducted i a T-maze, involving a discrete trial, paired-run, contingently reinforced alternation procedure (15,24). Briefly, in each pair of runs, rewarded, forced information runs provided left-right spatial information in the 2 arms of the T-maze. Next, after memory delays of 10, 90, or 180 sec, win-shift, or alternation, choice runs to the opposite arm location were rein-forced and served as measure of spatial memory. The test consisted of 12 pairs of trials per day across 3 days, totaling 12 trials for each of the 3 memory delays, or interrun intervals (IRIs). On each day, the 3 IRIs were randomized across 12 trials. The intertrial interval (ITI) was 30 sec. Th number of correct choice, win-shift, or alternation responses at each IRI memory delay, out of 12 trials, across 3 days, were a functional measure of spatial memory. Start, choice, and goal latencies, converted to speeds by reciprocals, served as measures of motivation and locomotor performance (15). For histological evaluations, the rats were overdosed with sodium pentobarbital and transcardially perfused with heparinized saline, followed by FAM (40% formaldehyde, glacial acetic acid, methanol; 1:1:8). Coronal sections, 6μm were cut from paraffin-embedded brain tissue, and every 15th section from -3.2 to -3.8mm Bregma (16) was stained with H&E. Because of possible anterior-to-mid-dorsal gradients of CA1 damage, and to determine more exactly the topographic location of CA1 neurons examined, the external skull marks of Bregma were supplemented by internal CA1 landmarks (16). CA1 cell layer damage was examined by: 1) a "tracing" method, 2) a standard neuropathological rating scale for cell necrosis, and 3) counting the number of remaining "viable" CA1 neurons, containing a nucleus and nucleolus (16). Quantitative data were subjected to analysis of variance (ANOVA) with BMDP2V, followed by Newman-Keuls' (N-K) tests for significance of differences among specific group means (15).

Results. Some effects of 4-VO ischemia on rat memory and CA1 damage have been published (15,26). A specific aim of this report was to demonstrate a correlation of memory loss and CA1 cell necrosis in the 4-VO model, with the 2 neural endpoints serving as therapeutic targets for remacemide.

Effects of 4-VO ischemia on memory and performance in the T-maze. In one study, the effects of 30 min of 4-VO ischemia on memory were compared between 8 sham surgery controls and eight 4-VO rats in the T-maze (15). In the pre-4-VO tests, no significant differences in memory or performance were found between the 2 groups subsequently assigned to surgery control or 4-VO ischemia. A mixed model, between and within subject, ANOVA of the 1-2 months post-4-VO effects on memory relative to the pre-4-VO baseline asymptotic memory scores indicated a significant impairment of memory in the 4-VO rats compared to controls [group x pre-post interaction: $F(1,15)=11.12$; $p \le 0.01$]. Three similar ANOVAs of pre-post 4-VO effects on the 3 performance measures indicated no significant effects on start, choice, or goal speeds. Figure 1 illustrates statistical comparisons of the significant pre-post 4-VO effects on memory between control and 4-VO rats at 10, 90, 180 sec IRI memory delays. Other studies have shown there was no recovery of the significant 4-VO memory impairment from 1 to 4 months post-ischemia (15). In histological evaluations, a significant bilateral CA1 cell necrosis was observed in the 4-VO rats tested for memory impairment from 1 to 2 months after ischemia, compared to controls [$F(1,27)=219.52$; $p \le 0.001$]. There was also a significant correlation between degree of 4-VO memory impairment and severity of CA1 damage (15).

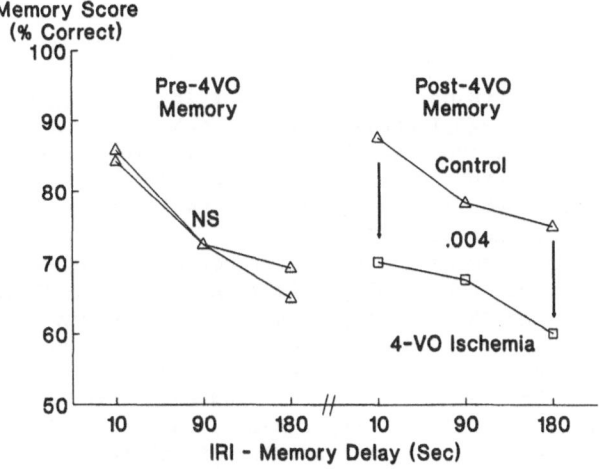

Fig. 1. 4-VO Impairment of Memory in T-Maze

Remacemide effects on memory of 4-VO rats in the T-maze. In this drug study, a group of rats was first trained to asymptotic memory, prior to being assigned to 4-VO saline or 4-VO remacemide groups. After 30 min of 4 VO ischemia, remacemide, at 20 mg/kg, i.p., or saline were administered 1 hour post-4-VO, and then once daily for 21 days, 1 hour after daily memory testing. Prior to 4-VO ischemia, there were no significant differences among the rats in memory or performance. According to the overall ANOVA, the treatment (saline-remacemide) effects were not significant. However, the pre-post by treatments interaction was highly significant [$F(1,5)=13.42$ $p \leq 0.01$]. This significant interaction clearly demonstrated that the memory scores of 4-VO remacemide-treated rats were comparable pre- and post-4-VO ischemia, whereas those of 4-VO saline rats were significantly impaired after ischemia. The neuroprotective effects of remacemide on memory in the pre-post 4-VO drug design are illustrated graphically in Fig. 2. In terms of selectivity of remacemide effects on memory independent of confounding interaction effects on performance, 3 ANOVAs indicated that there were no significant remacemide effects on start, choice, and goal speeds. Histological evaluations indicated a significant reduction of CA1 cell necrosis in 4-VO remacemide-treated rats, compared to those given saline, tested in the T-maze [$F(1,5)=10.14$; $p \leq 0.02$].

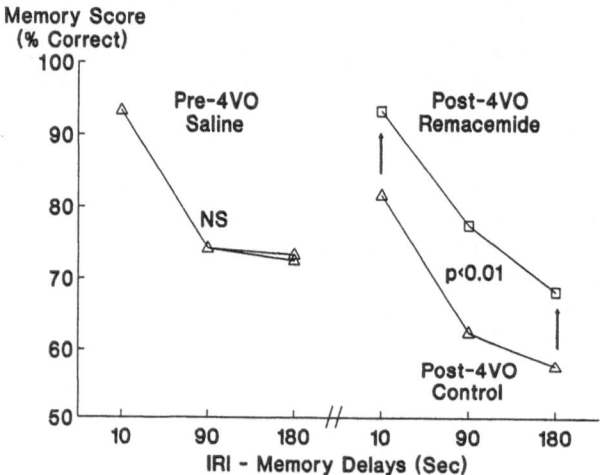

Fig. 2. Remacemide Efficacy on Memory in 4-VO Rats

Remacemide Effects on CA1 Damage After 15 Min. of Temperature-Controlled (37.5°C) 4-VO Ischemia. The effects of remacemide or saline were evaluated histologically on CA1 cell damage in nine 4-VO saline and ten 4-V remacemide-treated rats. Remacemide, at 20 mg/kg, i.p., or saline, 1 ml/kg i.p., were administered 1 hr after 4-VO ischemia, then once daily for 14 days. The significance of differences in % CA1 damage between the 4-VO saline and 4-VO remacemide-treated rats, between right and left CA1, and 8 sections per rat, was analyzed by a 2x2x8 ANOVA design (2 treatments x 2 L/ x 8 sect.). Percent CA1 cell layer damage was determined by length of damaged CA1 cell layer divided by the total length of CA1 x 100. The CA1 cell layer % damage values were first subjected to arc sine transformations for the ANOVA. The 4-VO rats that received remacemide had significantly less % left and right CA1 cell layer damage than the 4-VO saline rats [$F(1,17)=6.12$; $p \leq 0.02$]. The advisability of selecting 8 sections from each of the 19 4-VO rats from topographically exact dorsal CA1 locations from -3.2 to -3.8mm Bregma was apparent because there was a significant gradient

of CA1 cell layer damage across the 8 sections, [F(7,119)=23.5; p≤0.001].
However, the drug-saline treatments by 8 sections interaction was not
significant. The lack of a significant interaction indicated uniform
neuroprotection by remacemide across the selected sections of the CA1
subfield. Significant remacemide-saline differences across the 8 sections
of the dorsal CA1 subfield are illustrated graphically in Fig. 3.

Remacemide Effects on Body Temperature After 15 Min. Temperature-Controlled (37.5°C) 4-VO Ischemia. In the histological study, body (rectal)
temperature was not only maintained at 37.5°C during 15 min of ischemia and
15 min of reperfusion by a "feedback" heating unit, but also recorded: 1)
before ischemia, 2) after ischemia, 3) at the end of reperfusion, 4) one
hour post-ischemia, prior to saline or remacemide treatment, and 5) at the
end of 14 days of post-4-VO saline and remacemide treatments. A one-way
significant ANOVA [F(1,119)=18.28; p≤0.001], followed by N-K tests across
the 4 ischemia/reperfusion time intervals indicated a very small, but

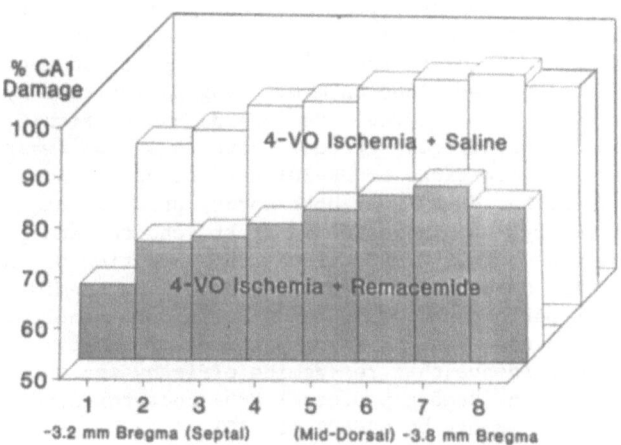

Fig. 3. Remacemide Efficacy on CA1 Damage in 4-VO Rats

statistically significant (0.8°C) increase from the end of ischemia to the
end of reperfusion (N-K: p≤0.01), and a similar decrease from the end of
reperfusion to recovery (N-K: p≤0.001). According to an overall ANOVA of
differences in body temperature between daily saline and remacemide treated
rats across 14 days, there were no significant remacemide effects on body
temperature. Studies have shown that reduction of brain temperature by
cooling (2.0°C) was necessary to show a significant reduction of CA1 cell
damage in the 4-VO model (3). Early reperfusion hypothermia by 2.0°C
protected CA1 neurons. When this hypothermia was delayed by 5 min after the
onset of reperfusion, no protective effects by hypothermia on CA1 neurons
could be demonstrated (3). Although there may be a dissociation between
brain and body temperature during reperfusion (3), because body temperature
was maintained at 37.5°C during ischemia and reperfusion, and remacemide was
administered 1 hr after the start of reperfusion, it may be concluded that
the neuroprotective effects of remacemide on CA1 neurons was not caused by
hypothermia in the rat 4-VO model of global ischemia.

Comparisons of Remacemide and MK-801 CNS Side Effects in Normal, Non-4-
VO Rats. From a clinical efficacy-safety standpoint, it seems remarkable

that, compared to the explosive number of *in vitro* and *in vivo* efficacy
studies reported with MK-801, the number of *in vivo* safety studies on CNS
side effects, or therapeutic safety margins has been surprisingly small (1)
Regardless of whether the putative neuroprotective effects of MK-801 may be
associated with antagonism of hippocampal NMDA receptors, non-competitive
NMDA antagonists, including MK-801, may block LTP and impair memory (13),
interact with PCP binding sites and produce disturbing psychotomimetic sign:
(27), as well as produce other adverse, possibly non-NMDA receptor mediated
effects on temperature, plasma glucose, and glucocorticoids, known to
modulate the severity of ischemic CA1 neuronal damage (8). The specific
aims of the CNS side effect studies were to compare the effects of
remacemide and MK-801 on: 1) memory and performance in the T-maze, 2) PCP-
like signs, 3) stereotypic behavior, 4) hyperactivity, 5) ataxia, 6) body
temperature, 7) plasma glucose, and 8) corticosterone. The therapeutic use
of such NMDA antagonists as MK-801, or any other NMDA antagonist proposed
for global ischemia would be "troublesome" if it blocked NMDA receptors, th
induction of LTP, and impaired memory in patients in whom ischemia had
already produced memory impairment and damaged CA1 neurons (1,4). The
effects of remacemide, at 10 and 20 mg/kg, i.p., and of MK-801, at 0.1 and
0.2 mg/kg, i.p., were compared on memory and performance in the T-maze, in
normal, non-4-VO rats. Each rat received randomly, over a 9 day schedule,
either saline or a dose of remacemide, 0.5 hrs prior to testing. The 3
treatment effects were examined on memory and performance by 4 within-
subject, repeated measure ANOVAs. Remacemide had no significant effects on
memory, start, choice, or goal speeds. After a 2 week "washout", the same
rats were used in evaluations of MK-801 effects on memory and performance.
According to 4 ANOVAs, MK-801, at both doses, produced significant
impairment of memory, and increases in start, choice, and goal speeds. It
remains unresolved whether MK-801 blocks the induction of NMDA-dependent LT
in vivo (11). The comparisons demonstrated that remacemide did not impair
memory or affect performance, whereas MK-801 significantly affected both
memory and performance. The ANOVA comparisons of remacemide and MK-801
provide credible evidence that remacemide efficacy and CNS side effects may
not be mediated by the same hippocampal NMDA receptor sites proposed for
blockade of LTP and memory by MK-801 (11,13).

There is considerable research on whether: 1) competitive and non-
competitive NMDA receptor antagonists represent PCP-like compounds, 2) these
effects are mediated through similar NMDA receptors, and 3) adverse CNS side
effects may seriously complicate or preclude therapeutic development of NMD,
antagonists for epilepsy, ischemia, and other neurodegenerative disorders
(27). PCP is known to produce psychotomimetic effects in man (27). A PCP-
like profile of head weaving, circling, hyperactivity, stereotypy, body
rolling, and ataxia has been reported for the rat (27). A PCP-like
behavioral rating scale was used to compare the effects of saline, 10, 20,
and 40 mg/kg of remacemide, and 0.1, 1.0, and 5.0 mg/kg of MK-801.
Experimental designs for each drug included 1 saline and 3 drug groups with
10 rats per group. Chi-square (X^2) analyses were used to test the
significance of differences in frequency and severity of PCP-like signs
among the 4 groups, from 15 to 30 min after treatment. X^2 analyses indicate
that remacemide produced no significant PCP-like effects at any dose. Ther
was some motor "atonia", or forelimb flaccidity at the 20 and 40 mg/kg
doses. In contrast to the lack of effects on PCP-like signs by remacemide,
MK-801 produced PCP-like signs and dose-dependent increases in stereotypy,
hyperactivity, and ataxia ($X^2=20.2$; df=1; $p \leq 0.001$), in the Opto-Varimex
Activity Monitor (see Table 1). Because of selective PCP binding at non-
competitive binding sites of the NMDA receptor complex, some non-competitiv
NMDA antagonists, including MK-801, have been characterized as PCP-like
compounds (27). Similar to the significant differences between remacemide
and MK-801 on memory and performance, the 2 drugs also differed
significantly in their effects on PCP-like behavior.

Table 1. Comparisons of Remacemide and MK-801 CNS Side Effects[1]

Tests	Remacemide (10, 20, 40)[2]	MK-801 (0.1, 0.2, 1.0, 5.0)[2]
A. Memory - Performance		
1. Memory, T-Maze	NS[3]	0.1, 0.2 - Impairment
2. Performance, T-Maze	10 - NS; 20 - Dec. Speed	0.1, 0.2 - Increased Speed
B. PCP-like Behavioral Profile		
3. PCP-like signs	NS	0.1 - NS; 1, 5 - PCP Signs
C. Opto-Varimex (O-V) Activity: Spontaneous, open-field exploration		
4. Distance Traversed	NS	0.1 - NS; 1, 5 - Ataxia
5. Stereotypic Time	NS	0.1 - NS; 1, 5 - Increased
D. Physiological Variables: Acute-single dose; Chronic-5 days		
1. Body Temp.- Acute	10, 20 - NS; 40 - Dec.	0.1, 1 - NS; 5 - Dec/Inc
2. Body Temp.- Chronic	10, 20 - NS; 40 - Dec.	0.1 - NS; 1, 5 - Increase
3. Plasma Glucose - Acute	NS	0.2 - NS; 1 - Increase
4. Plasma Glucose - Chronic	NS	0.1 - NS; 1, 5 - Increase
5. Corticosterone - Acute	NS	0.2 - NS; 1 - Increase

1) Variables analyzed by ANOVAs, followed by Newman-Keuls (N-K).
2) All drug doses - mg/kg, i.p. n=10 per group.
3) NS - not significantly different from saline controls. Significant N-K, decreases (Dec) or increases (Inc) were at $p \leq 0.01$ from controls.

Acute and chronic studies were also performed to compare the effects of remacemide and MK-801 on body temperature, plasma glucose, and corticosterone. At 10 and 20 mg/kg, remacemide produced no significant effects on temperature, plasma glucose, and corticosterone. The 0.1 and 1.0 mg/kg doses of MK-801 had no significant acute effects on body temperature. The 5.0 mg/kg dose produced a significant biphasic decrease (N-K: $p \leq 0.01$) followed by a significant increase in body temperature (N-K: $p \leq 0.01$). There was also a significant increase in body temperature after the 1.0 and 5.0 mg/kg doses in the 5 day chronic study (N-K: $p \leq 0.01$). The 5.0 mg/kg dose significantly increased plasma glucose in the chronic study (N-K: $p \leq 0.01$). MK-801, at 1.0 but not 0.2 mg/kg, also produced significant increases in plasma corticosterone (N-K: $p \leq 0.001$). Hyperthermia, and increases in plasma glucose and corticosterone are known to exacerbate CA1 damage in ischemia (8). Remacemide had no effects on these variables. MK-801 produced significant increases in all 3 variables and may exacerbate CA1 damage in global ischemia. Table 1 provides a summary of significant differences between remacemide and MK-801 on 10 different variables in which the effects of the 2 drugs appear to have acted on different sites in the NMDA receptor complex, or may have been mediated through different non-NMDA receptors.

Discussion. In view of the neuroprotective efficacy of remacemide on memory and CA1 neurons in the 4-VO model, it seems appropriate to make some brief comments concerning: 1) validity of the rationale for using combined evaluations of 2 neural endpoints for studies of the pathophysiology and pharmacotherapy in global ischemia, 2) putative neural mechanisms for remacemide efficacy and lack of NMDA receptor-mediated CNS side effects in context of current evidence for the glutamate "excitotoxic hypothesis" as the dominant strategy in clarification of neural mechanisms that may be involved in the pathophysiology and pharmacotherapy of global ischemia, 3) relations among duration of 4-VO ischemia, degree of memory loss, severity of CA1 damage, and the duration of a "therapeutic window" during which CA1 neurons may be ischemically compromised but still remain viable for drug treatment, and 4) advisability, and/or necessity of conducting concomitant efficacy-safety studies with neuroprotective drugs in appropriate *in vivo* models to help identify possible challenges in studies of therapeutic safety margins. Further research is needed to establish more fully whether circumscribed memory loss and selective CA1 cell damage represent distinct and predictable characteristics of global ischemia in man and in other animal models. Recently, selective loss of CA1 neurons, with minimal loss from cerebral cortex and thalamus have been reported in patients after cardiac arrest ischemia (21). An estimated 20% of cardiac arrest ischemia patients may suffer some cognitive impairments (25). Thus far, protection of CA1 cells has been used as a valid endpoint, or target for drug intervention, including evaluation of MK-801 in the 4-VO model (20). While this specific cellular endpoint may represent a focused and rapid approach for new drug treatment strategies, the ultimate clinical goal of drug therapy involves functional improvements, or benefits (4,25). From a neuroscience perspective, complex post-ischemic changes in bilateral cellular organization in hippocampal memory circuits, cell redundancy in hippocampal subfields, disturbed synaptic plasticity, or induction of LTP in learning and memory may seriously complicate predictions of functional drug efficacy from drug modification of hippocampal cellular morphology (15,24). The neuroprotective efficacy of remacemide on memory and CA1 cell damage in the 4-VO model as summarized in this report may represent a valid extension in the use of the 4-VO model from both clinical and neuroscience perspectives.

A role for glutamate in neurotoxicity of CA1 neurons in global ischemia appears convincing (7). Recent studies with MK-801, however, have shown that competitive and non-competitive NMDA receptors may play little or no role in the pathophysiology of CA1 neurons (20). In contrast to numerous previously reported positive studies, more recent temperature-controlled studies with gerbils (20), the 2-VO (14), and 4-VO models, have shown that non-specific, uncontrolled hypothermia effects of MK-801, rather than specific NMDA receptor-mediated effects, produced the previously reported protective effects on CA1 neurons (1,20). NMDA receptors have been studied most widely in ischemia. However, glutamate may interact with 3 subtypes of receptors: NMDA, quisqualate/AMPA, and kainate (KA) in global ischemia (22). Until recently, there were no specific antagonists to the AMPA and KA receptors for glutamate which were known to cross the blood-brain barrier. Recently, a novel, non-NMDA antagonist, a quinoxaline dione (NBQX), with no effects at NMDA or glycine sites, has been reported to reduce ischemia-induced CA1 cell damage in the gerbil (20), the 2-VO (14), and 4-VO models, independent of temperature effects (2). Specific molecular mechanisms whereby NBQX may have reduced CA1 cell damage in global ischemia remain to be clarified. Interestingly, the highest concentration of AMPA sites are localized in CA1 (6). Preliminary studies have shown that NBQX may inhibit Schaffer-collateral CA3 neurotransmission via post-synaptic blockade of AMPA or KA receptors (7). It is still widely believed that Ca++ permeability of non-NMDA receptors is low or absent (4). Recent studies, however, have reported that specific combinations of KA/AMPA receptor subunits, expressed in oocytes, were permeable to Ca++ (10). These observations suggest that NBQX blockade of AMPA receptors may represent a possible neuroprotective mechanism of CA1 cells in global ischemia (22). Thus far, however, *in vivo* studies of effects on memory, and on CNS side effects, or therapeutic safety margins have not been reported for NBQX (7,22).

Regarding putative neural mechanisms for the neuroprotective efficacy of remacemide on memory and CA1 cell damage in the 4-VO model, biochemical evaluations of remacemide and its desglycine metabolite, FPL 12495, have shown that remacemide is a weak NMDA receptor antagonist (Ki=70μM), whereas the values for FPL 12495 were 0.45μM, compared to 0.01μM for MK-801. Binding affinities were measured in rat cortical and hippocampal tissues in the presence of 10μM each of glutamate and glycine. *In vitro* electrophysiological studies with rat hippocampal slices, and receptor binding studies of remacemide at AMPA and KA sites may help determine to what extent, if any, remacemide may interact with different subtypes of glutamate receptors. Regarding the duration of a "therapeutic window" for remacemide in the 4-VO model, there is evidence on selective vulnerability, protracted delay in onset (24 hrs), and progression of post-ischemic CA1 cell necrosis from 24 to 72 hours (20). More research is needed on duration of 4-VO ischemia, degree of memory loss, severity of CA1 cell damage, and duration of a therapeutic window, during which CA1 neurons may remain viable and amenable to drug intervention (16). Remacemide was administered 1 hour after reperfusion, and once daily for 7, 14, or 21 days. Optimum post-ischemia dosing for remacemide remains to be determined.

Studies of the pharmacology of non-competitive NMDA receptor antagonists have suggested sufficient concern as to whether effects on LTP and memory (13), PCP-like side effects (27), and effects on temperature, glucose, and glucocorticoids (9) may seriously complicate, or even preclude therapeutic development of NMDA receptor antagonists (1,4). In studies summarized in this report, direct comparisons of remacemide with MK-801 on CNS side effects clearly indicated that the effects of remacemide differed significantly from those of MK-801 on memory and performance in the T-maze, PCP-like signs, stereotypic behavior, hyperactivity, ataxia, body temperature, plasma glucose, and corticosterone. There is currently no approved neuroprotective reference drug for ischemia. Some CNS and other side effects may be tolerable in hospital settings, if treatment duration is brief, and if therapeutic benefits outweigh adverse side effects (4). As with the demonstrated use of the 2 neural efficacy endpoints for remacemide in the 4-VO model, the concomitant efficacy-safety comparisons of remacemide with MK-801 present credible extensions in the use of the 4-VO model from both clinical and neuroscience perspectives.

References

1. Buchan, A.M. (1990) Do NMDA antagonists protect against cerebral ischemia: are clinical trials warranted? Cerebrovasc. Brain Metab. Rev. 2:1-26.
2. Buchan, A.M., Li, H., Cho, S., and Pulsinelli, W.A. (1991) Blockade of the AMPA receptor prevents CA1 hippocampal injury following severe but transient forebrain ischemia in adult rats. Submitted: J. Neurosci.
3. Busto, R., Dietrich, W.D., Globus, M.Y.-T., and Ginsberg, M.D. (1989) The importance of brain temperature of cerebral ischemic injury. Stroke. 20:1113-1114.
4. Choi, D.W. (1990) Cerebral hypoxia: some new approaches and unanswered questions. J. Neurosci. 10:2493-2501.
5. Cotman, C.W., and Monaghan, D.T. (1988) Excitatory amino acid neurotransmission: NMDA receptors and Hebb-type synaptic plasticity. Ann. Rev. Neurosci. 11:61-80.
6. Cotman, C.W., Monaghan, D.T., Ottersen, O.P., and Storm-Mathisen, J. (1987) Anatomical organization of excitatory amino acid receptors and their pathways. TINS. 10(7):273-280.
7. Diemer, N.H., Johansen, F.F., Sheardown, M., Honore, T., and Jorgensen, M.B. (1990) The role of excitatory mechanisms for the development of ischemia-induced damage in rat hippocampus. In: Krieglstein, J., and Oberpichler, H., eds. Pharmacology of Cerebral Ischemia. Wiss. Verl.-Gis. pp. 113-116.
8. Ginsberg, M.D., and Busto, R. (1989) Rodent models of cerebral ischemia. Stroke. 20:1627-1642.
9. Ginsberg, M.D., and Dietrich, W.D. (1989) Preface/Introduction. In: Ginsberg, M.D., and Dietrich, W.D., eds. Cerebrovascular Diseases. New York: Raven Press, pp. v-vii.
10. Hollmann, M., Hartley, M., and Heinemann, S. (1991) Ca^{2+} permeability of KA-AMPA-gated receptor channels depends on subunit composition. Science. 252:851-853.

11. Keith, J.R., and Rudy, J.W. (1990) Why NMDA-receptor-dependent long-term potentiation may not be a mechanism of learning and memory: reappraisal of the NMDA-receptor blockade stragety. Psychobiology. 18:251-257.

12. Meldrum, B. (1989) Excitotoxicity in ischemia: an overview. In: Ginsberg, M.D., and Dietrich, W.D., eds. Cerebrovascular Diseases. New York: Raven Press, pp. 47-60.

13. Morris, R.G.M. (1989) Synaptic plasticity and learning: selective impairment of learning in rats and blockade of long-term potentiation in vivo by the N-methyl-D-aspartate receptor antagonist AP5. J. Neurosci. 9:3040-3057.

14. Nellgard, B., and Wieloch, T. (1991) Post-ischemic blockade of AMPA but not NMDA receptors mitigate neuronal damage in the rat brain following transient but severe cerebral ischemia. J. Cereb. Blood Flow Metab. (in press).

15. Ordy, J.M., Thomas, G.J., Volpe, B.T., Dunlap, W.P., and Colombo, P.M. (1988) An animal model of human-type memory loss based on aging, lesion, forebrain ischemia, and drug studies with the rat. Neurobiol. Aging. 9:667-683.

16. Ordy, J.M., Wengenack, T.M., Bialobok, P., Coleman, P.D., Rodier, P., Baggs, R.B., Dunlap, W.P., and Kates, B. (1991) Selective vulnerability, early progression of hippocampal CA1 pyramidal cell degeneration and GFAP-positive astrocyte reactivity in the rat 4-VO model of transient global ischemia. Exp. Neurol. (in press).

17. Palmer, G.C., Stagnitto, M.L., Ordy, J.M., Griffith, R.C., Napier, J.J., Gentile, R.J., Woodhead, J.H., and Swinyard, E.A. (1991) Preclinical profile of stereoisomers of the anticonvulsant remacemide in mice. Epilepsy Res. 8:36-48.

18. Petito, C.K., Feldmann, E., Pulsinelli, W.A., and Plum, F. (1987) Delayed hippocampal damage in humans following cardiorespiratory arrest. Neurology. 37:1281-1286.

19. Pulsinelli, W.A., and Buchan, A.M. (1988) The 4-vessel occlusion rat model: method for complete occlusion of the vertebral arteries and control of the collateral circulation. Stroke. 19:913-914.

20. Pulsinelli, W.A., and Buchan, A.M. (1990) The NMDA receptor/ion channel its importance to in vivo ischemic injury to selectively vulnerable neurons. In: Krieglstein, J., and Oberpichler, H., eds. Pharmacology of Cerebral Ischemia. Wiss. Verl.-Ges. pp. 169-175.

21. Ross, D.T., and Graham, D.I. (1991) Selective loss and selective sparing of neurons in the thalamic reticular nucleus following human cardiac arrest. J. Cereb. Blood Flow Metab. Abstr. 11(suppl 2):S855.

22. Sheardown, M.J., Nielsen, E.O., Hansen, A.J., Jacobsen, P., and Honore, T. (1990) 2,3-Dihydroxy-6-nitro-7-sulfamoyl-benzo(F)quinoxaline: a neuroprotectant for cerebral ischemia. Science. 247:571-574.

23. Squire, L.R. (1986) Mechanisms of memory. Science. 232:1612-1619.

24. Thomas, G.J., and Ordy, J.M. (1991) Memory: a behavioristic and neuroscience approach. In: Squire, L.R., and Butters, N., eds. Neuropsychology of Memory. New York: Guilford Press, (in press).

25. Volpe, B.T., and Petito, C.K. (1985) Dementia with bilateral medial temporal lobe ischemia. Neurology. 35:1793-1797

26. Volpe, B.T., and Davis, H.P. (1989) An approach to the functional analysis of an animal model of ischemic injury. In: Ginsberg, M.D., and Dietrich, W.D., eds. Cerebrovascular Diseases. New York: Raven Press, pp. 335-340

27. Willets, J., Balster, R.L., and Leander, J.D. (1990) The behavioral pharmacology of NMDA receptor antagonists. TIPS. 11(10):423-428.

28. Zola-Morgan, S., Squire, L.R., and Amaral, D.G. (1986) Human amnesia an the medial temporal region: enduring memory impairment following a bilateral lesion limited to field CA1 of the hippocampus. J. Neurosci 6:2950-2967.

COMPARATIVE EFFECTS OF MAGNESIUM CHLORIDE AND MK-801 ON

INFARCT VOLUME AFTER MCA OCCLUSION IN FISCHER-344 RATS

E. Pinard, Y. Izumi, S. Roussel and J. Seylaz

C.N.R.S. U.A. 641,
Université Paris VII
Paris, France

INTRODUCTION

The endogenous divalent cation, magnesium, is a natural competitor of calcium and is essential in cellular metabolism, thus its deficiency is highly deleterious. The increasing interest in glutamate as a potential pathogenetic agent has led to extensive research on the N-methyl-D-aspartate (NMDA) receptor. Nowak et al. (1984) found that magnesium gates NMDA receptor-associated channel. MK-801 is an exogenous blocker of the NMDA receptor-associated channel. It inhibits calcium entry into cells via this channel once the magnesium blockade has been released by a certain level of membrane depolarization. Though magnesium has been shown to be neuroprotective in vitro (Kass et al., 1988) and in several pathological conditions, including brain trauma (Vink et al., 1988; McIntosh et al., 1989), spinal cord ischemia (Vacanti and Ames, 1984), quinolinate-induced lesions (Wolf et al., 1990) and global cerebral ischemia (Tsuda et al., 1989), no attempt has been made until now to determine whether it has beneficial effects on focal cerebral ischemia. In contrast, the ability of MK-801 to reduce the volume of infarction in normotensive rats and cats is well documented (Park et al., 1988; Ozyurt et al., 1988). We will therefore, in the present paper, use the neuroprotection exerted by MK-801 as a reference to try to elucidate the mechanisms by which magnesium protects against focal cerebral ischemia in similar experimental conditions.

MATERIALS AND METHODS

Fischer-344 rats were chosen as experimental animals. The middle cerebral artery (MCA) was occluded under halothane anesthesia using the method of Tamura et al. (1981). Temperature was regulated during the whole surgical procedure. Forty eight hours after MCA occlusion, triphenyl tetrazolium chloride (TTC) was infused via the heart to stain the healthy tissue. The brain was then removed and frozen after 45 minutes. The infarct areas were measured using an image analyser from black and white photographs taken every 0.5 mm on the brain cut in a cryomicrotome.

Systemic variables and brain and body temperatures were measured in additional groups of rats. In that case, a catheter was chronically inserted in a femoral artery the day before MCA occlusion and a thermoprobe was chronically implanted in the cerebral cortex.

There were 3 groups in the magnesium study: one control and two treated groups. The first treated group was given $MgCl_2$ alone, 1 mmol/kg i-p just after MCA occlusion and another mmol/kg one hour later. The second was given insulin (Actrapid insulin, 1 u/kg i-p) together with the first injection of $MgCl_2$ in order to prevent the hyperglycemic effect of magnesium.

There were 3 groups in the MK-801 study: one control and two treated groups. The first treated group was given 0.5 mg/kg and the second was given 5 mg/kg MK-801 i-p thirty minutes before MCA occlusion.

The Role of Neurotransmitters in Brain Injury, Edited by
M. Globus and W.D. Dietrich, Plenum Press, New York, 1992

RESULTS

MgCl$_2$ induced a significant increase in glycemia as shown in fig. 1. The arterial blood glucose content was doubled 2 hours after the first injection of magnesium and then returned to its control level within the following 24 hours. Insulin effectively compensated for the hyperglycemic effect of MgCl$_2$ since the blood glucose content of the control and the MgCl$_2$ + insulin-treated groups were not significantly different.

MgCl$_2$ treatment did not alter any of the systemic parameters, including arterial blood pressure, and did not influence brain and body temperatures compared to the control group. However, the MgCl$_2$-treated rats were less reactive than control rats when they recovered from anesthesia, but there was no marked difference in the behavior of the 3 groups a few hours later. MgCl$_2$ significantly reduced the volume of cortical infarction by 29% and the reduction was significantly enhanced when the rats remained normoglycemic due to insulin. In this case the reduction was by about 44%. The effects of both treatments on the areas of cortical infarction are shown in fig. 2. The infarcted areas in MgCl$_2$-treated rats and MgCl$_2$ + insulin-treated rats were significantly smaller than in control rats at most coronal planes. In addition, the rats given magnesium + insulin had smaller infarcted areas than magnesium-treated rats. Neither treatment had any significant effect on the striatum.

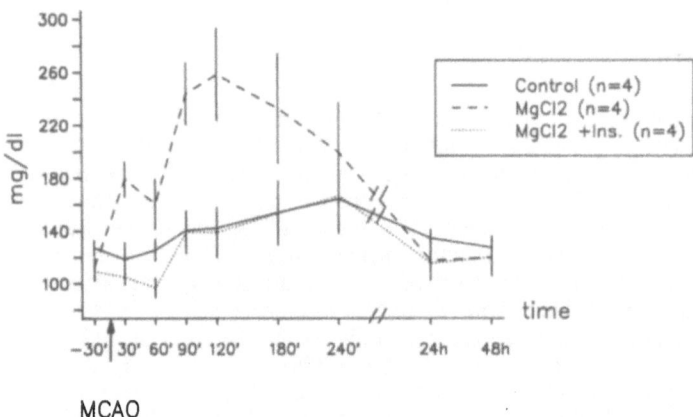

Fig. 1. Mean values (±SD) of arterial blood glucose content before and at various times after MCA occlusion for the 3 groups of the magnesium study.

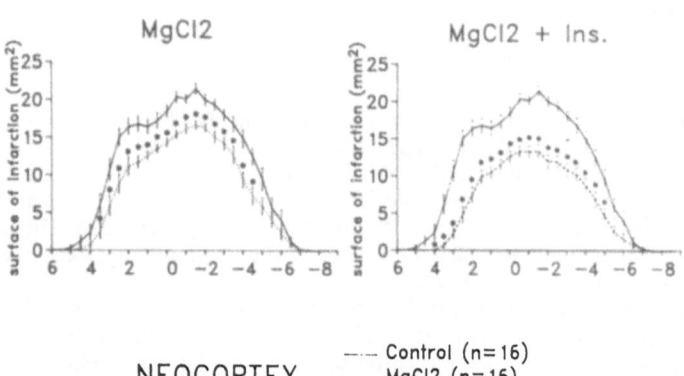

Fig. 2. Mean values (± SE) of infarction areas at various stereotaxic coronal levels. O indicates the bregma. Left: comparison between control group (solid line) and MgCl$_2$-treated group (dotted line). Right: comparison between control group (solid line) and MgCl$_2$ + insulin-treated group (dotted line). * indicates a significant difference at p < 0.05.

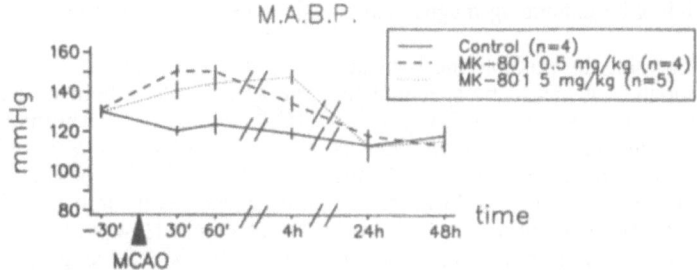

Fig. 3. Mean values (± SE) of MABP before and at various times after MCA occlusion.

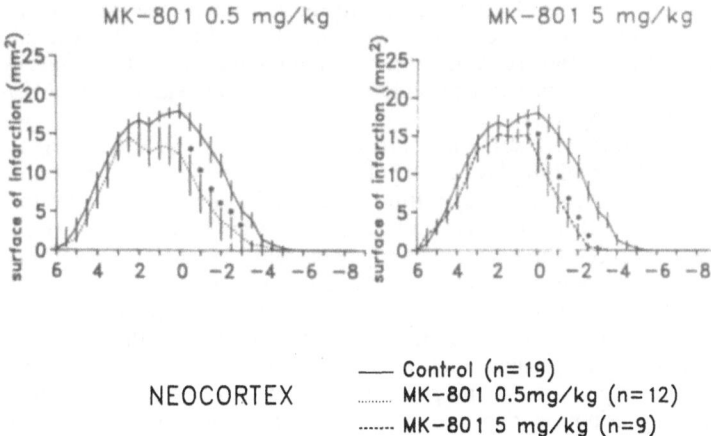

Fig. 4. Mean values (± SE) of infarction areas at various stereotaxic coronal levels. O indicates the bregma. Left: comparison between control group (solid line) and 0.5 mg/kg MK-801-treated group (dotted line). Right: comparison between control group (solid line) and 5 mg/kg MK-801-treated group (dotted line). * indicates a significant difference at p < 0.05.

Neither dose of MK-801 altered systemic variables, including glycemia, except arterial blood pressure. The comparison of MABP between the 3 groups (fig. 3) revealed that MK-801 induced a significant increase in arterial pressure as soon as the rats recovered from anesthesia. The pressure effect was not dose-dependent.

The rats given 0.5 mg/kg MK-801 had an increased locomotor activity, whereas the rats given 5 mg/kg were markedly less reactive, by comparison with control rats. The cortical volumes of infarction were significantly smaller in the treated groups than in the control group (reduction by about 32%), but they were not significantly different in the 2 treated groups. From the examination of the mean areas of infarction in each group (fig. 4), it is clear that only the caudally located areas were reduced in MK-801-treated rats. No significant neuroprotection was observed in the striatum.

DISCUSSION

In the present paper, we report that magnesium has beneficial effects on focal cerebral ischemia and the amplitude of the neuroprotection it produces is similar to that of MK-801. Furthermore, the reduction in infarct size induced by $MgCl_2$ is enhanced in rats given insulin. At the present time, it is not possible to conclude whether hyperglycemia is deleterious, as it is the case in global ischemia (Ginsberg et al., 1980), or whether insulin by itself is neuroprotective, as suggested by Auer and Voll (1991), or whether insulin increases magnesium influx by enhancing magnesium translocation in the cells (Lostroh and Krahl, 1974).

Before discussing the possible mechanisms involved in the neuroprotection afforded by the two compounds, it has to be specified that the systemically administered magnesium enters the brain mainly via the cerebrospinal fluid (Oppelt et al., 1963; Thurnau et al., 1987), but also by crossing the vascular endothelium as shown by Smith et al. (1983) with ^{28}Mg.

The generally proposed action mechanism of MK-801 is the blockade of NMDA receptor-associated channels and thus a reduction of calcium entry into cells. MK-801 also blocks the acetylcholine nicotinic receptor-associated channels. However, MK-801 has been shown to have central sympathomimetic effects mediated via catecholamine-dependent processes, likely involving dopamine (Clineschmidt et al., 1982; Liljequist et al., 1991). In addition, MK-801 blocks the spreading depression in the penumbral zone of focal ischemia (Gill et al., 1991). MK-801 induces an increase in cerebral oxidative metabolism and blood flow whose duration but not amplitude is dose-dependent (Pinard et al., 1991). Such a phenomenon does not occur when cell membranes are hyperpolarized by anesthesia as previously found by Nehls et al. (1989). It is difficult to estimate the extent to which each of these effects is involved, but interestingly enough MK-801 protects the caudal part of the infarct in rats, and this is the area where extension of infarct size occurs primarily (Ridenour et al., 1991) and where glucose consumption is mostly increased after MCA occlusion (Nedergaard et al., 1986).

By comparison with MK-801, there are even more ways in which magnesium might protect neurones. Magnesium plays an important role in stabilizing cellular and subcellular membranes. It is involved in the activation of about 300 enzymes, in the formation and utilization of energy-rich compounds and in the synthesis of electron carriers (Ebel and Günther, 1980). Extracellular magnesium competes with calcium (Iseri and French, 1984), inhibits the release of neurotransmitters, including glutamate (Ault et al., 1980), depresses cerebral metabolism and dilates cerebral blood vessels (Altura and Altura, 1984). Chi et al. (1990) have shown that blood flow is increased in the territory of the occluded MCA after systemic injection of magnesium. Such an effect, which is not suppressed by anesthesia, could be of major importance in the generalized reduction of infarct size. It must also be stressed that magnesium is depleted in brain injury following head trauma and the compensation for this decrease produces a significant neurological improvement (McIntosh et al., 1989). It is likely that a leak of magnesium also occurs in stroke.

Despite all these possible actions, the magnesium-induced neuroprotection is of the same order of magnitude as that of MK-801. One mechanism common to the 2 compounds is their blocking of calcium entry into cells, via the voltage-dependent channels for magnesium and via the NMDA-receptor associated channels for MK-801. This could also lead to a decrease in glutamate release by presynaptic effects. However, the protective action of magnesium is exerted all along the rostro-caudal axis of the infarct, whereas MK-801 only protects the caudal part of the infarct. Such a location of the MK-801 neuroprotection is in line with the results of Park et al. (1988) in Sprague Dawley rats. The differing locations of the neuroprotective effects of the 2 compounds may be due to differing mechanisms. MK-801 may block the spreading depression which seems to spread caudally, while magnesium may increase blood flow in the infarcted area, which would enable the whole penumbral zone to survive.

Though most in vivo investigations are based upon in vitro studies, their data can be far more difficult to interpret. The pathophysiologic evolution of stroke, including edema, spreading depression, collateral blood supply and glial processes have all to be taken into account, in addition to the cellular mechanisms of neuronal death. The pharmacological tools, even the more specific ones, all have side effects which may interfere with the phenomena studied. The surgical approach, together with the anesthetics used, may also be of some importance. Whatever the mechanisms involved, we can only observe the results of combined actions irrespective of their relative implication. It is thus somewhat illusive to extend the results of in vitro findings on neuronal death directly to in vivo ischemic conditions. However, comparison of the neuroprotective effects of several pharmacological compounds, not only in quantitative terms, but also in terms of location and therapeutical window, may be an indirect approach to a better understanding of the mechanism(s) by which neuroprotection is afforded. It should also be of interest to associate various approaches with combined therapies to overcome the deleterious consequences of cerebral ischemia, even though the infarct core cannot be reduced.

REFERENCES

Altura, B.T., and Altura, B.M., 1984, Interactions of Mg and K on cerebral vessels. Aspects in view of stroke, review of present status and new findings, Magnesium, 3:195-211.

Auer, R.N., and Voll, C.L., 1991, Insulin reduces ischemic brain damage by a direct CNS effect, J. Cereb. Blood Flow Metab., 11:S760.

Ault, B., Evans, R.H., Francis, A.A., Oakes, D.J., and Watkins, J.C., 1980, Selective depression of excitatory amino acid induced depolarizations by magnesium ions isolated spinal cord preparations, J. Physiol., 307:413-428.

Chi, O.Z., Pollak, P., and Weiss, H.R., 1990, Effects of magnesium sulfate and nifedipine on regional cerebral blood flow during middle cerebral artery ligation in the rat, Arch. Int. Pharmacodyn., 304:196-205.

Clineschmidt, B.V., Martin, G.E., Bunting, P.R., and Papp, N.L., 1982, Central sympathomimetic activity of (+)5-methyl-10,11-dihydro-5H-dibenzo [a,d] cyclohepten-5,10-imine (MK-801), a substance with potent anticonvulsant, central sympathomimetic, and apparent anxiolytic properties, Drug Dev. Res., 2:135-145.

Ebel, H., and Günther, T., 1980, Magnesium metabolism: A review, J. Clin. Chem. Clin. Biochem., 18:257-270.

Gill, R., Andiné, P., Hillered, L., Persson, L., and Hagberg, H., 1991, The effect of MK-801 on cortical spreading depression in the penumbral zone following focal ischaemia in the rat, J. Cereb. Blood Flow Metab., 11:S286.

Ginsberg, M.D., Welsh, F.A., and Budd, W.W., 1980, Deleterious effect of glucose pretreatment on recovery from diffuse cerebral ischemia in the cats. I. Local cerebral blood flow and glucose utilization, Stroke, 11:347-354.

Iseri, L.T., and French, J.H., 1984, Magnesium: Nature's physiologic calcium blocker, Am. Heart. J., 108:188-194.

Kass, I.S., Cottrell, J.E., and Chambers, G., 1988, Magnesium and cobalt, not nimodipine, protect neurons against anoxic damage in the rat hippocampal slice, Anesthesiology, 69:710-715.

Liljequist, S., Ossowska, K., Grabowska-Andén, M., and Andén, N.E., 1991, Effect of the NMDA receptor antagonist, MK-801, on locomotor activity and on the metabolism of dopamine in various brain areas of mice, European J. Pharmacol., 195:55-61.

Lostroh, A.J., and Krahl, M.E., 1974, Mg a second messenger for insulin: ion translocation coupled to transport activity, Adv. Enzyme Regul., 12:73-81.

McIntosh, T.K., Vink, R., Yamakami, I., and Faden, A.I., 1989, Magnesium protects against neurological deficit after brain injury, Brain Res., 482:252-260.

Nedergaard, M., Gjedde, A., and Diemer, N.H., 1986, Focal ischemia of the rat brain: autoradiographic determination of cerebral glucose utilization, glucose content and blood flow, J. Cereb. Blood Flow Metab., 6:414-424.

Nehls, D.G., Park, C.K., and McCulloch, J., 1989, Cerebral circulatory effects of dizocilpine (MK-801) in the rat, J. Cereb. Blood Flow Metab., 9:S376.

Nowak, L., Bregestovski, P., Ascher, P., Herbet, A., and Prochiantz, A., 1984, Magnesium gates glutamate-activated channels in mouse central neurones, Nature, 307:462-465.

Oppelt, W.W., MacIntyre, I., and Rall, D.P., 1963, Magnesium exchange between blood and cerebrospinal fluid, Am. J. Physiol., 205: 959-962.

Ozyurt, E., Graham, D.I., Woodruff, G.N., and McCulloch, J., 1988, Protective effect of the glutamate antagonist, MK-801 in focal cerebral ischemia in the cat, J. Cereb. Blood Flow Metab., 8:138-143.

Park, C.K., Nehls, D.G., Graham, D.I., Teasdale, G.M., and McCulloch, J., 1988, The glutamate antagonist MK-801 reduces focal ischemic brain damage in the rat, Ann. Neurol., 24:543-551.

Pinard, E., Roussel, S., Charbonné, R., and Seylaz, J., 1991, Dynamic cerebrovascular effects of MK-801 in anesthetized and conscious rats, J. Cereb. Blood Flow Metab., 11:S289.

Ridenour, T.R., Warner, D.S., Shou, J.G., Todd, M.M., and McAllister, A., 1991, Middle cerebral artery occlusion in the SHR rat: evaluation of infarct volume as a function of duration of ischemia, J. Cereb. Blood Flow Metab., 11:S322.

Smith, Q.R., Tai, C.Y., and Rapoport, S.I., 1983, Brain capillary permeability to inorganic ions, Soc. Neurosci. Abstr.

Tamura, A., Graham, D.I., McCulloch, J., and Teasdale, G.M., 1981, Focal cerebral ischaemia in the rat: 1. Description of technique and early neuropathological consequences following middle cerebral artery occlusion, J. Cereb. Blood Flow Metab., 1:53-60.

Thurnau, G.R., Kemp, D.B., and Jarvis, A., 1987, Cerebrospinal fluid levels of magnesium in patients with preeclampsia after treatment with intravenous magnesium sulfate: A preliminary report, Am. J. Obstet. Gynecol., 157:1435-1438.

Tsuda, T., Kogure, K., and Nishioka, K., 1989, Post-ischemic injury in CA1 neurons of the rat was prevented by Mg^{2+}., J. Cereb. Blood Flow Metab., 9(suppl 1):S752.

Vacanti, F.X., and Ames, III A., 1984, Mild hypothermia and Mg^{++} protect against irreversible damage during CNS ischemia, Stroke, 15:695-698.

Vink, R., McIntosh, T.K., Demediuk, P., Weiner, M.W., and Faden, A.I., 1988, Decline in intracellular free Mg^{2+} is associated with irreversible tissue injury after brain trauma, J. Biol. Chem., 263:757-761.

Wolf, G., Keilhoff, G., Fischer, S., and Hass, P., 1990, Subcutaneously applied magnesium protects reliably against quinolinate-induced N-methyl-D-aspartate (NMDA)- mediated neurodegeneration and convulsions in rats: are there therapeutical implications? Neurosci. Lett., 117:207-211.

GENERATION AND PROPAGATION OF CORTICAL SPREADING DEPRESSION IS MEDIATED

BY THE NMDA RECEPTOR

D. Scheller, U. Heister and F. Tegtmeier

Janssen Research Foundation, Raiffeisenstr. 8

4040 Neuss 21, FRG

INTRODUCTION

Cortical spreading depression (CSD), a transient and propagating suppression of the electrical activity of the brain (Leao, 1944), is supposed to be elicited during small traumatic injury (Gardner-Medwin, 1984; Mutch and Hansen, 1984; Rothner, 1989), during surgery (Gardner-Medwin, 1984), during local vascular· occlusions (Kempinsky, 1954) and within post-infarction penumbra zone (Nedergard and Astrup, 1986; Hakim, 1987). The corresponding short lasting (minutes), but dramatic electrolyte shifts (Nicholson and Kraig, 1981) are accompanied by a transient increase of local cortical blood flow (LCBF) followed by a long lasting hypoperfusion (hours, Lauritzen et al. 1982). Glutamate (glu) and aspartate (asp) have been described as possible endogenous mediators of CSD (Van Harreveld, 1959) indicating the involvement of the NMDA receptor (Marrannes et al. 1988, Scheller et al. 1990) in generation of CSD as a common final feature (Gardner-Medwin, 1984; Faden et al. 1989) for all of the known, different stimuli generating CSD.

With respect to propagation, no _in vivo_ studies have been performed so far in order to demonstrate a possible contribution of the NMDA receptor for the propagation of CSD.

The aim of this study was to measure the extracellular levels of aspartate (asp) and glutamate (glu) during generation and propagation of CSD by means of two microdialysis (MD) probes implanted in different cortical areas: one MD probe served for local application of KCl as well as to measure amino acids; the other MD probe was used to detect the changes of asp and glu at a remote site towards which the CSD was propagating.

In addition, NMDA antagonists were applied separately to one of the MD perfusates in order to selectively interfere with NMDA receptors at the site of CSD generation or at the site of CSD propagation.

METHODS

Rats were anaesthetized with urethane (1.75 g/kg) and fixed within a stereotaxic frame. Two holes (rostral (r) and caudal (c) to the

bregma) were drilled into the skull at a distance of 6 to 10 mm and the
dura was incised. Two microelectrodes (ME, rostral ME: MEr, caudal ME:
MEc) were inserted, each to a depth of 1000 μm and at a distance of
about 100 to 500 mm from one of the two MD probes (CMA/10 Carnegie
Medicin, Stockholm, Sweden) which were implanted separately within both
burr holes (rostral probe: MDr; caudal probe: MDc, see Fig.1). 4 or
10 μl samples were collected at a flow rate of 2 μl/min. Perfusate
consisted of artificial cerebrospinal fluid (CSF). Asp and glu were
identified by fluorescence detection after derivatization with OPA and
reversed phase HPLC.

To induce a single CSD, CSF was switched to high K^+-CSF (NaCl replaced
by KCl, K^+=128 mmol/l) for 2 min at either the caudal or the rostral MD
probe. High K^+-CSF was applied repetitively with recovery periods of
45 to 90 min between each application. In one series of experiments,
either Ketamine (1 or 20 mmol/l in perfusate) or 2-APV (0.1 mmol/l in
perfusate) was locally administered together with the K^+-stimulus. In a
second series of experiments, 2-APV (0.1 mmol/l in CSF) was added to
the perfusate of the MD probe remote to the K^+-stimulus.

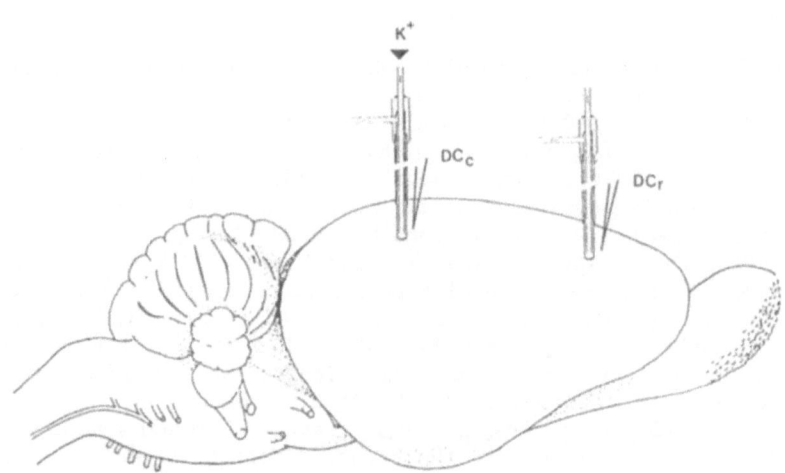

Fig.1. Ilustration of the experimental setup: two MEs are implanted
together with two MD probes. CSD is elicited at the rostral
or caudal MD probe and registered with two MEs. A the same
time, extracellular asp and glu are measured at both sites.
If required, NMDA antagonists are applied together with the
K^+-stimulus or at the site remote to the K^+-stimulus.

RESULTS

Local K^+-application induced a pronounced increase of
extracellular asp and glu and elicited CSD with a latency of 81.6 ±
20.16 s (Fig.2a). The CSD propagated towards the remote ME and MD probe
with a velocity of 3.5 ± 1.04 mm/min. However, changes of extracellular
asp and glu could hardly or not be detected when CSD approached the
remote ME (Fig.2b).

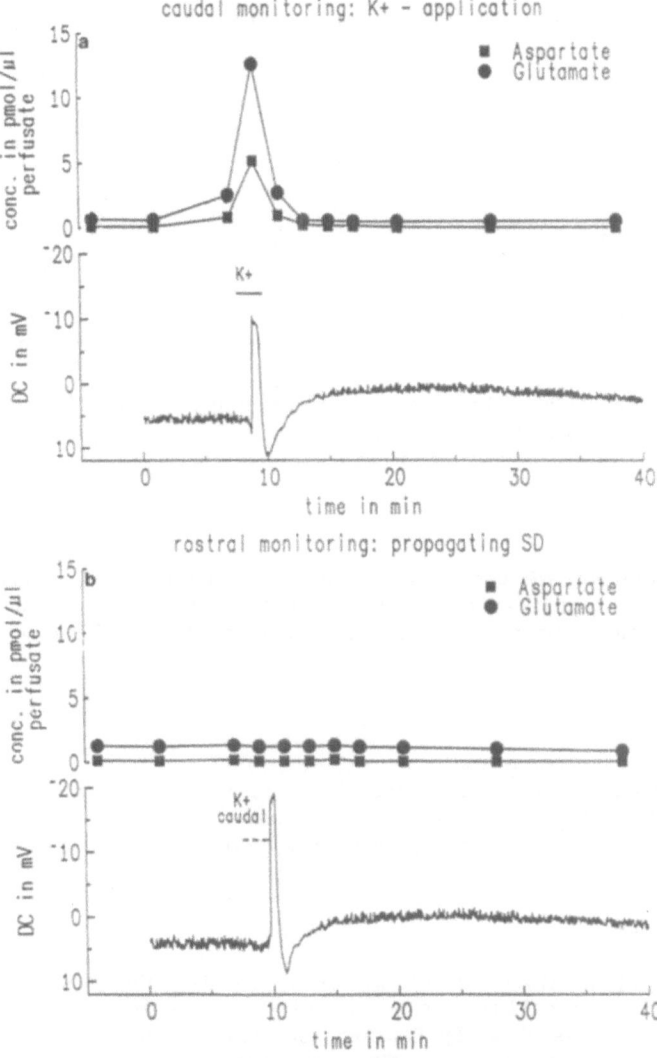

Fig.2 a: DC-recording (lower tracing) and dialysate concentrations of asp and glu (upper tracing) at the site of K⁺-application (CSD generation).

b: DC-recording (lower tracing) and dialysate concentrations of asp and glu (upper tracing) at the site remote to the K⁺-administration (CSD propagation) (see Fig.1).

Table 1. Summary of the effects of NMDA antagonists on
frequency of generation and propagation of CSD.

CSD Generation

inducing agent		Ketamine 1 mmol/l	Ketamine 20 mmol/l	2-APV 0.1 mmol/l
K$^+$	c	5/5	4/4	7/7
	t	0/5	0/4	1/7
NMDA	c	5/5	4/4	4/4
	t	3/5	0/4	1/4

CSD Propagation

inducing agent		2-APV 0.1 mmol/l
K$^+$	c	12/12
	t	6/12

(n/n: n CSDs induced / n applications of stimuli;
c = controls; t = treated)

Local application of either Ketamine or 2-APV blocked K$^+$-provoked CSD
(Tab.1). Local application of 2-APV at the MD probe towards which the
CSD was moving suppressed propagating CSD locally in 50 % of the cases
(Tab.1).

DISCUSSION

The present experiments were performed to study the contribution
of the NMDA receptor during generation as well as during propagation of
CSD _in vivo_ using microdialysis and microelectrode techniques in
combination.

CSD Generation

Local application of K$^+$ by means of the MD probe mediated an
increase of extracellular asp and glu due to increased transmitter
release (Fig.2a) as shown by Paulsen and Fonnum (1989). These transmitters are supposed to activate NMDA receptors (Rothman and Olney, 1987)
and to cause dramatic transmembranal electrolyte fluxes especially
after prolonged administration (Collingridge and Lester, 1990). Such
massive electrolyte fluxes accompany CSD's (Nicholson and Kraig, 1981)
which also can be identified as rapid negative DC shifts. The DC shifts
observed here can be identified as CSD's. It is important to note, that
they occure with a latency of about 1.5 min in regard to stimulus
(Fig.2a). Local application of the receptor agonist NMDA has been shown
to induce CSD (Curtis and Watkins, 1963; Scheller et al. 1990) with an
even shorter latency (of about 30 s). Such K$^+$ or NMDA induced CSD's (as
well as electrically provoked CSD's) could be blocked by either
Ketamine or 2-APV (Marrannes et al. 1988, Scheller et al. 1990, Tab.1).
Therfore, we conclude that local application of K$^+$ caused release of
aspartate and glutamate; these transmitters in turn elicited CSD; the
generation of these CSDs is mediated via the NMDA receptor as

illustrated further by its blockade with NMDA antagonists. These data suggest a central role of the NMDA receptor in generation of spreading depressions.

CSD Propagation

Locally generated CSD's spreaded across the hemisphere with a velocity of 3.5 mm/min and could be locally suppressed by 2-APV applied remote to the site of CSD generation. Although 2-APV was efficient in only 50 % of applications (possibly due to tissue heterogeneity and non-homogeneous diffusion of the antagonist in the ECS with respect to the location of CSD detecting electrode and/or contribution of non-NMDA receptors), the results indicate a participation of the NMDA receptor in propagation of CSD too. Accordingly, the endogenous agonists of the NMDA receptor mediating the spreading of CSD, asp and/or glu, should be detectable at the remote MD probe. However, only very tiny changes of both asp and glu were measurable with the MD technique at the site of CSD propagation – as opposed to the observation of a dramatic increase at the triggering site. This might be explained methodologically: at the triggering site, K^+ application not only induces neuronal transmitter release but also blocks re-uptake of transmitters (Erecinska, 1987). Only with the aid of this additional effect of K^+, the synaptic events may become detectable. At the remote site, such additional and the detectability improving effect of K^+ is missing. Moreover, the release of asp and glu during propagation of CSD might be less than during generation and, therefore, below the detection limit of the analytical procedure (0.05 pmol/µl perfusate). Alternatively, other endogenous agonists of the NMDA or Non-NMDA receptor as e.g. L-homocysteinic acid, quinolinic acid or N-acetyl-aspartylglutamate (Headley and Grillner, 1990) or the NMDA receptor modulator glycine (Collingridge and Lester, 1990) could have been involved (but were not measured).

CONCLUSION

The results support the idea that an excitatory mechanism, involving asp and/or glu and the NMDA- receptor, plays a key role in generation and propagation of CSD. CSDs are frequently found to be provoked during traumatic injury, local infarction or concussion. A CSD, therefore, could be a common final pathway of excitotoxicity during various pathological conditions. Consequently, NMDA antagonists, are to be expected to find a widespread application to prevent at least some of the excitotoxic damage after various types of cerebral injury.

REFERENCES

Collingridge, G.L. and Lester, R. A. J., 1990, Excitatory amino acid receptors in the vertebrate central nervous system, Pharmacol. Rev. 40(2): 143-210

Curtis, D. R. and Watkins, J. C. 1963, Acidic amino acids with strong excitatory actions on mammalian neurons, J. Physiol., 166: 1-14

Faden, A. I., Demediuk, P., Panter, S. and Vink, R., 1989, The role of excitatory amino acids and NMDA receptors in traumatic brain injury, Science, 244: 798-800

Erecinska, M., 1987, The neurotransmitter amino acid transport systems: a fresh outlook on an old problem, Biochem. Pharmacol., 36: 3547-3555

Gardner-Medwin, A. R. and Mutch, W. A. C., 1984, Experiments on
    spreading depression in relation to migraine and neurosurgery,
    An. Acad. Brasil. Cienc., 56: 423-430
Hakim, A. M., 1987, The cerebral ischemic penumbra, Can. J. Neurol.
    Sci., 14: 557-559
Headley, P. M. and Grillner, S., 1990, Excitatory amino acids and
    synaptic transmission: the evidence for a physiological function,
    TIPS,   11: 205-211
Kempinsky, W. H., 1954, Steady potential gradients in experimental
    cerebral vascular occlusion, EEG Clin. Neurophysiol., 6: 375-388
Lauritzen, M., Jorgensen, M. B., Diemer, N. H., Gjedde, A. and Hansen,
    A. J., 1982, Persistent oligemia of rat cerebral cortex in the
    wave of spreading depression, Ann.Neurol., 12: 469-474
Leao, A. A. P., 1944, Spreading depression of activity in the cerebral
    cortex, J.Neurophysiol., 8: 359-390
Marrannes, R., Willems, R., De Prins, E. and Wauquier, A., 1988,
    Evidence for a role of the N-methyl-D-aspartate (NMDA) receptor
    in cortical spreading depression in the rat, Brain Res., 457:
    226-240
Mutch, W. A. C. and Hansen, A. J., 1984, Extracellular pH changes
    during spreading depression and cerebral ischemia: mechanisms of
    brain pH regulation, J.Cerebr. Blood Flow Metab., 4: 17-27
Nedergaard, M. and Astrup, J., 1986, Infarct rim: effect of
    hyperglycemia   on   direct   current   potential   and $[^{14}C]2$-
    deoxyglucose phosphorylation, J.Cereb. Blood Flow Metab., 6:
    607-615
Nicholson, C. and Kraig, R. P., 1981, The behavior of extracellular
    ions during spreading depression, in: "The Application of
    Ion-Selective Microelectrodes", T. Zeuthen ed., Elsevier,
    Amsterdam
Paulsen, R. E. and Fonnum, F., 1989, Role of glial cells for the basal
    and Ca^{++}-dependent K^+-evoked release of transmitter amino acids
    investigated by microdialysis, J. Neurochem., 52: 1823-1829
Rothman, S. M. and Olney, J. W., 1987, Excitotoxicity and the NMDA
    receptor, TINS, 10: 299-302
Rothner, A. D., 1989, Minor head trauma: its relationship to migraine,
    Cephalagia, 9 (suppl. 10): 323
Scheller, D., Heister, U., Dengler, K, and Peters, T., 1990, Do the
    excitatory amino acids aspartate and glutamate generate spreading
    depressions in vivo, in: "Pharmacology of Cerebral Ischemia", J.
    Krieglstein,   ed.,   Proceedings   of   the   Third   International
    Symposium on Pharmacology of Cerebral Ischemia, Marburg 1990,
    Wissenschaftliche Verlagsgesellschaft, Stuttgart
Van Harreveld, A., 1959), Compounds of brain extracts causing spreading
    depression of cerebral cortical activity and contraction of
    crustacean muscle, J.Neurochem., 3: 300-315

EFFECTS OF KETAMINE ON RECOVERY OF AUDITORY BRAINSTEM RESPONSE AFTER TOTAL BRAIN ISCHEMIA

Yasuhiko Kushida, Koujirou Hirota, Yutaka Yoshita,
Kazuki Tohyama, Ken Yamamoto, Tsutomu Kobayashi, and
Seiitsu Murakami

Department of Anesthesiology and Intensive Care
Medicine, School of Medicine, Kanazawa University

Introduction

L-glutamate was first proposed as a neuroexitatory agent[1,2]. Progress has been particulary rapid in understanding the N-metyl-D-aspartate (NMDA) class of glutamate receptor[3] and ketamine has been known as an antagonist of NMDA receptor[4]. Prolonged stimulation of exitatory amino acid receptors of either the NMDA or non-NMDA types evertually results in the central neuronal injury. The exact mechanism of neuronal injury is complicated, since depolarization can lead to neural swelling, calcium influx, and other consequences[5].

In the present study, the effects of ketamine on recovery of auditory brainstem response (ABR) after induction of total brain ischemia were evaluated.

Method

Twenty-two adult mongrel dogs (weight range 8 ∼ 15 kg) were randomly divided into two groups of 11 dogs each. One group received ketamine administration (ketamine group) and in the other did not (control group). Total brain ischemia was induced by clamping the ascending aorta at its origin, SVC and IVC under general anesthesia (FiO2 1.0, Halothane 0.2%) during 15 minutes (Fig.1). Dogs in ketamine group were intravenously injected with 5 mg/kg of ketamine for 5 minutes; baseline values were measured, and total brain ischemia was induced 10 minutes after the injection (Fig.2). Auditory brainstem response (ABR) in response to stimulation of the right ear with clicking sounds at a sound pressure of 80 dBHL was measured with an apparatus for the measurement of evoked response (NicoLet Co). One electrode was placed in the cranial vault and a reference electrode was placed in the cranial process. ABR was measured with a filter band of

150-1,500 Hz, at an analysis time of 10 msec, and a mean summation of 1,000 times. ABR was determined to be present if ABR amplitude was over 0.2 microV and the same wave form in ABR configulation appeared again. ABR measurements were performed before brain ischemia as well as every 10 minutes for the first one hour, and at 90 minutes, and 2,3,4,6,9, and 12 hours after recirculation. To monitor systemic parameters, systemic blood pressure measurements and blood gas analysis were done via a femoral artery canula. Pulmonary artery pressure, central venous pressure, pulmonary artery wedge pressure, cardiac output and blood temperature were monitored using a pulmonary thermo-dilution catheter. Esophageal temperature and blood glucose level were also measured.

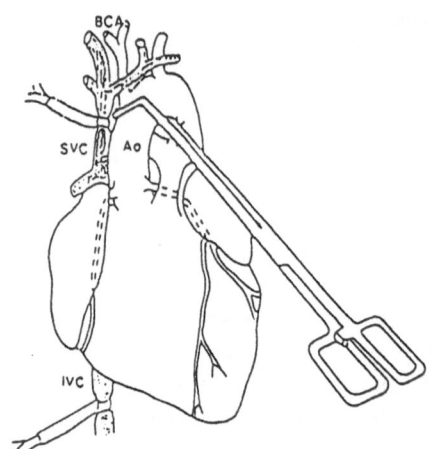

Fig. 1 Method for total brain ishemia.
Total brain ishemia was induced by occluding the aorta with a DeBakey's vascular clamp which was placed just proximal to the brachiocephalic trunk. The superior and inferior vena were also occluded with umbilical tapes.
(Ao:ascending aorta, BCA:brachiocephalic trunk, SVC:superior vena cava, IVC:Inferior vena cava).

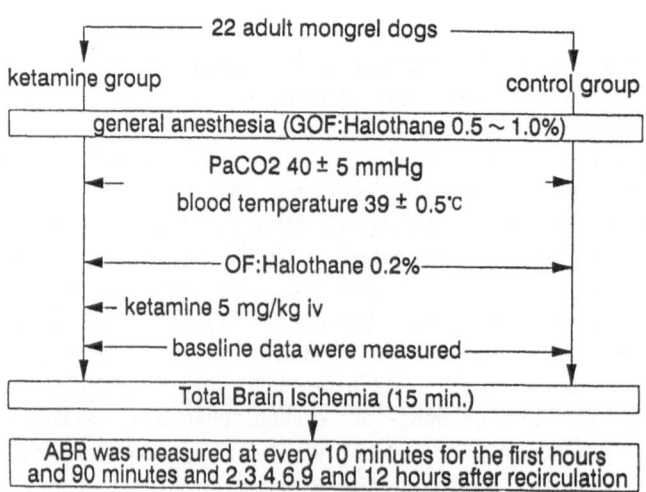

Fig 2. Experimental procedures.

Results

ABR disappeared immediately after clamping of the aorta in both groups(Fig 3). Analysis of ABR recovery after recirculation showed that incidences of Ⅲ wave at 20,30,40, and 50 minutes after recirculation (Fig 4), of Ⅳ wave at 30 and 40 minutes (Fig 5), and of V waves at 40 minutes (Fig 6) were significantly higher in the ketamine group than in those of control group. Analysis of inter peak latencies (IPL) of I - Ⅲ, Ⅲ - V, and I - V after recirculation showed that IPL's of Ⅲ - V at 60 minutes and at 4 hours were significantly shorter (but not shorter than the baseline value)in the ketamine group than in those of control group (Fig 7),as was that of I - V at 60 and 90 minutes and at 2,4 hours (Fig 8). IPL's of I - Ⅲ did not differ significantly between both groups.

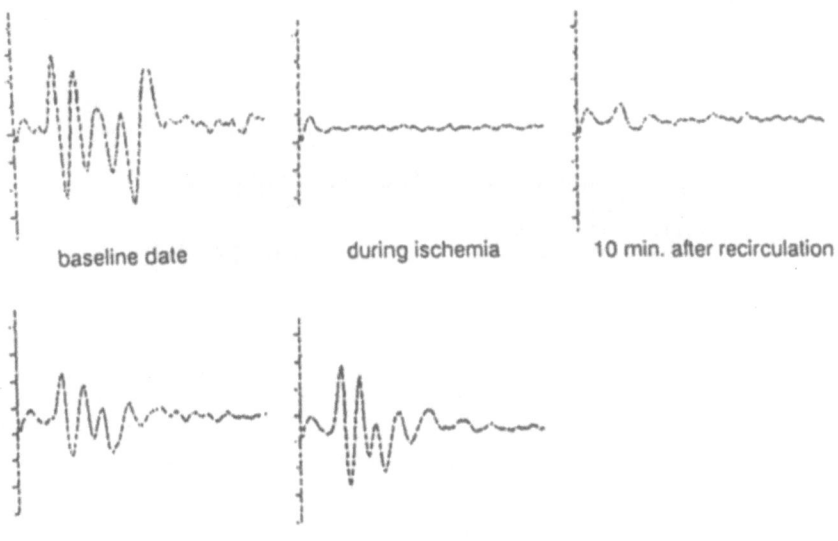

baseline date   during ischemia   10 min. after recirculation

20 min. after recirculation   30 min. after recirculation

Fig. 3 An example of ABR recovery.

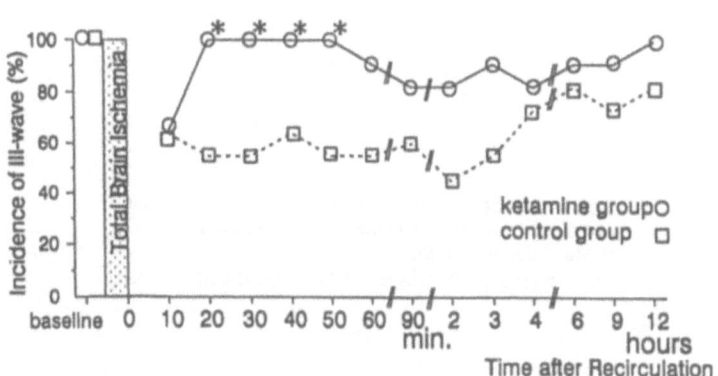

Fig.4 Incidences of III wave at 20,30,40 and 50 minutes after recirculation were significantly higher in the ketamine group than in those of control group.
(Fisher's exact probability test ＊ :p < 0.05 vs control group)

107

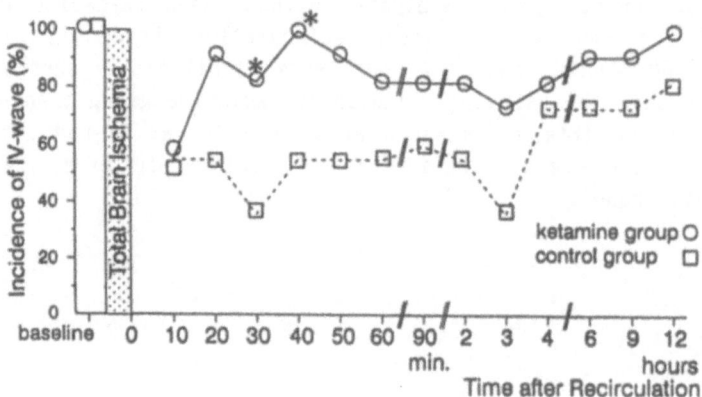

Fig.5 Incidences of IV wave at 30 and 40 minutes after recirculation were significantly higher in the ketamine group than in those control group.
(Fisher's exact probability test *:p< 0.05 vs control group)

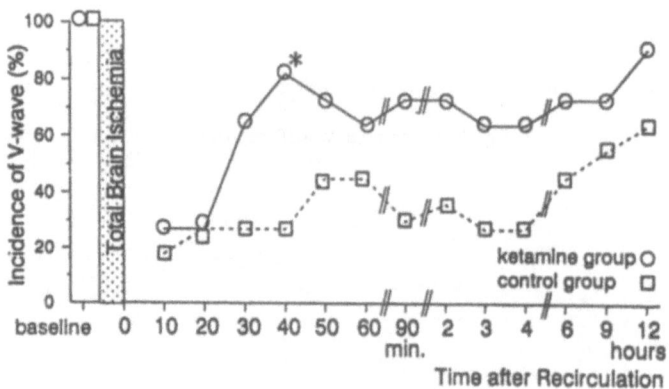

Fig.6 Incidence of V wave at 40 minutes after recirculation was significantly higher in the ketamine group than those of control group.
(Fisher's exact probability test * :p< 0.05 vs control group)

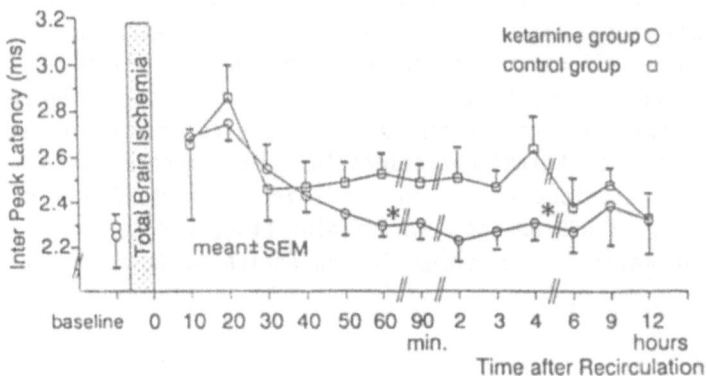

Fig.7 IPL's of III-V at 60 minutes and at 4 hours were
significantly shorter in the ketamine group in those
of control group
(nonpaired t-test ＊:p＜0.05 vs control group)

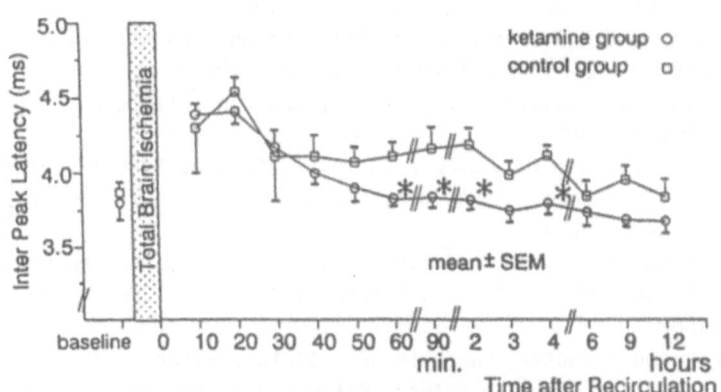

Fig.8 IPL's of I-V at 60 and 90 minutes and 2,4 hours
were significantly shorter in the ketamine group
in those of control group.
(nonpaired t-test ＊:p＜ 0.05 vs control group)

Discussion

We compared ABR recoveries after total brain ischemia of the ketamine group with those of the control group. ABR recovery was greater and the prolongation of IPL's were shorter in the ketamine group compared to controls. Recently, there has been interest in the protective effect on the brain NMDA antagonists such as MK-801[6, 7, 8]. Ketamine has been also reported to have an essentially equivalent pharmacological effect and to reduce delayed neural injury[9, 10]. ABR is one of the assessment techniques for brainstem damage, however good recovery of ABR does not necessarily good cerebral cortex recovery. The probable action of ketamine accelerating the cerebral blood flow may indicate a deleterious effect on the cerebral cortex. The clear and rapid recovery of ABR observed in ketamine group may be caused from the reduction of early neural damage for an action of ketamine as an NMDA antagonist.

Conculusion

The administration of ketamine before induction of total brain ischemia in dogs accelerates the subsequent recovery of ABR.

References

1. T.Hayashi, Chemical Physiology of Exitation in Muscle and Nerve, Nagayama-Shoten
2. D.R.Curtis and J.C.Watkins, The exicitation and depression of spinal neurons by structually related amino acids, J.Neurochem. 6:117 (1960)
3. C.W.Cotman and L.L.Iversen, Exitatory amino acids in the brain-focus on NMDA receptors, Trends Neurosci. 10:263 (1987)
4. J.C.Watkins and H.J.Olverman, Agonists and antagonists for excitatory amino acid receptors, Trends Neurosci. 10:265 (1987)
5. S.M.Rothman and J.W.Olney, Exitotoxity and the NMDA receptor, Trends Neurosci. 10:299 (1987)
6. R.Gill and A.C.Foster, Systemic administration of MK-801 protects against ischemia-induced hippocampal neurodegeneration in the gerbil, J Neurosci. 7:3343 (1987)
7. A.I.Faden and J.A.Nobel, Effects of competitive and non-competitive NMDA receptor antagonists in spinal cord injury, Eur J Pharmacol. 175:765 (1990)
8. A.I.Faden and P.Panter, The role of exitatory amino acids and NMDA receptors in traumatic brain injury, Science 244:798 (1989)
9. F.W.Marcoux and J.E.Goodrich, Ketamine prevents ischemic neuronal injury, Brain Research, 452:329 (1988)
10. J.Weiss and M.P.Goldberg, Ketamine protects cultured neocortical neurons from hypoxic injury, Brain Research, 380:186 (1986)

Chapter 3

The Role of Monoamines in Brain Injury

ROLE OF NIGROSTRIATAL PROJECTIONS IN THE VULNERABILITY OF THE STRIATUM

A. Buisson, V. Pateau, M. Plotkine, and R.G. Boulu

Laboratoire de Pharmacologie, Faculté des Sciences

Pharmaceutiques et Biologiques, Université rené

Descartes, 75006 Paris, France

INTRODUCTION

The striatum, a structure particularly susceptible to transient ischemia, is innervated by interacting corticostriatal glutamatergic and nigrostriatal dopaminergic projections. GLOBUS et al.[3] demonstrated that dopamine (DA) depletion due to lesioning of the subtantia nigra (SN) protects the striatum against ischemic damage in the 4-vessels occlusion model in rats. As glutamate receptor activation is thought to play a major role in ischemic damage[2], we have evaluated the role of the nigro-striatal pathway in the development of neuronal damage following injection of quinolinic acid (QA), an excitotoxin that mimics the neurotoxic action of glutamate, into the striatum and following of a focal cerebral ischemia induced by cauterisation of the middle cerebral artery (MCA) in rats.

MATERIALS AND METHODS

SN LESIONING

Male Sprague Dawley rats (280-320 g), were pretreated with desipramine (25mg/kg,i.p.), anesthetized with chloral hydrate (300mg/kg), and given unilateral injection of 6-hydroxydopamine (12µg in 2µl 0.2 % ascorbate) into the pars compacta of the left SN : coordinates 2.2 mm lateral, 5.3 mm posterior to the bregma, and 5.8 mm ventral to the surface of the skull. Eight days after SN lesioning, the rats were injected with apomorphine (0.5 mg/ kg s.c.)and their rotational behavior checked 30 min later; only rats exhibiting a marked contralateral turning indicating over 90 % depletion of the DA in the ipsilateral striatum[4], were used. Eight to fourteen days after 6-OHDA injection, the rats were injected with QA or subjected to middle cerebral artery occlusion.

QUINOLINIC ACID INJECTION

A group of 9 rats was anesthetized with chloral hydrate(300 mg/ kg) and QA (150 nmol in 1 µl, dissolved in 1N NaOH and adjusted to pH 7) was infused stereotaxically into the left striatum (coordinates: 3 mm lateral, 0.2 mm posterior to the bregma and 5.5 mm below to the surface of the skull). A control group (11 rats) with intact SN were given the same QA injection. All the rats were sacrificed three days later. Their brains were removed and frozen in isopentane (-40°C). Left striatum damage was determined on coronal slices (30 μm) stained with cresyl violet. The extent of neuronal damage was assessed by the volume of the striatal lesions, determined by measuring the surface of necrotic tissue on each striatal section using an image analyser.

MIDDLE CEREBRAL ARTERY OCCLUSION

Six SN lesioned rats and 6 control rats (280-320 g) were anesthetized with chloral hydrate and the left MCA was cauterized[5] at its proximal site. Rectal and temporal temperature were maintained at 37°C.

MICRODIALYSIS TECHNIQUE AND BIOCHEMICAL ANALYSIS

The change in the extracellular concentration of dopamine and glutamate during MCA occlusion, were monitored by horizontal transtriatal microdialysis. The microdialysis fibres (Hospal AN 69) were inserted into the striatum (at the bregma level and 4.5 mm below to the surface of the skull) 24 hours before ischemia. The length of dialysis section was 3 mm on the left side, the rest of the fibre was coated with epoxy glue. The fibre was perfused with Ringer solution (147 mM NaCl, 4 mM KCl, 2.4 mM CaCl$_2$) at a constant flow rate (2.5 μl/min). The rats were sacrificed 2 days later. Their brains were removed and frozen in isopentane. The extent of ischemic damage was assessed by the volume of the infarctus, determined by measuring the surface of necrotic tissue on selected coronal brain sections with an image analyzer. The correct implantation of the fibre was checked on cryostat sections

Each 20 min period yielded a sample volume of 50 μl, of which 25 μl was analyzed for DA, and 25 μl for glutamate. The DA concentration was determined by HPLC with electrochemical detection and the glutamate concentration by HPLC with fluorescence detection and automated pre-column derivatization with o-phtaldialdehyd.

All values are means ± SEM. The concentrations of glutamate and DA were compared using Student's t test. The null hypothesis was rejected when $P \leq 0.05$.

RESULTS

QUINOLINIC ACID INJECTION

QA injection produced a highly reproducible, well delineated necrosis in the control rats. The necrotic area was less intensely stained, with a maximum (17.7 ± 1.3 mm^2) at 0.5mm anterior to the injection site (Fig 1).

The volume of necrosis was 34.9 ± 2.3 mm^3 in control rats (n= 11) (fig 2). Rats with unilateral SN lesions (n= 9) had much smaller volumes of necrotic tissue due to QA injection (9.1 ± 1.4 mm^3, P<0.001)(fig 2). Each striatal section showed a marked reduction in the excitotoxic damage, with maximum necrosis (7.1 ± 2.5 mm^2) at the injection plane. The striatal sections at 11.2 and 10.7 mm showed no damage.

MCA OCCLUSION

MCA occlusion induced a highly reproducible infarctus, affecting both the cortex and striatum of control rats. In SN-

112

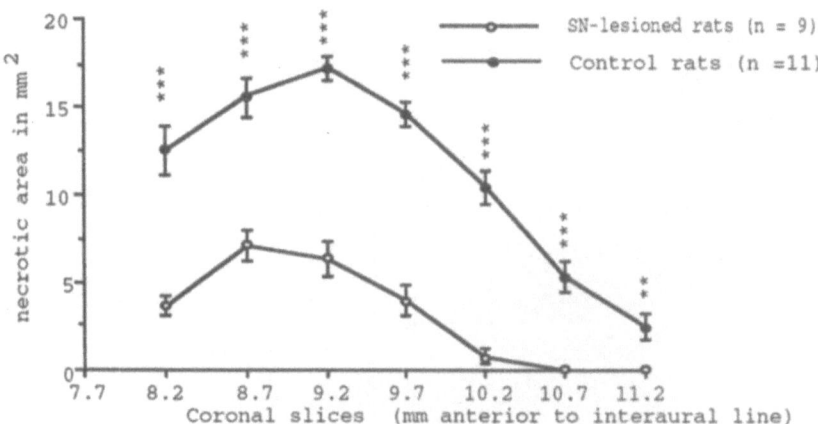

Figure 1. Effect of SN lesioning on neuronal degeneration, expressed as the necrotic area, following intrastriatal injection of quinolinic acid (150 nmol). Neuronal damage was measured on sections (30 μm) corresponding to 8.2 - 11.2 mm planes anterior to the interaural line. Values are means ± S.E.M. for each group.
(** : P<0.01; *** : P<0.001)

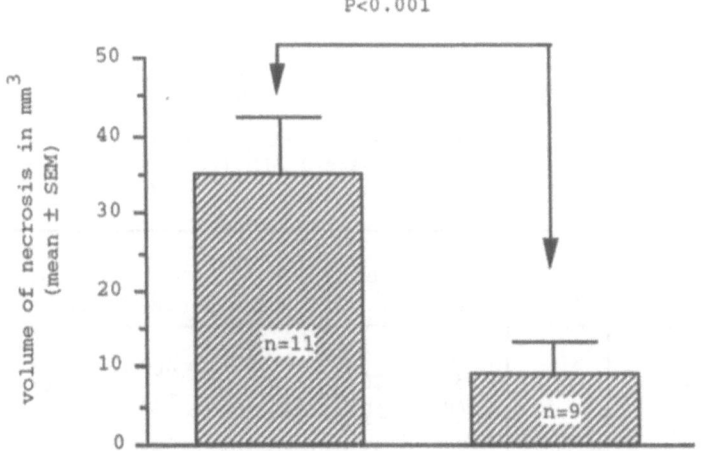

Figure 2. Volume of neuronal damage induced by an intrastriatal injection of quinolinic acid (150 nmol) in non-lesioned and SN-lesioned rats. Values are means ± S.E.M.

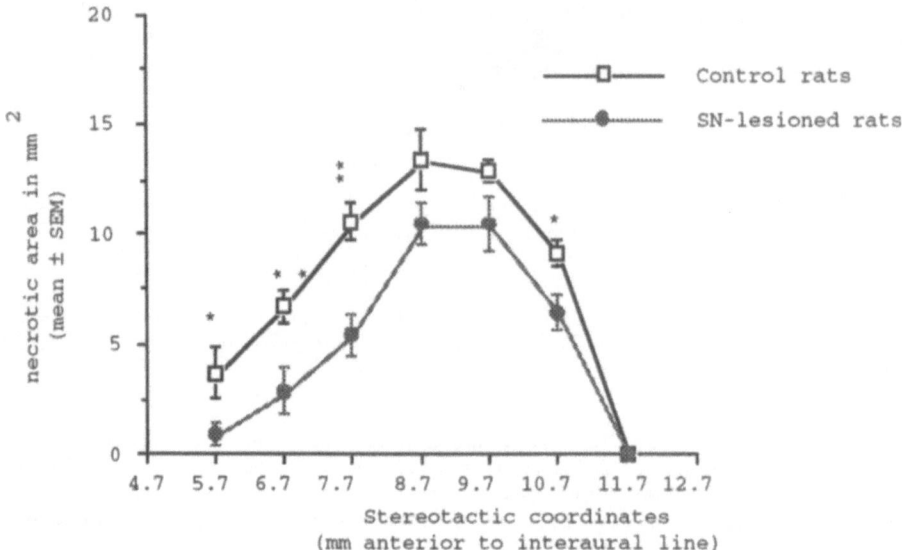

Figure 3. Effect of SN lesioning on neuronal degeneration, expressed as the necrotic area, following MCA occlusion. Neuronal damage was measured on sections (30 μm) corresponding to 5.7 - 11.7 mm planes anterior to the interaural line. Values are means ± S.E.M. for each group. (* : P<0.05; ** : P<0.01)

Table 1. Effect of MCA occlusion on the perfusate concentration of dopamine (nmole/l). Values are means ± SEM. (a : P<0.05 versus basal level; * : P<0.05 control group versus SN-lesioned group)

	basal level	20 min	40 min	60 min	80 min	100 min	120 min
control n = 6	0.3 ±0.1	353 [a] ±148	253 [a] ±80	103 [a] ±32	66 [a] ±23	55 [a] ±17	37 [a] ±15
SN-lesioned n = 6	0.1 ±0.2	29 [a] ±21 *	14 [a] ±9 *	4 [a] ±2 *	9 [a] ±5 *	3 [a] ±1*	2 [a] ±1 *

Table 2. Effect of MCA occlusion on the perfusate concentration of glutamate (μmole/l). Values are means ± SEM.(a : P<0.05 versus basal level; * : P<0.05 control group versus SN-lesioned group).

	basal level	20 min	40 min	60 min	80 min	100 min	120 min
control n = 6	3,6 ±0.6	28.2^a ±4.5	27.2^a ±4.1	18.8^a ±3.5	12.5^a ±2.8	10.1^a ±2.1	8.2 ±2.8
SN-lesioned n = 6	8.3 ±4.3	10.8 ±4.7 *	12 ±6.7 *	8.2 ±5.7	6.6 ±4.6	7.3 ±4.9	6.9 ±3.2

lesioned rats there was a marked reduction of the volume of infarction in the DA-depleted striatum from 50.8 ± 3.3 to 31.5 ± 3.2 mm^3 (P<0.01). Four out of seven striatal sections showed a reduced necrotic area after SN lesioning (fig 3). The infarct volume in the cortex was reduced after SN lesion, but the difference from the control did not reach statistical significance (controls : 126.8 ± 12.1, SN lesioned : 70.8 ± 24.5.mm^3; P = 0.06).

In the control group, the extracellular striatal concentration of DA increases immediatly after MCA occlusion and remained very high 2 hours later. MCA occlusion also induces a marked increase in glutamate concentration, which remains significantly above to the basal level 100 minutes later.In the SN lesioned rats, the release of dopamine after ischemia was dramatically reduced, indicating that our procedure almost totally suppressed striatal dopaminergic innervation (table 1). The glutamate concentration before ischemia tended to increase towards basal levels, but MCA occlusion did not induce any significant change in the extracellular concentration of glutamate (table 2).

DISCUSSION

These results demonstrate that SN lesioning in rats markedly reduces the vulnerability of the striatum to both quinolinic acid injection and focal cerebral ischemia. These results are in agreement with the data of CHAPMAN et al.[6] showing that lesioning of the nigrostriatal dopaminergic pathway decreases the excitotoxic effect of striatal injections of N-methyl-D-aspartate and kainate, assessed by measuring the glutamate acid decarboxylase (GAD) and the choline acetyltransferase (ChAT) activities. Similarly, CLEMENT and PHEBUS[7] have demonstrated that unilateral SN lesioning markedly attenuates the degree of ischemic neuronal damage in the ipsilateral striatum in a 4-vessel occlusion ischemia model.Taken together, these data indicate that the protection observed by CLEMENT AND PHEBUS is probably the result of reducing the excitotoxic insult due to the glutamate released during ischemia. However, WIELOCH et al.[8] found no difference in the densities of striatal damage in DA-lesioned and sham-operated animals. They suggest that the lack of effect might be due to the severity of the energy depletion occuring during the insult in the 2-vessel occlusion ischemia model used.

The mechanism by which SN lesioning reduces the vulnerability of the striatum to excitotoxin is not yet known. Microdialysis has revealed that the extracellular striatal concentration of dopamine increases 1100-fold in the controls but not in the SN lesioned rats during cerebral focal ischemia. The deleterious effect of DA might be due to the free radicals formed as a consequence of dopamine oxidation[9]. Ischemic-induced glutamate release is also prevented in the DA-depleted striatum. These findings suggest that the protective effect could be linked to reduction in the release of DA or glutamate, or both, during ischemia.

Although ARBUTNOTT et al.[10] found that iontophoretic applied DA only inhibited striatal cells, others [11] have reported that a few of these cells responds to DA with increased activity. Similarly, KNAPP and DOWLING[12] observed that DA enhanced amino acid-gated conductance in cultured retinal cells. Thus DA may, under certain circumstances, potentiate the consequences of glutamatergic receptor overstimulation.

On the other hand, there is good evidence that the nigrostriatal dopaminergic input exerts presynaptic control over glutamatergic corticostriatal transmission[13]. TOSSMAN et al.[14] measured the extracellular concentration of glutamate in the rat striatum after a unilateral 6-OHDA lesion of the dopaminergic system, and found a marked increase in the basal and evoked overflow of glutamate in the ipsilateral striatum. These results raise the question of the activity of the corticostriatal glutamatergic pathway after SN lesioning. The precise mechanism underlying the reduced vulnerability of the striatum to exitotoxic insult following lesioning of the SN remains to be established.

ACKNOWLEDGEMENT

This work was supported by a grant from the Institut Scientifique Roussel.

REFERENCES

1. W.A. Pulsinelli, J.B. Brierley, and F. Plum, Temporal profile of neuronal damage in a model of transient forebrain ischemia, Ann. Neurol., 11 : 491-498 (1982).
2. H. Benveniste, J. Drejer, A. Schousboe, and N.H. Diemer, Elevation of the extracellular concentration of glutamate and aspartate in rat hippocampus during transient cerebral ischemia monitored by intracerebral microdialysis, J. Neurochem., 43 : 1369-1374 (1984).
3. M.Y.T. Globus, R. Busto, W.D. Dietrich, P. Scheinberg, and M.D. Ginsberg, Substantia nigra lesion protects against ischemic damage in the striatum, Neurosci. Lett.,80 : 251-256 (1987).
4. F Hefti, E. Melamed, and R.J. Wurtman, Partial lesion of the dopaminergic nigrostriatal system in rat brain : biochemical characterisation, Brain Res., 195 : 123-137 (1980).
5. A. Tamura, D.I. Graham, J. Mc Culloch, and G.M. Teasdale, Focal cerebral ischemia in the rat : 1. Description of technique and early neuropathological consequences following middle cerebral artery occlusion, J. Cereb. Blood Flow Metab., 1 : 53-60 (1981).
6. A.G. Chapman, N. Dürmuller, G.J. Lee, and B.S Meldrum, Excitotoxicity of NMDA and kainic acid is modulated by nigrostriatal dopaminergic fibres, .Neurosci. Lett.,107 : 256-260 (1989).
7. J.A. Clement, and L.A. Phebus, Dopamine depletion protects striatal neurons from ischemia-induced cells death, Life Sci.,42 : 707-713 (1988).
8. T. Wieloch, Y. Miyauchi, and O. Lindvall, Neuronal damage in the striatum following forebrain ischemia : lack of effect of selective lesions of mesostriatal dopamine neurons, Exp. Brain Res., 83 : 159-163 (1990).
9. A. Slivka, and G. Cohen, Hydroxyl radical attack on dopamine, J.Biol. Chem., 260 : 15466-15472 (1985).

10. G.W. Arbuthnott, J.R. Brown, N.K. MacLeod, R. Mitchell, and A.K. Wright, The action of dopamine on synaptic transmission through the striatum, in Neurotransmitter interactions in the basal ganglia, M. Sandler et al. (Eds.), Raven Press, New York, (1987).

11. K. Hirata, C. Y. Yim, and G.J. Mogenson, Excitatory input from sensory motor cortex to neostriatum and its modification by conditioning stimulation of the substantia nigra, Brain Res, 321 : 1-8 (1984).

12. A.G. Knapp, and J.E. Dowling, Dopamine enhances excitatory amino acid-gated conductances in cultured retinal horizontal cells, Nature (Lond.), 325 : 437-439 (1987).

13. L. Kerkerian, N. Dusticier, and A. Nieoullon, Modulatory effect of dopamine on high-affinity glutamate uptake in the rat striatum, J. Neurochem., 48 : 1301-1306 (1987).

14. U. Tossman, J. Segovia, and U. Ungerstedt, Extracellular levels of amino acids in striatum and globus pallidus of 6-hydroxy dopamine-lesioned rats measured with microdialysis, Acta Physiol. Scand.,127 : 547-551 (1986).

NORADRENERGIC MODULATION OF EXCITOTOXICITY

B. Arvin, E. Le Peillet, N. Dürmüller,
A.G. Chapman and B.S. Meldrum

Department of Neurology, Institute of
Psychiatry, London (U.K.)

INTRODUCTION

The noradrenergic system originating in neurons of the locus coeruleus diffusely innervates forebrain structures. It is thought to modulate synaptic activity in many ways; noradrenaline has presynaptic actions on auto- and hetero-receptors and post-synaptic actions on several classes of receptor that influence second messenger systems and ionic conductances. Several specific modulatory effects on excitatory and inhibitory synaptic potentials have been identified (see Discussion below). Additionally noradrenaline influences local blood flow and the movement of ions between plasma and brain tissue. Thus it is not surprising that reports of the action on ischemic brain damage of agents modifying noradrenergic transmission or of lesions of the ascending noradrenergic system describe a diversity of effects that are difficult to integrate into a consistent explanatory scheme. We have therefore attempted to isolate one aspect of the problem by studying the effects of lesioning the ascending noradrenergic system on the excitotoxic action of kainate injected focally into the rat hippocampus. We have used two different methods of lesioning the pathway unilaterally (electrolytic lesions of the locus coeruleus (LC), and 6-OH-dopamine lesions of the medial forebrain bundle (MFB)), and have assessed the completeness and specificity of the lesion by tissue measurements of monoamine content and by studying evoked release of monoamines by intrahippocampal microdialysis.

METHOD

Male Wistar rats (250-300g) were anaesthetised with pentobarbitone (60mg/kg i.p) for LC lesioning and for MFB and KA injections and with chloral hydrate (300 mg/kg, i.p.) for implantation of dialysis probes . For all procedures the rats were placed in a small stereotaxic frame. The coordinates (from I.A.) were as follows:- Unilateral LC

The Role of Neurotransmitters in Brain Injury, Edited by
M. Globus and W.D. Dietrich, Plenum Press, New York, 1992

lesion: AP:5.5, L:1.1, V:2.1 (from the skull at an angle of 20° to the vertical); Unilateral MFB lesion: AP:5.2, L:1.2, V:1.6; Bilateral KA lesion: AP:5.2, L:2.4, V:6.7.

LC lesions were carried out using a monopolar stimulating electrode with a current of 0.3mA for 10sec. KA (1.1nmol/1µl/10min dissolved in phosphate buffered saline, pH 7.4) was injected bilaterally into dorsal hippocampus one week after the LC lesions. Following the injection the cannulae was left in place for a further 10 min.

Medial forebrain bundle (MFB) lesions were performed unilaterally using 6-hydroxy dopamine H-Br (8.8 µg 6-OHDA/2µl saline with 0.2% ascorbic acid/5min).

Determinations of tissue monoamine levels were carried out one week after the electrolytic lesion. The striatum and hippocampus were dissected on ice, weighed, homogenized and monoamine levels determined using acid extraction and HPLC analysis with coloumetric detection.

Dialysis probes were implanted 7 days after the MFB lesion. A transverse dialysis probe (Filtrial AN 69 HF; 200µm I.D. supported by a fine stainless steel wire) was inserted through a drilled hole in the left temporal bone and recovered from a hole in the right side. The fibres were secured on the surface of the skull using cannulae and dental acrylate. After a 3 hour recovery period 3 x 20 min (rate of 1µl/min) basal samples the perfusion medium (120 mM $NaCl_2$, 12 mM glucose, 1 mM $Cacl_2$, in 25 mM NaH PO_4 buffer pH 7.2) was changed to one containing 500 µM KA for 20 min and 3 further samples were collected. All samples were assayed for catecholamines using an HPLC-electrochemical detector employing reverse-phase liquid chromatography and citric acetate buffer pH 5.0 in 2% methanol. Estimation of percentage recovery of monoamines was done in two fibers using a standard solution of monoamines (0.1µM at 37°C) and was found to be between 15-23%.

Histology was performed 7 days after the KA injections and 5 days after microdialysis. The animals were perfusion-fixed (10% formalin for focal injections, and formaldehyde: acetic acid: methanol, 1:1:8 in the dialysis experiments). Subsequently 40 µm and 7 µm sections were cut and stained with cresyl fast violet and luxol fast blue/cresyl fast violet respectively. Hippocampal sections 160 µm anterior and posterior to the injection site and 1 mm either side of the dialysis probe were used to give a score for each hippocampus zone (CA1-CA4 and DG cells). A six point scoring system was adopted with values ranging from 0-5 (corresponding to cell losses of approximately 0-10%, 10-30%, 30-50%, 50-70%, 70-90%, 90-100%).

RESULTS

Seven days following LC lesion DA and NA levels in the hippocampal tissue were reduced by 80% and 60% respectively. The 5HT content was unaltered. DA, NA and 5HT content in the striatal tissue also remained unchanged after LC lesion. In the microdialysis experiments basal extracellular levels of

NA were found to be 400 nM. KA (500 μM for 20 min through
the probe) induced a rise in extracellular NA to 520 nM the
values remained high until the end of the experiment (sample
8). On the MFB lesioned side the basal NA levels remained
the same (Table 1) but the KA-stimulated rise in NA was
absent. Extracellular DA levels were extremely low (2nM)
and in some cases beyond the detection scope of our system.

Basal and post KA-stimulation values of HIAA and 5HT
were similar in both control and lesioned rats (see table
1).

For assessment of histology in the LC lesioned animals
three groups were examined. (I) Non-operated (n=20), (II)
sham-operated (n=4) and (III) LC lesioned (n=8). Bilateral
KA injection (1.1nmol) in groups I and II produced
widespread damage with 90-100% damage of cells in CA1, CA2,
CA3 and CA4 cells and 70-90% damage in DG cells. In group
III, KA toxicity was reduced in both the ipsilateral (0-30%
damage) and contralateral side (50-70% damage in CA1 and DG

Table 1. Hippocampal Monoamines in Sham
and 6OHDA treated rats

		Basal nM	20'post KA nM
NA	-SHAM	420±30	525±45
	-6OHDA	408±32	410±38
DA	-SHAM	~2	~2
	-6OHDA		
DOPAC	-SHAM	31±2	17±2
	-6OHDA	28±4	20±2
HIAA	-SHAM	140±10	100±10
	-6OHDA	190±30	130±30
5HT	-SHAM	~20	~20
	-6OHDA		

Seven days after a unilateral MFB lesion a
dialysis probe was implanted into the
hippocampus (exposed only in the side
ipsilateral to MFB lesion) and after 3h 3 x
20min samples were collected. Following this
500 μM KA was infused through the probe for 20
min and 3 further samples were collected. All
samples were assayed for catecholamines using
HPLC-electrochemical detection. The values in
the table represent mean values for basal
levels 20 min prior to and mean values 20 min
after KA. Statistical analysis was done using
Student's t-test.**= <0.01.n/s not significant.

cells). In the ipsilateral side there was no protection in
CA3 or CA4 cells and in the contralateral side no protection
in CA2, CA3 and CA4. For microdialyis studies, infusion of
KA (500µM through the probe) in one hippocampus for 20
minutes in control animals resulted in total loss of
hippocampal pyramidal cells and a 50% loss of DG cells
(assessed 1mm either side of the probe). In animals which
were injected with 6-OHDA in MFB KA failed to produce damage
in CA1 and CA2 and granule cells whilst little or no
protection was seen in CA3 and CA4 neurons.

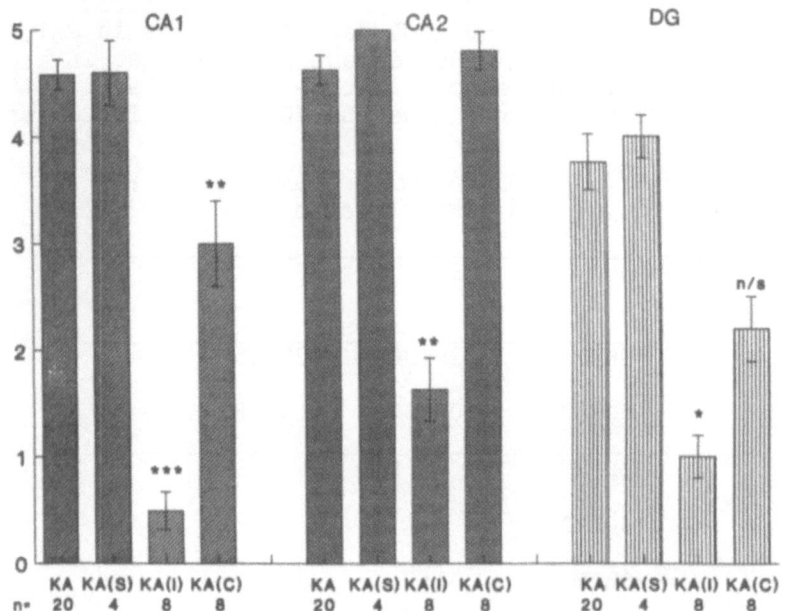

Fig. 1. Seven days after unilateral LC lesion bilateral
        injections of KA (1.1nmol) were carried out in
        control (KA, n=20), sham operated (KA(S), n=4) and
        LC lesioned ipilateral side KA(I) and contralateral
        side KA(C), n=8). Sections were cut (40 µm) and
        stained for Nissl substance. Damage was assessed
        160 µm anterior and posterior to the injection site.
        A six point scoring system was used with values from
        0 to 5:- 0=0-10%, 1=10-30%, 2=30-50%, 3=50-70%,
        4=70-90% and 5=90-100% neuronal loss.
        Statistical analysis were done using Mann-Whitney
        test.   *** = <0.001 ** = <0.01 * = <0.1
        n/s = not significant.

DISCUSSION

    We have previously shown that lesioning of the nigro-
striatal dopaminergic pathway protects against excitotoxic
damage induced by kainate or N-Methyl-D-aspartate (NMDA) in
the rat striatum (Chapman et al., 1989). A similar

protective effect against the toxicity of quinolinate has recently been reported (Buisson et al., 1991) . Our current results suggest that the noradrenergic input to the hippocampus facilitates excitotoxic damage in a manner parallel to that in which the dopaminergic input to the striatum facilitates excitotoxic damage. Noradrenaline has been shown to enhance glutamate mediated excitation and decrease synaptic inhibition in several electrophysiological studies but until now there have been no studies investigating how NA modulation affects excitotoxicity. Electrophysiological studies have shown that NA enhances glutamate-induced excitation in rat neocortical neurons (Dodt et al., 1991), rat cortical slices (Mouradian et al., 1991), rat hippocampus (Sara and Bergis., 1991) and frog spinal cord (Wohlberg et al., 1987). In contrast Lehmenkuehler et al., 1991 reported that NA decreases NMDA-induced responses in rat motorcortex. In one study NA has been shown to decrease synaptic inhibition in the hippocampus, possibly by acting on GABAergic interneurons (Madison and Nicoll., 1988).Another study shows that NA (via β_1 receptors) excites inhibitory interneurons and enhances ortho- and anti-dromically evoked IPSPs (Andreasen & Lambert., 1991).

Our results show that LC lesions (which result in a decrease in the tissue content of NA) and MFB lesions (which abolish the KA-induced increase in extracellular NA) protect against KA toxicity in the hippocampus. Since LC lesions led to a reduction in tissue level of DA as well as NA and also since in the MFB lesions the extracellular DA were not accurately measurable, a contribution of DA to KA-induced toxicity cannot be ruled out. However, noradrenergic inputs to the hippocampus are 1-2 orders of magnitude greater than the dopaminergic inputs; the role of NA in enhancing excitotoxicity in the hippocampus is likely to be predominant.

The pattern of protection against KA toxicity in both LC and MFB lesions (CA1 and CA2 and not in CA3 and CA4) would argue against a direct interaction between KA and NA receptors. This pattern of protection against KA toxicity has also been reported with NMDA antagonists (Jarrard and Meldrum., 1991) and non-NMDA antagonists (Arvin et al., 1991) and therefore may be related to a generalized reduction in excitatory activity.

ACKNOWLEDGEMENTS

We thank the Medical Research Council the Bethlem-Maudsley Research fund and the Wellcome Trust for financial support.

REFERENCES

Andreasen, M. and Lambert, J.D.C., 1991, Noradrenaline receptors participate in the regulation of GABAergic inhibition in area CA1 of the rat hippocampus. J. Physiol., 439:649-669.

Arvin, B., Chapman, A.G., and Meldrum, B.S., 1991,
    Monoaminergic activity and excitotoxicity, in:
    "Excitatory Amino Acids", B.S. Meldrum, F. Moroni,
    R.P. Simon and J.H. Woods, eds., Raven Press, New
    York, p.627-634.

Arvin, B., Moncada, C., Le Peillet, E. and Meldrum, B.S.,
    1991, The non-NMDA antagonists NBQX and GYKI 52466
    protect against kainic acid toxicity but not AMPA
    toxicity in the rat hippocampus. XVth International
    Symposium on Cerebral Blood Flow and Metabolism
    (BRAIN-91), The Role of Neurotransmitters in Brain
    Injury, Key West, Florida, June 7-9.

Buisson, A., Pateau, V., Plotkine, M. and Boulu, R.G., 1991,
    Role of nigrostriatal projections in the vulnerability
    of striatum to quinolinic caid. XVth International
    Symposium on Cerebral Blood Flow and Metabolism (BRAIN
    91), The Role of Neurotransmitters In Brain Injury,
    Key West Florida, June 7-9.

Dodt, H., Pawelzik, H., and Zieglgaensberger., 1991, Action
    of noradrenaline on neocortical neurons in vitro.
    Brain Res., 545:307-311.

Chapman, A.G., Dürmüller, N., Lees, G.J. and Meldrum, B.S.,
    1989, Excitotoxicity of NMDA and kainic acid is
    modulated by nigrostriatal dopaminergic fibers.
    Neurosci. Lett., 107: 256-260.

Jarrard, L.E. & Meldrum, B.S., 1991, Selective excitotoxic
    pathology in the rat hippocampus. The Hippocampus.,
    submitted.

Loy,R., Koziell, D., Lindsey, J. and Moore, R.Y., 1980,
    Noradrenergic innervation of the adult rat hippocampal
    formation. J. Comp. Neurol., 189:699-710.

Lehmenkuehler, C., Walden, J., and Speckmenn, E.J., 1991,
    Decrease of N-methyl-D-aspartate responses by
    noradrenaline in the rat motocortex in vivo, Neurosci.
    Lett., 121:5-8.

Madison, D.V. and Nicoll, R.A., 1988, Norepinephrine
    decreases synaptic inhibition in the rat hippocampus.
    Brain Res., 442:131-138.

Mouradian, R.D., Sessler, F.M., and Watrehouse, B.D., 1991,
    Noradrenergic potentiation of excitatory transmitter
    action in cerebrocortical slices: evidence for
    mediation by an alpha 1 receptor-linked second
    messenger pathway, Brain Res., 546:1991) 83-95.

Sara, S.J., and Bergis, O., 1991, Enhancement of
    excitability and inhibitory processes in hippocampal
    dentate gyrus by Noradrenaline: A pharmacological
    study in awake, free moving rats, Neurosci. Lett.,
    126:1-5.

Wohlberg, C.J., Heckman., J.C and Davidoff, R.A., 1987,
    Epinephrine and norepinephrine modulate neuronal
    responses to excitatory amino acids and agonists in
    frog spinal cord., Synapse 1:202-207.

EFFECT OF INJURY ON α_1-ADRENORECEPTORS

IN RAT BRAIN *IN VIVO*

Hanna M. Pappius, Suzan Dyve*,
Michael McHugh and Albert Gjedde*

Goad Unit of the Donner Laboratory
of Experimental Neurochemistry, and
*Positron Imaging Laboratories
Montreal Neurological Institute
McGill University
Montreal, Quebec, Canada H3A 2B4

INTRODUCTION

We have used a freezing lesion in the rat as a model of cerebral injury (Pappius, 1981) and the deoxyglucose method of Sokoloff et al. (1977) to assess the functional state of traumatized brain. These studies indicated that with time after a focal lesion a widespread but not uniform depression of local cerebral glucose utilization (LCGU) developed, measurable within 4 hours and persisting for at least 5 days. Quantitation of CMRgluc in various structures showed that the cortical areas throughout the lesioned hemisphere were the most affected, the average cortical glucose utilization falling to about 50% of normal 3 days following the lesion. The effect was not restricted to areas surrounding the lesion or overlying edematous white matter, but involved the whole hemisphere from frontal to visual cortex. Heat lesions had a similar effect (Pappius, 1981).

Since we also demonstrated that blood flow was not significantly altered under these conditions, we interpreted the demonstrated depression of cortical glucose metabolism as reflecting a depression of cortical activity. In other words, we believe that the decreased cortical glucose use in the traumatized hemisphere is not due to decreased energy supply but rather is the result of diminished need for energy because of diminished cortical functional activity. Such an interpretation is supported by the significant correlation of some somatosensory deficits with the extent of depression of cortical glucose utilization in the traumatized hemisphere (Colle et al., 1986) and the finding that the reduction of cerebral metabolism by pentobarbital and isoflurane in lesioned brain was limited by the metabolic depression which had already occurred as a result of injury (Archer et al., 1990). Furthermore, our recent studies showed that the energy status and substrate supply in the cortical areas of the injured hemisphere had not been compromised. On the contrary, we found an enriched metabolic profile in the very cortical areas in which glucose use was depressed (Buczek et al., 1991).

Subsequent studies implicated changes in both the serotonergic and the noradrenergic neurotransmitter systems in the mechanisms underlying the functional depression in the freezing lesion model of brain injury (Pappius and Dadoun, 1986, 1987; Pappius et al., 1988; Inoue et al., 1991). Most recently, using a method developed by Dyve et al. (1989) and subsequently modified by Gjedde et al. (1991), we showed that in normal rat brain specific binding of a selective α_1-adrenoreceptor ligand, [^{125}I]HEAT,([^{125}I]-iodo-2-[ß-(4-hydroxyphenyl)-ethyl-amino methyl]tetralone) occurs *in vivo*. The anatomical distribution of the specific binding of labelled HEAT *in vivo*, in excess of physical solubility and non-specific binding, agreed closely with that defined previously *in vitro* (Jones et al., 1985). Furthermore, we were able to demonstrate that the *in vivo* binding occurs at two different sites with different affinities (Gjedde et al., 1991) similar to those reported for α_{1A}- and α_{1B}-adrenoreceptor subtypes (Morrow and Creese, 1986; Johnson and Minneman, 1987; Minneman, 1988). We then applied this method to study the effects of injury.

The data summarized below provide evidence that changes in low-affinity α_1-adrenoreceptors, usually designated α_{1B}, may contribute to the functional depression in injured brain.

MATERIALS AND METHODS

Standardized freezing lesions were made on the exposed dura in the left parietal region of 33 halothane anaesthetized Sprague-Dawley male rats (280-320g) as described previously (e.g. Pappius et al., 1981, 1988). Three days later the ligand ([^{125}I]HEAT-80μCi on production date with unlabelled HEAT 0 to 250 nmol) was injected into awake, partially restrained animals. Arterial blood samples were taken and the animals were decapitated 60 minutes later. Tissue radioactivity was determined by quantitative autoradiography using [^{125}I]HEAT standards prepared from each batch of the ligand. Blood radioactivity was converted to HEAT concentration using specific activity in the injectate. For each animal the total volume of distribution of the tracer [^{125}I]HEAT was estimated for each region of interest from tissue and blood levels of HEAT at 60 minutes of tracer circulation. Non-specific binding was corrected for by subtraction of the volume of distribution for the corresponding cerebellar cortex where little or no specific binding was previously demonstrated (Dyve et al., 1989). Sixteen normal animals were similarily studied. The kinetic constants for the two groups of animals, normal and injured, were calculated as described (Gjedde et al., 1991). Briefly, the B_{max} for the two adrenoreceptor subtypes were calculated in three stages. The apparently single Michaelis constant Kd was first shown to be concentration dependent signifying more than one binding site. The Michaelis constants K_d^L and K_d^H for low and high affinity binding for all regions combined were then obtained by non-linear regression. These were found not to be significantly different in normal and injured brain. Finally, using the above constants and non-linear regression, B_{max} values were calculated for low and high affinity binding for individual regions of interest.

RESULTS AND DISCUSSION

Results summarized in Table 1 show that three days after a freezing lesion changes in the B_{max} of the *in vivo* low-affinity binding of [^{125}I]HEAT were restricted to the cortical areas of the traumatized hemisphere.

B_{max} of the low-affinity binding was highly significantly increased in the lesioned left hemisphere in the three cortical areas examined, namely sensorimotor, auditory and visual, as compared to both the normal and the non-lesioned side. At the same time in representative sub-cortical structures, medial geniculate and lateral thalamus, there was no significant effect of injury on the B_{max} of the low-affinity binding of [^{125}I]HEAT. The areas of increased low-affinity binding correlated closely with areas of diminished glucose utilization, suggesting that the demonstrated changes in the density of α_{1B}-adrenoreceptors may be of functional importance in injured brain.

Table 1.   Effect of Injury on Low-Affinity Binding of [I^{125}]HEAT in Rat Brain *In Vivo*

B_{max}[a]
(pmol g^{-1})

Region	Normal	Three days after injury Hemisphere	
		Left (lesioned)	Right
Sensory Motor Ctx	185 ± 42	442 ± 60*‡	157 ± 31
Auditory Ctx	339 ± 23	468 ± 16*‡	217 ± 31
Visual Ctx	316 ± 33	525 ± 72*†	238 ± 35
Medial Geniculate	689 ± 163	534 ± 47	561 ± 98
Lateral Thalamus	467 ± 144	473 ± 55	645 ± 60

[a] Estimates by non-linear regression $\pm$ SE. N (number of different plasma concentrations of HEAT at which binding was determined) 8 for normal, 12 for lesioned; 2-6 animals were studied at each plasma concentration.
* $p < 0.01$ from right hemisphere
‡ $p < 0.01$ from normal
† $p < 0.05$ from normal

Such a conclusion is also strongly supported by the effects of prazosin, another specific α_1-adrenergic receptor blocker, on cortical glucose utilization in injured brain (Inoue et al., 1991). Prazosin given 30 min before lesioning in a single 1 mg/kg dose which was without effect in normal animals, significantly improved cortical glucose utilization three days later, at a time when average CMRgluc in untreated animals was 50% of normal. Prazosin 3 mg/kg/day, started before the lesion and given for 3 days, produced further normalization of the depressed glucose utilization specifically in areas affected by injury, despite the fact that this regimen induced significant global decrease in glucose utilization in normal animals. Prazosin given acutely 3 days after the lesion just before the deoxyglucose study was without effect, at least in the dose used in our studies.

In contrast, results presented in Table 2 show that three days after injury the B_{max} of the *in vivo* high-affinity binding of $[^{125}I]$HEAT was significantly decreased bilaterally in most of the cortical and subcortical structures in which it was determined, as compared to the values obtained in normal animals.

This effect may be related to the bilateral activation of the norepinephrine metabolism in injured brain which we have demonstrated previously. Norepinephrine (NE) content was shown to be decreased in the cortex of both hemispheres for up to 8 days after lesioning (Pappius and Dadoun, 1986) while in other studies a bilateral increase in 3-methoxy-4-hydroxy-phenylglycol sulfate (MHPG-SO$_4$), the principal NE metabolite in rat brain, was demonstrated (Pappius, 1991).

Table 2.   Effect of Injury on High-Affinity Binding of $[I^{125}]$HEAT in Rat Brain *In Vivo*

Region	Normal	Three days after injury Hemisphere	
		Left (lesioned)	Right
Sensory Motor Ctx	9 ± 1	5 ± 1[†]	4 ± 1[‡]
Auditory Ctx	14 ± 1	9 ± 2[†]	8 ± 1[‡]
Visual Ctx	5 ± 1	7 ± 1	5 ± 1
Medial Geniculate	41 ± 4	24 ± 1[‡]	25 ± 1[‡]
Lateral Thalamus	36 ± 3	18 ± 1[‡]	18 ± 1[‡]

B_{max}[a]
(pmol g^{-1})

[a] Estimates ± SE, see Table 1.
[†] $p < 0.05$ from normal
[‡] $p < 0.01$ from normal

The distribution of the apparently generalized changes in the B_{max} of the high-affinity binding of $[^{125}I]$HEAT in lesioned brain was not correlated in any way with the distribution of the unilateral, mostly cortical, decrease of glucose utilization. Thus there is no evidence that the decreased α_{1A} binding contributes to the cortical dysfunction resulting from injury, as delineated by unilateral decrease in cortical CMRgluc.

CONCLUSIONS

On the basis of these studies we conclude that the noradrenergic system plays a role in functional consequences of brain injury and that this effect is, at least in part, mediated by α_1-adrenergic receptors. From the changes in the B_{max} of the low-affinity adrenoreceptor binding in lesioned brain it appears that mediation by α_{1B} adrenoreceptors is most likely involved.

SUMMARY

We have shown that three days following a focal cortical freezing lesion the B_{max} of the low-affinity α_1-adrenoreceptor binding (α_{1B}) is significantly increased in the cortical areas of the injured hemisphere while B_{max} of the high-affinity α_1-adrenoreceptor binding (α_{1A}) is globally decreased.

Areas of the injury-induced increase in the cortical low-affinity α_1-adrenoreceptor binding (α_{1B}) correspond to the areas of functional cortical depression as delineated by decreased CMR_{gluc}.

Together with the previous finding that prazosin, a specific α_1-adrenoreceptor blocker, prevents the development of cortical dysfunction as reflected by decreased CMR_{gluc}, these results suggest that the demonstrated increase in cortical low-affinity α_1-adrenoreceptor density throughout the cortex of the lesioned hemisphere may be of functional importance in injured brain.

REFERENCES

Archer, D.P., Elphinstone, M.G., and Pappius, H.M., 1990, The effect of pentobarbital and isoflurane on glucose metabolism in thermally injured rat brain, J. Cereb. Blood Flow Metab., 10:624.

Colle, L.M., Holmes, L.J., and Pappius, H.M., 1986, Correlation between behavioral status and cerebral glucose utilization in rats following freezing lesion, Brain Res., 397:27.

Dyve, S., Gjedde, A., Diksic, M., Sherwin, A., Hakim, A., 1989, In vivo quantification of blood-brain transfer and binding of [^{125}I]HEAT, an α_1 adrenoceptor antagonist, Synapse, 3:205.

Gjedde, A., Dyve, S., Yang, Y-J., McHugh, M., and Pappius, H.M., 1991, Bi-affinity alpha-1 adrenoceptor binding sites in normal rat brain in vivo, Synapse, (in press).

Inoue, M., McHugh, M., and Pappius, H.M., 1991, The effect of alpha adrenergic receptor blockers prazosin and yohimbine on cerebral metabolism and biogenic amine content of traumatized brain, J. Cereb. Blood Flow Metab., 11:242.

Johnson, R.D., and Minneman, K.P., 1987, Differentiation of α_1-adrenergic receptors linked to phosphatidylinositol turnover and cyclic AMP accumulation in rat brain, Mol. Pharmacol., 31:239.

Jones, L.S., Gauger, L.L., and Davis, J.N., 1985, Anatomy of brain alpha$_1$-adrenergic receptors: In vitro autoradiography with [^{125}I]-HEAT, J. Comp. Neurol., 231:190.

Minneman, K.P., 1988, Alpha 1-adrenergic receptor subtypes, inositol phosphates, and sources of cell Ca^{2+}, Pharmacol. Rev., 40:87.

Morrow, A.L., and Creese, I., 1986, Characterization of α_1-adrenergic receptor subtypes in rat brain: A reevaluation of [^{3}H]WB4104 and [^{3}H]prazosin binding, Mol. Pharmacol., 29:321.

Pappius, H.M., 1981, Local cerebral glucose utilization in thermally traumatized rat brain, Ann. Neurol., 9:484.

Pappius, H.M., 1988, Significance of biogenic amines in functional disturbances resulting from brain injury, Metabolic Brain Dis., 3:303.

Pappius, H.M., 1991, Brain injury: New insights into neurotransmitter and receptor mechanisms, Neurochem. Res., 16:491.

Pappius, H.M., and Dadoun, R., 1986, Biogenic amines in injured brain, Trans. Am. Soc. Neurochem., 17:298.

Pappius, H.M., and Dadoun, R., 1987, The effects of injury on the indoleamines in cerebral cortex, J. Neurochem., 49:321.

Sokoloff, L., Reivich, M., Kennedy, D., DesRosiers, M.H., Patlak, C.S., Pettigrew, K.D., Sakurada, O., Shinohara, M., 1977, The (^{14}C)deoxyglucose method for the measurement of local cerebral glucose utilization: theory, procedure and normal values in the conscious and anesthetized albino rat, J. Neurochem., 28:897.

MODULATION OF BRAIN PROSTAGLANDIN SYNTHESIS BY THE NORADRENERGIC SYSTEM

[1]E. Shohami, and [2]J. Weidenfeld

[1]Department of Pharmacology, Hebrew University-Hadassah Medical
School, and [2]Department of Neurology, Hadassah University
Hospital and Laboratory of Endocrinology, Bikur Holim Hospital
Jerusalem, Israel

INTRODUCTION

The presence of prostaglandin (PG) in the central nervous system was
demonstrated more than 25 years ago (Samuelsson, 1964), however their
precise roles in normal and pathological conditions are not known to date.
The levels of free arachidonic acid (AA) and its metabolites in the CNS are
very low (Anggard, 1988; Wolfe, 1982). However various stimuli induce AA
release from membrane phospholipids (e.g. ischemia, trauma, electroconvulsive shock) resulting in the accumulation of PGs in the brain (Gaudet et al.
1980; Shohami et al., 1987; Ellis et al. 1984; Bazan et al., 1986). Little
is known about the endogenous factors that regulate brain PG synthesis, but
it is likely that both, hormones and neurotransmitters are involved in this
mechanism. Phospholipase A_2 (PLA_2) releases AA from membrane phospholipids
(Kunze and Wogt, 1971), and interactions of hormones or neurotransmitters
with specific receptors are among the factors that affect its activity.

The effect of reducing NE levels by α-MPT, on the levels of arachidonic
acid in mouse brain after electroconvulsion shock was studied by Aveldano de
Calderoni & Bazan (1979). They found that NE inhibition in vivo resulted in
reduced levels of AA, implying modulatory effect of catecholamines (CA) on
the eicosanoid cascade. A role for CA in regulating PG synthesis in neural
tissue in-vitro was also demonstrated (Wolfe et al. 1976). Noradrenaline,
dopamine and adrenaline, when added to rat brain synaptosomes significantly
stimulated the generation of PGE_2 (Hillier et al. 1976).

There is evidence that PGs are elevated as a response to immunological
challenge by bacterial endotoxin, lipopolysaccharide (LPs), (Cook et al.
1987). In the present study we investigate the role of endogenous CA in
regulating brain PG levels under basal conditions and following the activation of PG synthesis by LPS. We examined the effect of depleting endogenous
CA content by three different pharmacological manipulations, on the biosynthesis of PGs.

The Role of Neurotransmitters in Brain Injury, Edited by
M. Globus and W.D. Dietrich, Plenum Press, New York, 1992

MATERIALS AND METHODS

Experimental protocol

Rats (200g, strain of the Hebrew University) were anesthetized with sodium pentobarbital (35mg/kg) and injected into the lateral ventricle (icv) with 300μg/kg bw 6-OHDA (Sigma, St. Louis, Mo., USA) as described in detail elsewhere (Weidenfeld et al. 1989a). Six days later, rats were sacrificed to evaluate either their CA content (Felice et al. 1978) or PG synthetic capacity (Shohami & Gross 1985). Five days after 6-OHDA or vehicle administration, other groups of rats were injected i.p. with 500μg/kg bw LPS in 1ml saline or with saline alone. Twenty-four hours later, the rats were decapitated and the production of PG in both groups was determined.

In order to investigate the effect of inhibition of CA production at different steps of their biosynthesis, two other drugs were used: 1) α-Methyl-p-tyrosine (α-MPT) (Sigma) a specific inhibitor of the enzyme tyrosine-hydroxylase was injected i.p. 250mg/kg bw in saline and 20h later the rats were sacrificed. 2) FLA-63 [bis(4-methyl-1 homo piperazinyl thio-carbonyl)]disulfide (Regis USA) an inhibitor of dopamine-ß-hydroxylase was given i.p. 50mg/kg bw in saline and 4h later the rats were sacrificed. In another group of rats the drug was given 20h after LPS administration. Upon sacrifice, brain tissue was taken from these rats to measure CA content or PG rate of synthesis.

RESULTS

Table 1 depicts NE levels in the cortex and hypothalamus after 6-OHDA, α-MPT or FLA-63 as compared to their respective controls.

Table 1. The effect of 6-OHDA, α-MPT and FLA-63 on the content
of NE in brain tissue (ng/mg protein)

Treatment	Frontal cortex	Hypothalamus
Vehicle (i.c.v.)	4.1±0.5	32±8
6-OHDA (i.c.v.)	0.82±0.2*	9.1±3*
vehicle (i.p.)	4.6±0.4	31±6
α-MPT (i.p.)	0.3±0.02*	2.1±0.2*
FLA-63 (i.p.)	0.4±0.02*	2.5±0.3*

Values represent the mean ± SEM obtained from 8-10 rats.
*P < 0.05 as compared to their respective vehicle control, by student t-test.
6-OHDA or vehicle were injected icv (300μg/kg). NE content was measured 6 days later.
α-MPT (250mg/kg) and FLA-63 (50mg/kg) were injected i.p. and NE content was measured 20h or 4h later.

As expected, icv injection of 6-OHDA significantly reduced NE content by 80 and 72% in the cortex and hypothalamus, respectively. The administration of TH inhibitor, α-MPT or DBH inhibitor, FLA-63 reduced NE content by more than 90% in both cortex and hypothalamus.

The ex-vivo release of eicosanoids from brain slices following icv injection of 6-OHDA is shown in Fig. 1.

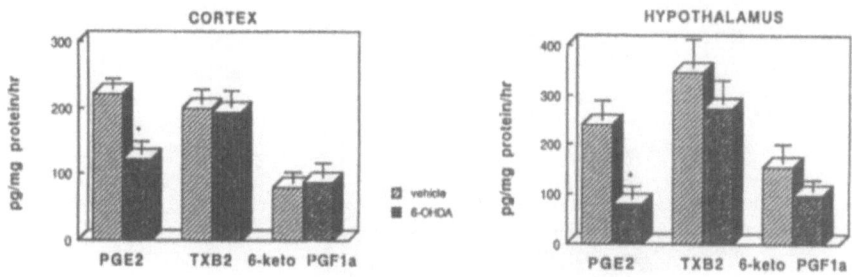

Rats were injected with 6-OHDA (300µg/kg, icv) or vehicle and
sacrificed 6 days later. Brain tissue was studied ex-vivo for
eicosanoids release.
*$P<0.05$ as compared to the respective vehicle control, by student
t-test.

Fig. 1. Release of eicosanoids (pg/mg protein/h) from various
brain regions taken from vehicle or 6-OHDA treated rats

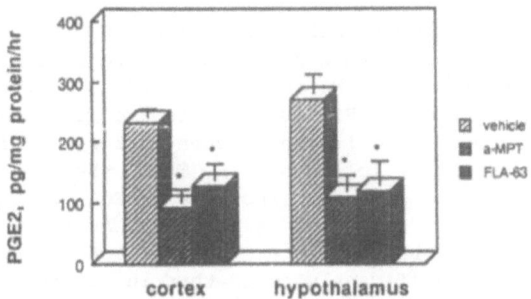

α-MPT (250mg/kg) and FLA-63 (50mg/kg) were injected i.p. and PGE_2
release was measured 20h or 4h later.
*$P<0.05$ as compared to control, by student t-test.

Fig. 2. Effect of α-MPT and FLA-63 on PGE_2 release (pg/mg
protein/h)

6-OHDA significantly reduced the release of PGE_2 from cortical slices
by 45%, and from the hypothalamus by 66%. The administration of 6-OHDA did
not affect the release of TXB_2 and 6-keto-$PGE_{1\alpha}$ in the cortex. In the hypothalamus, 6-OHDA slightly reduced TXB_2 and 6-keto-$PGE_{1\alpha}$ release but this
effect did not reach statistical significance.

In the following experiments, we did not measure TXB_2 and 6-keto-$PGE_{1\alpha}$
since the depletion of NE by 6-OHDA did not affect the release of these
eicosanoids.

Fig. 2 shows the effect of α-MPT and FLA-63 on the release of PGE_2 from
slices of cortex, and hypothalamus.

Both drugs significantly reduced PGE_2 release by 45-60% from the
frontal cortex and the hypothalamus.

Administration of LPS enhanced PGE_2 release by 2-2.5-fold as compared
to control rats in both brain areas tested. Fig. 3 summarizes the effect of
NE depletion on the LPS stimulated production of PGE_2.

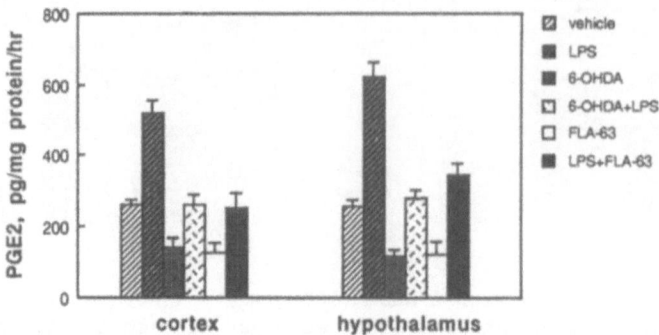

LPS (500µg/kg) was given i.p. to control (vehicle treated) rats or to
rats treated with either 6-OHDA (300µg/kg) or with FLA-63 (50mg/kg).
*P < 0.05 as compared to the respective control, by student's t-test.

Fig. 3. Release of PGE2 from brain slices after LPS administration

6-OHDA pretreated animals, which exhibited reduced basal PG synthesis,
also responded to LPS with a 2-2.5 fold increase in synthesis, which is
similar to the extent of the increase found in control animals. Similar
results were found when LPS was given in a combination with FLA-63. From the
results shown in Fig. 3, it can be seen that FLA-63 failed to affect the
induced PGE_2 synthesis following LPS. Thus, the combined treatment of LPS
with FLA-63 resulted in 2-fold increase in PGE_2 as compared to FLA-63 treatment alone, the same extent of stimulation was found when control rats were
given LPS. However, the absolute levels of PGE_2 after LPS challenge in CA
depleted rat is significantly lower than in control LPS treated rats.

DISCUSSION

In the present study we have reduced the content of brain CA, and
specifically of NE, by three different pharmacological manipulations. The
three drugs, namely, 6-OHDA, FLA-63 and α-MPT, effectively reduced NE
content in the cortex and hypothalamus. At the same time, a marked reductio
in the basal synthesis of PGE_2 was found in the cortex (by 45%) and hypothalamus (by 66%). It is interesting to note that in the brain structures i
which PGE_2 production was reduced following NE depletion, the production of
TXB_2 and $6-keto-PGE_{1\alpha}$ remained unchaged, similar to control rats. This
finding may be attributed to the fact that TXB_2 and $6-keto-PGE_{1\alpha}$ are synthesized mainly by cell types in the CNS (namely astroglia and endothelial
cells) which are relatively less sensitive to changes in NE levels (Murphy
et al 1988).

Our results indicate that, at least part of the tonic production of
PGE_2 in the cortex and the hypothalamus is dependent upon intact CA neural
input. The specific involvement of NE in regulating PGE_2 synthesis is based
on the finding that administration of FLA-63, a specific DBH inhibitor whic
selectively reduces brain NE and epinephrine, resulted in about 50% reduction of PGE_2 production. The possibility that the inhibitory effect of
FLA-63 is due to a direct effect of the drug on PGE_2 synthesis is unlikely,
since it did not reduce PGE_2 production in the liver (data not shown), whic
is devoid of NE synthetic capacity.

In order to investigate the effect of NE depletion on induced PG production, we injected the bacterial endotoxin LPS. This endotoxin was chosen

as a stimulant to PG synthesis since the well known effect of LPS on brain eicosanoid synthesis plays an important role in host defence mechanisms e.g. production of fever. Following LPS administration, 2-2.5 fold increase was found in PGE_2 and TXB_2 production in both cortex and hypothalamus. The reduced noradrenergic input following 6-OHDA did not prevent the PGE_2 response to LPS stimulation. Although the basal PGE_2 levels were lower, the degree of its increased induced production was similar to that found in control rats. When the enzyme DBH was inhibited by FLA-63 in rats pretreated with LPS, the level of PGE_2 measured 4h later was significantly higher (about 2-fold) as compared to rats treated with FLA-63 alone. It is therefore possible to assume that DBH inhibition affected only basal PGE_2 synthesis and not the LPS induced production.

These findings may suggest that brain PG synthesis is differentially regulated by the adrenergic system under basal conditions and following LPS stimulation. Specifically, the central adrenergic system appears to exert a facilitatory effect on basal PG synthesis. In contrast, LPS-induced PG synthesis does not require an intact adrenergic system. The notion of separate regulatory systems for basal and induced PG synthesis is in accordance with previous studies suggesting that endogenous glucocorticoids attenuate stress-induced brain PG but do not affect basal synthesis (Weidenfeld et al 1989b).

Although the present study has examined the effect of CA levels on PG synthesis, previous studies have provided evidence that PG inhibit NE release from central and peripheral nerve endings (Otorii et al 1985, Hedquist 1976). Thus, the data of the present study showing that the central CA system facilitates basal PG synthesis may lead one to speculate on the existence of a feedback mechanism between the central CA and PG systems.

Several modes of brain injury (i.e. ischemia, trauma) are known to enhance brain CA synthesis. In view of our present results, it may be speculated that the increase in PG production after brain injury is mediated at least in part by the increase in CA.

REFERENCES

Anggard E. (1988) Biosynthesis and metabolism in the brain. In: Prostaglandins: Biology and Chemistry of Prostaglandins and Related Eicosanoids (Curtis-Prior P.B., ed.) pp. 381-385. Churchill Livingstone, Edinburgh.

Aveldano de Caldironi M.I. and Bazan N.G. (1979) α-methyl-p-tyrosine inhibits the production of free arachidonic acid and diacylglycerols in brain after a single electroconvulsive shock. Neurochem. Res. 4: 213-2.

Bazan N.G., Birkle D.L., Tang W., and Reddy T.S. (1986) The accumulation of free arachidonic acid and the formation of prostaglandins and lipoxygenase reaction products in the brain during experimental epilepsy. In: Advances in Neurology, vol. 44; Basic Mechanisms of the epilepsies. Molecular and cellular approaches (Delgado-Escueta A.V., Ward A.A. and Woodbury D.M. eds) pp 879-902. Raven Press, New York.

Cook J.A., Tempel G.E., Ball H.A., Wise W.C., Matera G., Reines H.D. and Halushka P.V. (1987) Eicosanoids in sepsis and its sequelae. In: Eicosanoids in the Cardiovascular and Renal Systems, MTP Press Limited. International Publishers pp. 92-127.

Ellis E.F., Wright K.F. and Wei E.P. (1984) Cyclooxygenase products of arachidonic acid metabolism in cat cerebral cortex after experimental concussive bain injury. J. Neurochem. 37, 892-896.

Felice W.L., Felice J.D. and Kissinger P.T. (1978) Determination of catecholamines in rat brain by reversed phase ion-pair liquid chromatography. J. Neurochem. 31, 1461-1466.

Gaudet R.J., Alam I. and Levine L. (1980) Accumulation of cyclooxygenase products of arachidonic acid metabolism in gerbil brain during reperfusion after bilateral common carotid occlusion. J. Neurochem. 35, 653-658.

Hedqvist P. (1976) Effects of prostaglandins on autonomic transmission. In: Advances in Prostaglandin Research. Prostagladins: Physiological, Pharmacological and Pathological Aspects. (S.M.M. Karim ed) MTP Press Ltd., Lancaster, p. 37.

Hillier K., Roberts P.J. and Woollard P. (1976) Catecholamine stimulated prostaglandin synthesis in rat brain synaptosomes. Br. J. Pharmac. 58, 426P-427P.

Kunze H. and Wogt W. (1971) Significance of phospholipase A for prostaglandin formation. Ann. N.Y. Acad. Sci. 180, 123-125.

Murphy S., Pearce B., Jeremy J. and Dandona P. (1988) Astrocytes as eicosanoid-producing cells. Glia 1, 241-245.

Otorri T., Ohkubo K. and Suzuki K. (1985) Central noradrenergic neurons and the hypertensive effects of intracerebroventricularly administered prostaglandin E_2 in anesthetized rabbits. Prostaglandins 29, 25-33.

Samuelsson B. (1964) Identification of a smooth muscle-stimulating factor i: bovine brain prostaglandins and related factors. Biochim. Biophys. Acta 84: 218-219.

Shohami E., and Gross J. (1985) An ex-vivo method for evaluation prostaglandin synthetase activity in cortical slices of mouse brain. J. Neurochem 45, 132-136.

Shohami E., Shapira Y., Sidi A. and Cotev S. (1987) Head injury induces increased prostaglandin synthesis in rat brain. J. Cereb. Blood Flow and Metab. 7, 58-63.

Weidenfeld J., Abramsky O., and Ovadia H. (1989a) Evidence for the involvement of the central adrenergic system in interleukin 1 induced adrenocortical response. Neuropharmacol. 28: 1411-1414.

Weidenfeld J., Abu-Amer Y., and Shohami E. (1989b) Inhbition of rat brain prostaglandin synthesis by dexamethasone: lack of effect of dexamethasone phosphate ester and various hormonal steroids. Neuropharmacol. 27, 1295-1299.

Wolfe L.S. (1982) Eicosanoids: prostaglandins, thromboxane and other derivatives of carbon-20 unsaturated fatty acids. J. Neurochem. 38, 1-14.

Wolfe L.S., Rostworowski K. and Pappius H.M. (1976) The endogenous biosynthesis of prostaglandins by brain tissue in vitro. Can. J. Biochem. 54, 629-640.

POTENTIAL ROLE OF 5-HYDROXYTRYPTAMINE₁ₐ RECEPTORS IN CEREBRAL ISCHEMIA

Jochen H. M. Prehn and Josef Krieglstein

Institut für Pharmakologie und Toxikologie
Philipps-Universität,
Ketzerbach 63, 3550 Marburg/L., Germany

INTRODUCTION

In the course of cerebral ischemia, there is a massive increase in the concentration of both excitatory and inhibitory neurotransmitters within the extracellular space (Benveniste et al., 1984; Hagberg et al., 1987; Globus et al., 1988; Sarna et al., 1990). An overexcitation of neurons caused by excitatory amino acid neurotransmitters has been suggested to play a major role in the pathogenesis of ischemic neuronal damage (Jørgensen and Diemer, 1982; Simon et al., 1984). By acting on N-methyl-D-aspartate (NMDA) and non-NMDA receptors, glutamate causes an influx of Ca^{2+} and Na^+ into the neuron. The neuronal membrane strongly depolarizes and may allow Ca^{2+} to enter the cell via additional routes, e.g. voltage-sensitive calcium channels. These events may lead to an intracellular Ca^{2+} accumulation, which is neurotoxic (for an overview see Choi, 1988). Beside antagonists of glutamate, agents that produce a hyperpolarisation of the neuronal membrane may be capable of reducing the influx of Ca^{2+} through these ionophores and may exert neuroprotective effects.

5-Hydroxytryptamine₁ₐ (5-HT₁ₐ) receptors mediate an inhibitory, hyper-polarizing effect on cortical and hippocampal neurons via a Ca^{2+}-independent K^+-conductance (Beck et al., 1985; Andrade et al., 1986; Colino and Halliwell, 1987, Davies et al., 1987). Agonists of 5-HT₁ₐ receptors have been shown to mimick the hyperpolarizing action of 5-HT on the resting membrane potential, to increase the firing threshold and to decrease the firing rate of hippocampal CA1, cortical, and dorsal raphe neurons (Basse-Tomusk and Rebec, 1986; Andrade and Nicoll, 1987; Colino and Halliwell, 1987; Davies et al., 1987; Rowan and Anwyl, 1987).

EFFECTS OF 5-HT₁ₐ RECEPTOR AGONISTS IN MODELS OF GLOBAL CEREBRAL ISCHEMIA

Beside high densities of 5-HT₁ₐ receptors within the lateral septum and the dorsal raphe nuclei, Pazos and Palacios (1985a) observed a high density of those receptors in the external layers of the entorhinal cortex. Since the layer II of the entorhinal cortex projects to the hippocampus via the perforant path, and since there are commissural projections of this cortical area, a modulation of the neuronal activity in cells of this layer can influence the neuronal activity along the entire intrinsic hippocampal circuit. Within the hippocampal formation, a high density of 5-HT₁ₐ receptors

Table 1. Effects of 5-HT₁ₐ agonists on ischemic hippocampal damage.

Drug	Species	Ischemia	Effect	Reference
8-OH-DPAT	gerbil	BCA-O, 5 min	protection	Bode-Greuel et al., 1990a
BAY R 1531	gerbil	BCA-O, 5 min	protection	Bode-Greuel et al., 1990b
CM 57493	rat	2-VO, 10 min	protection	Nuglisch et al., 1990
Gepirone	gerbil	BCA-O, 5 min	no effect	Bode-Greuel et al., 1990b
Ipsapirone	gerbil	BCA-O, 5 min	protection	Bode-Greuel et al., 1990b

has also been observed in the molecular layer of the dentate gyrus, which might be similarily important, since this layer contains all the dendrites of the granule cells. Furthermore, a high density of 5-HT₁ₐ receptors has been found in the CA1 subfield. It therefore is of interest, to test 5-HT₁ₐ agonists for their neuroprotective activity in models of transient forebrain ischemia, which causes extensive neuronal damage of the CA1 pyramidal neurons (Kirino, 1982; Pulsinelli et al., 1982).

Bode-Greuel et al. (1990a, b) have explored the neuroprotective activity of 8-hydroxy-2-(di-n-propylamino)tetralin (8-OH-DPAT), ipsapirone, gepirone, and BAY R 1531 (6-methoxy-4-(di-n-propylamino)-1,3,4,5-tetrahydrobenz(c,d)indole) in a model of transient forebrain ischemia in the Mongolian gerbil (Table 1). The authors observed a pronounced neuroprotection of the CA1 neurons by 8-OH-DPAT (94 % protection at 5 mg/kg) and by BAY R 1531 (100 % protection at 3 mg/kg). Ipsapirone exhibited a protection of 53 % at the dose of 3 mg/kg. Nuglisch et al. (1990) have investigated the neuroprotective potency of CM 57493 (4-(3-trifluoromethylphenyl)-1-(2-cyanoethyl)-1,2,3,6-tetrahydropyridine), using a 2-vessel occlusion model in the rat (Table 1). These authors observed a 10 % reduction of the hippocampal neuronal damage at a dose of 1 mg/kg.

Small differences in cerebral temperature during and after ischemia have been demonstrated to influence the histopathological outcome after transient forebrain ischemia (Busto et al., 1987, 1989). Since 5-HT₁ₐ receptor agonists produce a dose-dependent hypothermia (Gudelsky et al., 1986), careful monitoring of the temperature is essential if 5-HT₁ₐ agonists are tested in experiments of cerebral ischemia. In the studies of Bode-Greuel et al. (1990a, b), the animals were prepared on a heating pad being adjusted at 37 °C. The rectal temperature of the gerbils varied in the range of ± 1.7 °C during the surgery. In the study of Nuglisch et al.(1990), the environmental temperature was adjusted at 30 °C both in the intraischemic and in the postischemic period. Measurement of the rectal and cerebral temperature revealed no differences in the temperature between drug-treated and vehicle-treated animals up to 4 h after ischemia (Prehn et al., submitted for publication, see also Fig. 1).

EFFECTS OF 5-HT₁ₐ RECEPTOR AGONISTS IN MODELS OF FOCAL CEREBRAL ISCHEMIA

The neuroprotective potency of 5-HT₁ₐ agonists have also been determined in models of focal cerebral ischemia. Bielenberg and Burkhardt (1990) have demonstrated neuroprotective properties of the 5-HT₁ₐ agonists 8-OH-DPAT, BAY R 1531, buspirone, ipsapirone, and gepirone in models of permanent focal cerebral ischemia in mice and rats after preischemic application of the drugs (Table 2). In the latter study, postischemic application of ipsa-

Table 2. Effect of 5-HT₁ₐ agonists on infarct size after permanent middle cerebral artery-occlusion.

Drug	Species	Effect	Reference
8-OH-DPAT	mouse, rat	reduction	Bielenberg and Burkhardt, 1990
BAY R 1531	rat	reduction	Bielenberg and Burkhardt, 1990
Buspirone	mouse, rat	reduction	Bielenberg and Burkhardt, 1990
CM 57493	mouse, rat	reduction	Nuglisch et al., 1990
Gepirone	mouse, rat	reduction	Bielenberg and Burkhardt, 1990
Ipsapirone	mouse, rat	reduction	Bielenberg and Burkhardt, 1990
Roxindole	mouse	reduction	Ausmeier and Krieglstein, to be published
Urapidil	mouse	reduction	Prehn et al., submitted for publication

pirone (30 mg/kg) 1 h after middle cerebral artery-occlusion (MCA-O) in rats also led to an infarct reduction of 50 %. The uncertain feature of this study was that the brain temperature of the animals had apparently not been controlled. Nuglisch et al. (1990) reported about neuroprotective properties of the 5-HT₁ₐ agonist CM 57493 after MCA-O in mice and rats. In addition, a neuroprotective activity of urapidil was found after MCA-O in mice (Prehn et al., submitted for publication). In the latter two studies, the environmental temperature was adjusted at 30 °C during the surgical procedure as well as after the vessel occlusion. Postischemic administration of CM 57493 (10 mg/kg) immediately after vessel occlusion also reduced the infarct size in mice (Prehn et al., submitted for publication), whereas application of CM 57493 (10 mg/kg) 1 h after the occlusion produced no neuroprotection (Karkoutly and Krieglstein, personal communication).

DORSAL-VENTRAL GRADIENT IN VULNERABILITY OF THE HIPPOCAMPUS TO GLOBAL CEREBRAL ISCHEMIA IS INFLUENCED BY 5-HT₁ₐ RECEPTOR ACTIVATION

Smith et al. (1984) found that in rats having been subjected to transient forebrain ischemia the dorsal hippocampal portion was more vulnerable to ischemia than the ventral hippocampal portion, although there was no significant difference in local cerebral blood flow between these two regions during the ischemic period. On the base of these observations, the question arose, whether differences in the distribution and density of excitatory or inhibitory neurotransmitter receptors or differences in the extracellular concentration of those neurotransmitters during ischemia could explain this descending vulnerability of the CA1 neurons along the dorsal-ventral axis.

It has been shown that the density of 5-HT innervation increases within all hippocampal layers successively from dorsal to ventral levels (Köhler, 1984; Oleskevich and Descarries, 1990). By using autoradiographic techniques, Köhler (1984) observed high densities of 5-HT₁ₐ receptors labeled with [³H]5-HT at all septotemperal levels of the hippocampal region, with higher density at ventral levels. In addition, Gage et al. (1978) demonstrated a

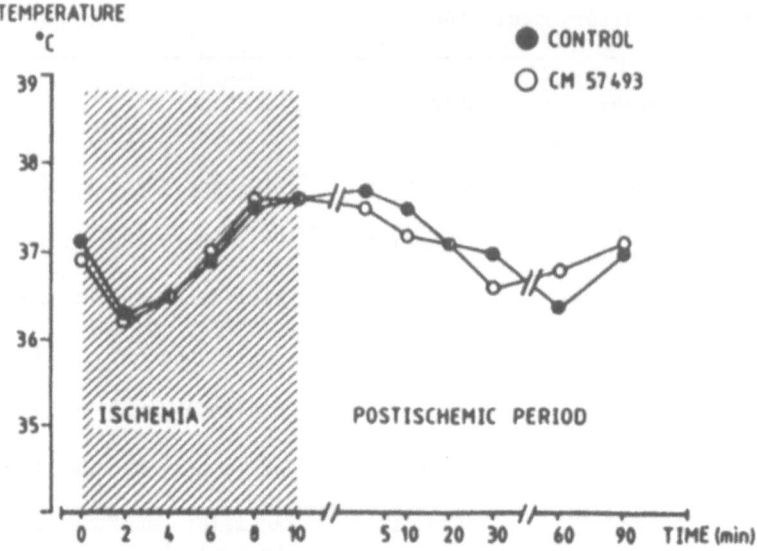

Fig. 1. Recording of the cerebral temperature during and after 10 min of forebrain ischemia in the rat. The environmental temperature was adjusted at 30 °C. Neither the intraischemic nor the postischemic curves of CM 57493 (1 mg/kg) were statistically different from the control curves. The values are given as means from three experiments.

higher concentration of the neurotransmitter 5-HT in the ventral than the dorsal hippocampal portion.

In order to determine, whether $5-HT_{1A}$ receptor activation can influence the pattern of ischemic hippocampal damage after global cerebral ischemia, transient forebrain ischemia of 10 min duration was performed in male Wistar rats by 2-vessel occlusion combined with hypotension. Thirteen animals received the $5-HT_{1A}$ agonist CM 57493 at a dose of 1 mg/kg (i.p.) 30 min before the induction of ischemia; 13 control animals received saline. The environmental temperature was adjusted at 30 °C both during ischemia and up to 2 h afterwards. The rectal temperature of the rats was controlled up to 4 h after ischemia. Additionally, the cerebral temperature was taken from 3 seperate drug-treated and saline-treated animals during the ischemic and postischemic period. For this purpose, a temperature probe was inserted into the brain between both hemispheres. The temperatures of the control animals and the drug-treated animals did not differ, neither in the intraischemic, nor in the postischemic period (Fig. 1).

Seven days after ischemia, the animals were perfused with a phosphate-buffered (pH 7.4) formalin-solution (4%). Slices were taken from each brain at 16 horizontal stereotaxic planes and were stained with a mixture of 1% celestine blue and 1% acid fuchsin.

The histological evaluation demonstrated a dorsal-ventral gradient in the vulnerability of the hippocampus both in $5-HT_{1A}$ agonist-treated and in saline-treated animals. However, in case of drug-treatment, this gradient was more distinct (Fig. 2). Within each CA1 subfield, the fimbrial end of CA1 showed a higher grade of damage than the subicular end of CA1. The $5-HT_{1A}$ agonist CM 57493 produced a statistically significant neuroprotection of the hippocampus, beginning at the level 5.2 mm above the interaural line (Fig. 3). The most pronounced neuroprotection was observed at the level 4.1 mm above the interaural line (30.3 ± 5.5 % CA1 and CA2 neurons were damaged in controls, and 13.4 ± 2.9 % neurons were damaged in

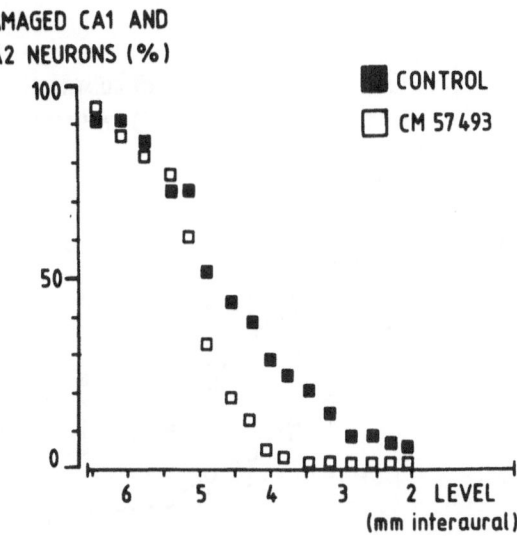

Fig. 2. Dorsal-ventral gradient in vulnerability of hippocampal CA1 and CA2 neurons. Seven days after ischemia, the percentage of the neuronal damage was determined at 16 stereotaxic horizontal planes, both in a representative control rat and in a representative CM 57493-treated rat (1 mg/kg). Note that the gradient in vulnerability was more pronounced after treatment with the 5-HT$_{1A}$ agonist.

CM 57493-treated rats; $p < 0.01$; Mann-Whitney U-test, Fig. 4). At this level, the neuronal injury was restricted to the CA2 subfield in most of the CM 57493-treated rats. It may be of interest, that the CA2 subfield, which is known to be early affected by global ischemia (Smith et al., 1984), has been shown to be rather poor of 5-HT$_{1A}$ receptors (Pazos and Palacios, 1985a).

DISCUSSION

The complexicity of the role of 5-HT in cerebral ischemia is probably due to the multiplicity of 5-HT receptors and their different distribution and densities within the brain. As above mentioned, 5-HT$_{1A}$ receptors mediate an inhibitory effect on neurons. Via 5-HT$_2$ receptors, however, 5-HT also stimulates hippocampal and cortical neurons (Colino and Halliwell, 1987; Davies et al., 1987).

In case of global cerebral ischemia, the concentration of 5-HT increases 8- to 10-fold within the hippocampal extracellular space during the ischemic period (Sarna et al., 1990). In contrast to the high density of 5-HT$_2$ receptors within the rat cerebral cortex (Pazos and Palacios, 1985b), there is a rather low density of those receptors within the hippocampal formation of the rat (Köhler, 1984; Pazos and Palacios, 1985b). 5-HT, being released in the course of cerebral ischemia, may therefore stimulate predominantly 5-HT$_{1A}$ receptors within the hippocampus and may inhibit neuronal activity. Indeed, our observations suggest that there could be a correlation between the density of 5-HT$_{1A}$ receptors and the pattern of ischemic neuronal damage within the hippocampal formation.

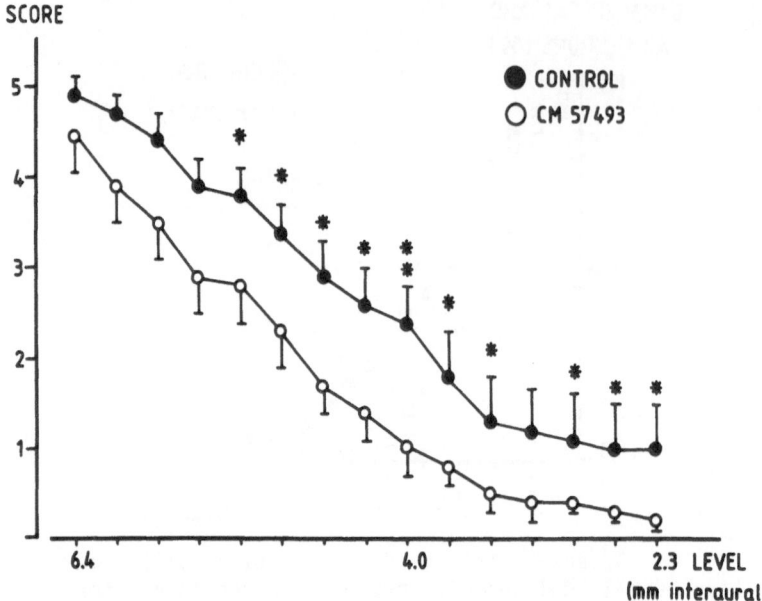

Fig. 3. The neuroprotective effect of CM 57493 (1 mg/kg) is more pronouced at the ventral part of the hippocampus. Seven days after transient forebrain ischemia, the hippocampal neuronal damage was scored on a five-point scale, corresponding to 0 - 100 % neuronal damage. The values of the semi-quantitative analysis are given as means ± S.E.M. from 13 experiments. * : P < 0.05; ** : P < 0.01 (Mann-Whitney U-test).

A bundle of 5-HT$_{1A}$ agonists have been tested for their neuroprotective activity in various rodent models of cerebral ischemia and were found to be capable of reducing ischemic neuronal injury. Although, at least in some experiments, the hypothermic effect of 5-HT$_{1A}$ agonists might have contributed to the resulting neuroprotection, the results of these studies give evidence for a role of 5-HT$_{1A}$ receptors in the pathogenesis of ischemic neuronal damage. The exact mechanism of the neuroprotective activity of 5-HT$_{1A}$ agonists, however, warrants further investigation.

5-HT$_{1A}$ receptors are located postsynaptically e. g. in the hippocampus and entorhinal cortex, but function presynaptically as autoreceptors in the neurons of the dorsal raphe nuclei (Hall et al., 1985). Since the cell bodies of the dorsal raphe neurons project to the neocortex and the hippocampus (Segal, 1975; Steinbusch, 1984), an inhibition of the activity of dorsal raphe neurons via presynaptic 5-HT$_{1A}$ receptors reduces the release of 5-HT (Hutson and Curzon, 1989). This fact might explain why 5-HT$_{1A}$ agonists show a protective effect on the cortical tissue after focal cerebral ischemia, although the density of 5-HT$_{1A}$ receptors is rather low in the neocortex. The density of 5-HT$_2$ receptors, however, is high within this structure (Pazos and Palacios, 1985b), and an inhibition of the release of 5-HT could therefore effect neuroprotection. Accordingly, 5-HT$_2$ antagonists have also been shown to reduce neuronal injury after focal cerebral ischemia (Nakayama et al., 1988). Interestingly, neuroprotective effects of 5-HT$_2$ antagonists have been reported in models of global cerebral ischemia, too (Fujikura et al., 1989; Krieglstein et al., 1989; Bode-Greuel et al., 1990a). A combined treatment using a 5-HT$_{1A}$ agonist and a 5-HT$_2$ antagonist could exert a pronounced neuroprotective effect in cerebral ischemia. Indeed, Bode-Greuel et al. (1990a) demonstrated an 83 % protection of the CA1 neurons by the drug-combination ipsapirone and ketanserin after tran-

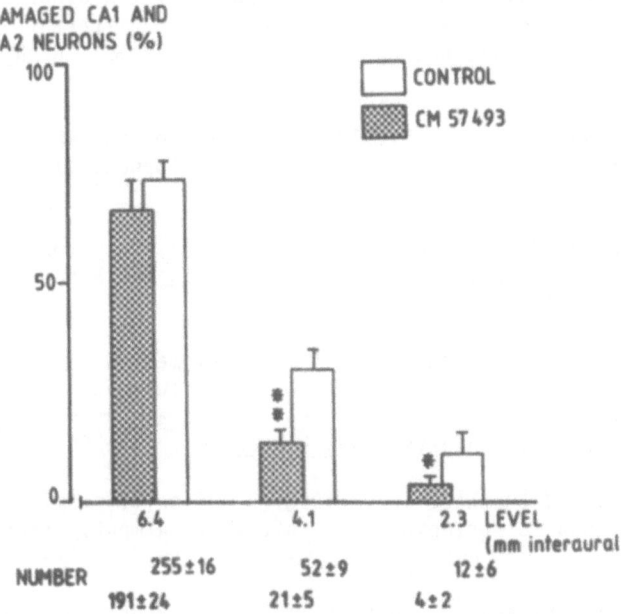

Fig. 4. CM 57493 (1 mg/kg) reduces neuronal damage caused by transient forebrain ischemia in the rat. The percentage of damaged CA1 and CA2 neurons was determined at three distinct hippocampal levels. Under each columne, the number of damaged CA1 and CA2 neurons is given. The values are given as means ± S.E.M. * : P < 0.05; ** : P < 0.01 (Mann-Whitney U-test).

sient forebrain ischemia in the gerbil. However, since both receptor types have been shown to be involved in thermoregulation (Gudelsky et al., 1986), this finding should be verified in animals, which show a less susceptibility to hypothermia than gerbils (Ginsberg and Busto, 1989). To our knowledge, a similar drug combination in a model of focal cerebral ischemia has not been published yet, but could produce a pronounced neuroprotection, too.

In conclusion, 5-HT$_{1A}$ receptors may play a beneficial role in the patho-physiology of cerebral ischemia. Thus, 5-HT$_{1A}$ agonists could become effective tools for the treatment of cerebral ischemic disorders.

Acknowledgment

The authors are grateful to Ms. S. Engel for her excellent technical assistance and to Dr. A. Rami for his helpful discussion. These experiments were supported by the Deutsche Forschungsgemeinschaft, Bonn-Bad Godesberg (Kr 354/13-3).

REFERENCES

Andrade, R., Malenka, R. C., and Nicoll, R. A., 1986, A G protein couples serotonin and GABA$_B$ receptors to the same channel in hippocampus, Science, 234:1261.

Andrade, R. and Nicoll, R. A., 1987, Novel anxiolytics discriminate between postsynaptic serotonin receptors mediating different physiological responses on single neurons of rat hippocampus, Arch. Pharmacol., 336:5.

Basse-Tomusk, A. and Rebec, G. V., 1986, Ipsapirone depresses neuronal activity in the dorsal raphe nucleus and the hippocampal formation, Eur. J. Pharmacol., 130:141.

Beck, S. G., Clarke, W. P., and Goldfarb, J., 1985, Spiperone differentiates multiple 5-hydroxytryptamine responses in rat hippocampal slices in vitro, Eur. J. Pharmacol., 116:195.

Benveniste H., Drejer, J., Schousboe, A., and Diemer, N. H., 1984 Elevation of the extracellular concentrations of glutamate and aspartate in rat hippocampus during transient cerebral ischemia monitored by intracerebral microdialysis, J. Neurochem., 43:1369.

Bielenberg, G. W. and Burkhardt, M., 1990, 5-Hydroxytryptamine$_{1A}$ agonists. A new therapeutic principle for stroke treatment, Stroke, 21 (suppl. IV):IV161.

Bode-Greuel, K. M., Klisch, J., Glaser, T., and Traber, J., 1990a, Serotonin (5-HT)$_{1A}$ receptor agonists as neuroprotective agents in cerebral ischemia, in: "Pharmacology of Cerebral Ischemia 1990", J. Krieglstein and H. Oberpichler, eds., Wissenschaftliche Verlagsgesellschaft, Stuttgart.

Bode-Greuel, K. M., Klisch, J., Horváth, E., Glaser, T., and Traber, J., 1990b, Effects of 5-hydroxytryptamine$_{1A}$-receptor agonists on hippocampal damage after transient forebrain ischemia in the Mongolian gerbil, Stroke, 21(suppl IV):IV164.

Busto, R. W., Dietrich, W. D., Globus, M. Y.-T., Valdés, I., Scheinberg, P., and Ginsberg, M. D., 1987, Small differences in intraischemic brain temperature critically determine the extent of ischemic neuronal injury, J. Cereb. Blood Flow Metab., 7:729.

Busto, R. W., Dietrich, W. D., and Globus, M. Y.-T., 1989, Postischemic moderate hypothermia inhibits CA1 hippocampal injury, Neurosci. Lett., 101:299.

Choi, D. W., 1988, Glutamate neurotoxicity and diseases of the nervous system, Neuron, 1:623.

Colino, A. and Halliwell, J. V., 1987, Differential modulation of three seperate K-conductances in hippocampal CA1 neurons by serotonin, Nature, 328:73.

Davies, M. F., Deisz, R. A., Prince, D. A., and Peroutka, S. J., 1987, Two distinct effects of 5-hydroxytryptamine on single cortical neurons, Brain Res., 423:347.

Fujikura, H., Kato, H., Nakano, S., and Kogure, K., 1989, A serotonin S$_2$ antagonist, naftidrofuryl, exhibited a protective effect on ischemic neuronal damage in the gerbil, Brain Res., 494:387.

Gage, R. H., Thompson, R. G., and Valdes, J. J., 1978, Endogenous norepinephrine and serotonin within the hippocampal formation during development and recovery from septal hyperreactivity, Pharmac. Biochem. Behav., 9:359.

Ginsberg, M. D. and Busto, R., 1989, Rodent models of cerebral ischemia, Stroke, 20:1627.

Globus, M. Y.-T., Busto, R., Dietrich, W. D., Martinez, E., Valdes, I., and Ginsberg, M. D., 1988, Effect of ischemia on the in vivo release of striatal dopamine, glutamate, and gamma-aminobutyric acid studied by intracerebral microdialysis, J. Neurochem., 51:1455.

Gudelsky, G. A., Koenig, J. I., and Meltzer, H. Y., 1986, Thermoregulatory responses to serotonin (5-HT) receptor stimulation in the rat: Evidence for opposing roles of $5-HT_2$ and $5-HT_{1A}$ receptors, Neuropharmacology, 125:1307.

Hagberg, H., Andersson, P., Lacarewicz, J., Jacobson, I., Butcher, S., and Sandberg, M., 1987, Extracellular adenosine, inosine, hypoxanthine, and xanthine in relation to tissue nucleotides and purines in rat striatum during transient ischemia, J. Neurochem., 49:227.

Hall, M. D., El Mestikawy, S., Emerit, M. B., Pichat, L., Hamon, M., and Gozlan, H., 1985, $[^3H]$8-hydroxy-2-(di-n-propylamino)tetralin binding to pre- and postsynaptic 5-hydroxytryptamine sites in various regions of the rat brain, J. Neurochem., 44:1685.

Hutson, P. H. and Curzon, G., 1989, Hippocampal 5-HT synthesis and release in vivo is decreased by infusion of 8-OH-DPAT into the nucleus raphe dorsalis, Neurosci. Lett., 100:276.

Jørgensen, M. B. and Diemer, N. H., 1982, Selective neuronal loss after cerebral ischemia in the rat: Possible role of transmitter glutamate, Acta Neurol. Scand., 66:536.

Kirino, T., 1982, Delayed neuronal death in the gerbil hippocampus following ischemia, Brain Res., 510:57.

Köhler, C., 1984, The distribution of serotonin binding sites in the hippocampal region of the rat brain. An autoradiographic study, Neuroscience, 13:667.

Krieglstein, J., Sauer, D., Nuglisch, J., Roßberg, C., Beck, T., Bielenberg, G. W., and Mennel, H. D., 1989, Naftidrofuryl protects neurons against ischemic damage, Eur. Neurol., 29:224.

Nakayama, H., Ginsberg, M. D., and Dietrich, W. D., 1988, (S)-Emopamil, a novel calcium channel blocker and serotonin S2 antagonist, markedly reduces infarct size following middle cerebral artery occlusion in the rat, Neurology, 38:1667.

Nuglisch, J., Karkoutly, C., Peruche, B., Prehn, J. H. M., Welsch, M., Mennel, H. D., Roßberg, C., and Krieglstein, J., 1990, Effect of the $5-HT_{1A}$-agonist CM 57493 on infarct area, infarct volume and hippocampal neuronal damage after focal and global cerebral ischemia in mice and in rats, in: "Pharmacology of Cerebral Ischemia 1990", J. Krieglstein and H. Oberpichler, eds., Wissenschaftliche Verlagsgesellschaft Stuttgart.

Oleskevich, S. and Descarries, L., 1990, Quantified distribution of the serotonin innervation in adult rat hippocampus, Neuroscience, 34:19.

Pazos, A. and Palacios, J. M., 1985a, Quantitative autoradiographic mapping of serotonin receptors in the rat brain. I. Serotonin-1 receptors, Brain Res., 346:205.

Pazos, A. and Palacios, J. M., 1985b, Quantitative autoradiographic mapping of serotonin receptors in the rat brain. II. Serotonin-2 receptors, Brain Res., 346:231.

Pulsinelli, W. A., Brierley, J. B., and Plum F., 1982, Temporal profile of neuronal damage in a model of transient forebrain ischemia, Ann. Neurol., 11:491.

Rowan, M. J. and Anwyl, R., 1987, Neurophysiological effects of buspirone and isapirone in hippocampus: Comparison with 5-hydroxytryptamine, Eur. J. Pharmacol., 132:93.

Sarna, G. S., Obrenovitch, T. P., Matsumoto, T., Symon, L., and Curzon, G., 1990, Effect of transient cerebral ischemia and cardiac arrest on brain extracellular dopamine and serotonin as determined by in vivo dialysis in the rat, J. Neurochem., 55:937.

Segal, M., 1975, Physiological and pharmacological evidence for a serotonergic projection to the hippocampus, Brain Res., 94:115.

Simon, R. P., Swan, J. H., Griffiths, T., and Meldrum, B. S., 1984, Blockade of N-methyl-D-aspartate receptors may protect against ischemic damage in the brain, Science, 226:850.

Smith, M.-L., Auer, R. N., and Siesjö, B. K., 1984, The density and distribution of ischemic brain injury in the rat following 2 - 10 min of forebrain ischemia, Acta Neuropathol. (Berl), 64:319.

Steinbusch, H. W. M., 1984, Serotonin-immunoreactive neurons and their projections in the CNS, in: "Handbook of Chemical Neuroanatomy, Vol. 3, Part II: Classical Transmitters and Transmitter Receptors in the CNS", Bjorklund, A., Hokfelt, T., and Kuhar, M. J., eds., Elsevier, Amsterdam.

ROLE OF SEROTONIN IN TRAUMATIC BRAIN INJURY: AN EXPERIMENTAL STUDY IN THE RAT*

Hari Shanker Sharma[1,2], Jorge Cervos-Navarro[1], Georg Gosztonyi[1] and Prasanta Kumar Dey[2]

[1]Institute of Neuropathology, Free University Berlin, 1000 Berlin 45, Federal Republic of Germany, and [2]Neurophysiology Research Unit, Department of Physiology, Institute of Medical Sciences, Banaras Hindu University, Varanasi-221 005, India

INTRODUCTION

Edema is a serious complication in many brain diseases including traumatic injury. The pathogenesis of traumatic brain edema is complex and includes physical destruction of microvessels, microcirculatory alterations in and around the primary injury and permeability changes of the vessel walls leading to a leakage of plasma constituents into the tissue (Long 1990, Reulen et al 1990). There are reasons to believe that many of these events are influenced by a number of chemical mediators which are released or become activated in and around the primary lesion like biogenic amines, arachidonic acid, leucotrienes, histamine and free radicals (Wahl et al 1988).

However, role of serotonin (5-hydroxytryptamine, 5-HT) in traumatic brain edema is not well understood. Several lines of recent evidences indicate a presumptive role of this amine in various neurological diseases and in pathological conditions. Thus, marked changes in serotonin metabolism occur in such important brain injuries as stroke, ischemia and trauma as well as in experimental cold injury lesions and various other neurological diseases (Essman 1978). Increased serotonin content in the walls of the cerebral vessels, cerebrospinal fluid and brain occurs following traumatic insults to brain (cf Pappius et al 1988). Abnormal levels of serotonin in blood and brain have been described in a wide variety of psychiatric illnesses and mental abnormalities (McEntee and Crook 1991). There are serotonergic receptors in the cerebral vessels and intracarotid, intravenous or intracerebroventricular infusion of serotonin markedly affects the cerebral circulation and metabolism as well as increases the permeability of the blood-brain barrier (BBB) (Wahl et al 1988, Olsson et al 1990, cf Sharma et al 1990).

Thus, it seems likely that 5-HT could play an important role in edema formation and cellular changes following traumatic insults to brain. Therefore, in present study we examined the role of endogenous 5-HT in BBB breakdown, edema formation and *early* cellular changes in experimental model of traumatic brain injury using a pharmacological approach.

*Supported by Grants from Alexander von Humboldt-Stiftung, Bonn, Germany and University Grants Commission, New Delhi, India to HSS.

MATERIALS AND METHODS

Experiments were carried out on 42 inbred male Wistar rats (200-250 g). Under urethane anesthesia (1.5 g/kg, i.p.) a burr hole (about 4 mm^2) was made in the right parietal bone and the underlying cerebral cortex was exposed. The dura was carefully removed and a stab wound about 3 mm deep and 3 mm long was inflicted under stereotaxic guidance using a sharp sterile scalpel blade (n=10) (Dey and Sharma 1983). The lesion was thus mainly in the cortex and the superficial parts of the subcortical white matter. At the end of 5 h period following trauma the animals were decapitated and the whole brain was immediately removed from the cranium (n=5). The blood clots, if any, on the injured hemisphere were removed. The right and left hemispheres were separated, placed on preweighed filter papers and immediately reweighed. The edema was determined from the changes in the water content of the brain (Dey and Sharma 1983). The volume swelling was calculated from the changes in brain water content according to Elliott and Jasper (1949).

In a separate group of rats, the BBB permeability (n=5) was measured using Evans blue albumin (EBA, 0.3 ml/100 g of a 2 % solution) injected into the right external branch of the jugular vein 5 min before sacrifice. The brain was removed after a brief saline rinse, the extravasation of dye was examined and measured colorimetrically (Sharma et al 1990). The mean arterial blood pressure (MABP) in these animals was continuously monitored through an indwelling catheter placed in the common carotid artery that was connected to a strain gauge pressure transducer (Statham P 23, USA) and a chart recorder. Timed samples of arterial blood were withdrawn in heparinised vials for determinations of blood gases and pH using a Radiometer apparatus (Copenhagen). Urethane anesthetised intact animals served as controls (n=10).

Separate groups of animals were treated with p-chlorophenylalanine (p-CPA, a 5-HT synthesis inhibitor drug obtained from Sigma Chemical Co., USA) in a dose of 100 mg/kg, i.p. for 3 consecutive days (n=10) (Koe and Wiessman 1966, Olsson et al 1990). The traumatic injury was made in these animals on the 4th day and brain edema (n=5) and the BBB permeability (n=5) were examined after 5 h trauma.

The structural changes of the perifocal parietal cerebral cortex (about 5 mm away from the primary injury site) was examined in samples obtained from injured rats with or without pretreatment with p-CPA and a survival period of 5 h. For this purpose, the animals were perfused transcardially at a pressure of 100 torr with 100 ml of fixative (2.5 % glutaraldehyde, 2 % paraformaldehyde in 0.1 M sodium phosphate buffer, pH 7.4 containing 2.5 % lanthanum chloride and 2 % picric acid) (Olsson et al 1990) preceded with a 50 ml 0.9 % saline rinse at room temperature. The brain was dissected and the tissue pieces from both the ipsilateral and contralateral parietal cerebral cortex were processed for routine light (n=3) and electron microscopy (n=3) according to commercial protocol.

The unpaired Student's t-test was used to evaluate the statistical significance of the quantitative data obtained. A p-value less than 0.05 was considered to be significant.

RESULTS

Trauma to right parietal cerebral cortex in untreated animals resulted in a marked visible swelling in the ipsilateral cerebral hemisphere after 5 h survival period. However, the increase in water content was evident in

Table I. Effect of p-CPA on water content and Evans blue extravasation in 5 h traumatized animals.

Type of Expt	Brain water content %		Evans blue dye mg %	
	Right half[a]	Left half	Right half[a]	Left half
Control n=5	78.56±0.56	78.64±0.32	0.28±0.06	0.30±0.04
5 h injury n=5	80.81±0.64* (+12)[b]	79.87±0.36* (+6)[b]	0.87±0.12* (+210 %)	0.66±0.12* (+120 %)
p-CPA+ 5 h injury n=5	78.89±0.54 (+2)[b]	78.77±0.43 (+1)[b]	0.56±0.11* (+100 %)	0.42±0.14 (+40 %)

Figures in parentheses indicate % change from control group.
Values are mean±SD, * =P <0.01 Student's unpaired t-test.
a= injured, b= volume swelling (Elliott & Jasper 1949).

both ipsilateral (2.86%) and contralateral (1.56%) hemisphere at this time period as compared to control (Table I).

A deep extravasation of Evans blue was noted around the border zone of the trauma which was diffusely extended to the perifocal injury site. However, a significant increase in EBA staining was noted in both ipsilateral and contralateral hemisphere as compared to control (Table I). Morphological examination showed profound cellular changes that were more prominent in the ipsilateral hemisphere in the vicinity of the lesion (Table II). Thus, a general expansion and sponginess around the perifocal lesion was evident. Perineuronal swelling, chromatolysis and appearance of dark neurones were frequent. Vacuolation in gray and white matter was much common.

Table II. Effect of p-CPA on cellular changes in 5 h traumatized animals. These qualitative changes were examined in blind fashion by at least two independent workers.

Type of Expt	Cellular changes				
	Dark neuron	Peri- vascular edema	Vacuo- lation	Myelin disrup- tion	Lanthanum infiltration[b]
A. Untreated injured 5 h (n=3)					
Right half[a]	++	+++	++++	+++	+++
Left half	-/+	+	+	+/-	+
B. p-CPA treated +5 h injury (n=3)					
Right half[a]	-/+	+	+	+	+
Left half	-	-	-	-/+	-

a=injured, b=across endothelium.
+ = mild, ++ = moderate, +++ = intense, ++++ = severe, - = nil.

149

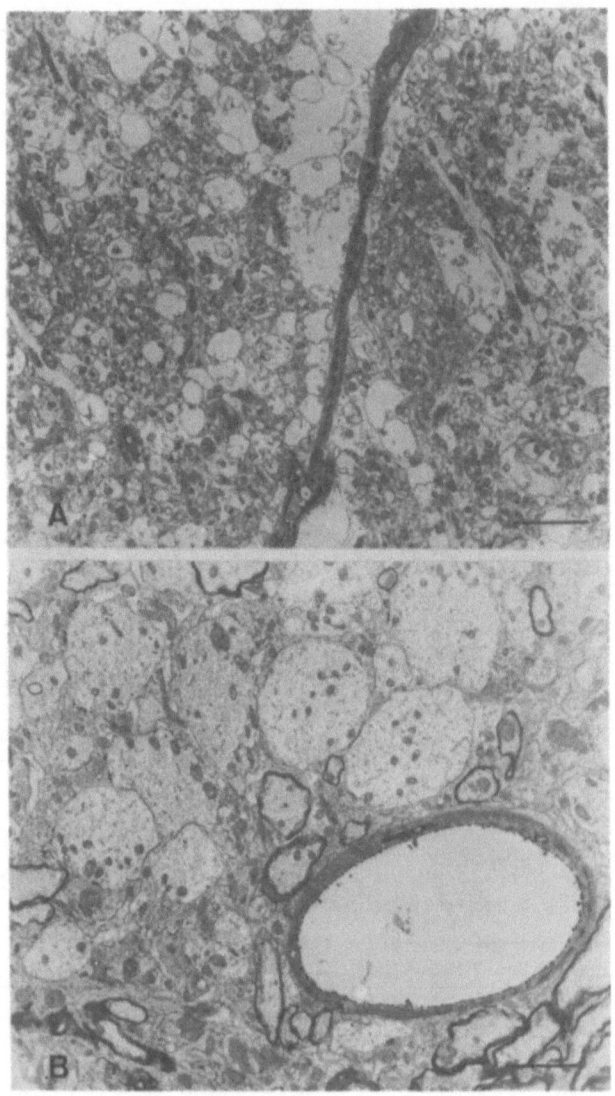

Figure 1. Electron micrograph from right parietal cerebral cortex of an untreated traumatized animal. The injury was made in right parietal cerebral cortex and the tissue piece was taken from about 5 mm in the periphery of the primary lesion. (a). Collapse of a longitudinally sectioned microvessel, vacuolation and edema is prominent. (b). Picture taken from similar area of a 5 h traumatized animal but pretreated with p-CPA (for details see text). Please note that the above cellular changes were mainly prevented by this drug treatment. (bars = 2.5 μm).

At ultrastructutral level, collapsed microvessels, perivascular edema, vacuolation (Figure 1a), and vesiculation of myelin were prominent. A diffuse extravasation of ionic lanthanum across the endothelium was seen in several areas. However, the tight junctions were mainly intact. The magnitude of such changes were mild in the contralateral hemisphere. The MABP at the end of 5 h trauma showed a mild hypotension (20±4 torr). The PaO_2 showed a mild increase, whereas the $PaCO_2$ and arterial pH were not significantly altered from the intact control group at this time period.

Pretreatment with p-CPA markedly prevented the increase of water content in both hemispheres following 5 h trauma (Table I). The visible swelling of the traumatized hemisphere was absent. The intensity of Evans blue staining was much less and the extravasation of EBA in both hemispheres was significantly reduced as compared to untreated traumatized group (Table I). The cellular changes were markedly preserved in this drug treated animals. Thus the appearance of dark neurones, general expansion and sponginess were less frequent (Table II). The magnitude of vacuolation, edema and cellular damage were considerably less at ultrastructural level (Figure 1b). Diffuse infiltration of lanthanum across cerebral endothelium, collapse of microvessels and perivascular edema were mainly absent. However, the MABP and blood gases did not differ significantly in this group as compared to untreated animals.

DISCUSSION

Our results demonstrate that endogenous depletion of 5-HT prior to traumatic insult to brain markedly thwarts the edema development and *early* cellular changes indicating an important role of this amine in the pathophysiology of traumatic brain injury.

Physical trauma to brain will obviously start a cascade of interrelated biochemical and structural events in and around the primary injury (Long 1990). It seems likely that edema is one of those secondary events which might aggravate a primary injury and the present results show that serotonin might be involved in the microvascular reactions causing edema.

Serotonin is present in several neuronal pathways emanating from the dorsal raphe nuclei and in mast cells of the leptomeninges and in platelets of blood (Essman 1978, McEntee and Crook 1991). During the progression of the injury changes in the concentration of serotonin most likely occur. Additional amounts of serotonin may be brought in from the blood or from neurones at the same time as edema is formed. Biochemical determinations, however, indicate that the net result is an increase in the serotonin content of traumatized brain (Dey and Sharma 1983). Serotonin is a potent chemical mediator of the microvascular response leading to edema of the nervous system (cf Wahl et al 1988). Our results with p-CPA pretreatment before the induction of trauma are in line with the assumption that serotonin plays a role as a compound with vascular permeability-increasing capacity leading to an early edema (cf Sharma et al 1990, Olsson et al 1990). This is further evident with the fact that the degree of cell changes in the periphery of the initial lesion is lower in rats pretreated with p-CPA as compared to animals with the same form of injury but without any reduction in serotonin by the drug. Obviously, the improved preservation of the tissue might be a consequence of a reduced edema. However, the release of high concentrations of serotonin in itself might also contribute to the development of early cellular changes (Sharma and Olsson 1990). It seems likely that p-CPA in the doses used in our study will induce a long-lasting inhibition of serotonin synthesis in brain and spinal cord (Koe and Wissman 1966, Olsson et al 1990). There are thus reasons to beleive that reduction of serotonin content in the nervous tissue is somehow reflected in the reduced swelling and the milder cell changes in the brain of our traumatized rats. However, apart from effects caused by serotonin, released arachidonic acid, prostaglandins, thromboxane can synergistically contribute to the formation of edema. The clarification of the role of these mediators requires additional investigation.

Acknowledgements. Elisabeth Scherer, Franziska Drum, Katja Deparade, Hanna Plückhan, Gabriele Kluge, Reyes Esparza assisted with the light and electron microscopical works and Aruna Misra gave us secretarial assistance.

REFERENCES

Dey P K, Sharma H S (1983) Ambient temperature and development of brain edema in anaesthetized animals. Indian J Med Res 77:554-563.

Elliott K A C, Jasper H (1949) Measurement of experimentally induced swelling and shrinkage. Am J Physiol 157: 1096-1121.

Essman W (1978) Serotonin in Health and Disease, Vols. I-V, Spectrum Publications, New York.

Koe B K, Weissman A (1966) p-Chlorophenylalanine: a specific depletor of brain serotonin. J Pharmacol Exp Ther 154:499-516.

Olsson Y, Sharma H S, Pettersson A (1990) Effects of p-chlorophenylalanine on microvascular permeability changes in spinal cord trauma. An experimental study in the rat using ^{131}I-sodium and lanthanum tracers. Acta Neuropathol 79: 595-603.

Long D M (1990) Brain Edema. Pathogenesis, imaging and therapy. Adv Neurol 52: 1-538, Academic Press, New York.

McEntee W J, Crook T H (1991) Serotonin, memory, and the aging brain. Psychopharmacology 103: 143-149.

Pappius H M, Dadoun R, McHugh M (1988) The effect of p-chlorophenylalanine on cerebral metabolism and biogenic amine content of traumatized brain. J Cereb Blood Flow Metab 8: 324-334.

Reulen H -J, Baethmann A, Fenstermacher J D, Marmarou A, Spatz M (1990) Brain Edema VIII, Acta Neurochir (Wien) Suppl 51: 1-417.

Sharma H S, Olsson Y, Dey P K (1990) Changes in blood-brain barrier and cerebral blood flow following elevation of circulating serotonin level in anesthetized rats. Brain Res 517:215-223.

Sharma H S, Olsson Y (1990) Edema formation and cellular alterations following spinal cord injury in the rat and their modification with p-chlorophenylalanine. Acta Neuropathol 79: 604-610.

Wahl M, Unterberg A, Baethmann A, Schilling L (1988) Mediators of blood-brain barrier dysfunction and formation of vasogenic brain edema. J Cereb Blood Flow Metab 8:621-634.

MEASUREMENT OF SEROTONIN IN PLASMA BY IN VIVO MICRODIALYSIS DURING

PHOTOCHEMICALLY INDUCED THROMBOSIS - METHODOLOGICAL ASPECTS

Per Wester, Ricardo Prado, Brant D. Watson, Mordecai Y.-T. Globus, Hans Leistra, W. Dalton Dietrich

Cerebrovascular Disease Research Center, Department of Neurology D4-5, University of Miami School of Medicine, P.O. Box 016960 Miami, FL 33101, U.S.A.

INTRODUCTION

The microdialysis technique represents a newly developed method that enables one to study biochemical alterations in extracellular fluid (ECF) as a function of time (Ungerstedt, 1984). The microdialysis setup consists of a narrow U-shaped probe with inlet and outlet tubes and a dialysis membrane where molecules diffuse from the ECF into the probe. The method was originally developed for monitoring brain ECF (Ungerstedt, 1984) and has become a routine tool used in pharmacokinetic studies for intravenous recordings. However, intra-arterial microdialysis applications are so far lacking.

It has long been postulated that serotonin may be closely related to the pathogenesis and progression of ischemic stroke (Welch et al, 1972; 1980; Wurtman and Zervas, 1974; Jellinger and Riederer, 1983; Shah et al, 1985; Wester et al, 1987a; Dietrich et al, 1989). In order to mimic large vessel thrombosis and TIA, a method of non-occlusive common carotid artery thrombosis (CCAT) in the rat was developed (Watson et al, 1987; Dietrich et al, 1988, 1991 in press). With non-occlusive CCAT, acute blood brain barrier leakage (Dietrich et al, 1988) and a heterogeneous pattern of abnormal local cerebral blood flow (Dietrich et al, 1991, in press) have been documented previously. Both these consequences are likely to be related to blood-borne substances released from the site of vascular thrombosis. The purpose of this study was twofold; firstly, to develop appropriate conditions for intra-arterial in vivo microdialysis measurements of monoamines and secondly to apply this technique to the direct characterization of serotonin (5-hydroxytryptamine) measurements in downstream plasma in response to non-occlusive common carotid artery thrombosis.

MATERIALS AND METHODS

Basic set up

Experiments were performed on fasted adult male Wistar rats weighing between 225 and 325 g. Anaesthesia was induced with 4% halothane for 5 minutes. Rats were then maintained on 1.5 % halothane and a mixture of 70% nitrous oxide and 30% oxygen delivered by a closely fitting face mask. Femoral and venous polyethylene catheters (PE-50) were inserted for measurements of arterial blood pressure, blood gases and for rose bengal administration. The rats were then intubated with PE-240 tubings and mechanically ventilated with 0.5% halothane, 70% nitrous oxide and 30% oxygen. Respiratory adjustments were made as needed to ensure normal arterial blood gases. Pancuronium bromide, 0.6mg/kg, was injected intravenously and additional doses of 0.2 mg/kg were administered to immobilize the animal. Rectal temperature was maintained at

36.7 °C by means of a rectal thermistor probe and a thermostatically regulated heating pad (CMA/140, Carnegie Medicine, Sweden).

Exposure of the carotid arterial system

With the rat in the supine position, a midline incision was made in the ventral aspect of the neck from the symphysis in the mandible to the sternal notch. The common carotid artery (CCA) was exposed by blunt dissection and retraction of the sternomastoideus and sternohyoideus muscles. The CCA was then carefully dissected from the vagus nerve and all surrounding connective tissues to the level of the bifurcation of the external (ECA) and internal (ICA) carotid arteries. The ICA distal to the bifurcation was carefully isolated from the surrounding tissues (see below). A small segment of the ECA between the lingual and superior thyroid artery was exposed by retracting the digastricus muscle as well as the hyoid bone. A CMA/11 guide (Carnegie Medicine, Stockholm, Sweden) was inserted in a retrograde fashion to the level of the bifurcation, but not beyond. Heparin (50 IU) was administered intra-arterially to prevent thrombosis at the cannula tip before any incisions were made into the external carotid artery. Atraumatic micro-aneurysm clips were then placed on the ICA and CCA for the insertion of a 3mm CMA/11 microdialysis probe (Carnegie Medicine, Stockholm, Sweden). The clips were then removed in the same order they were placed. Interruption of blood flow through the carotid artery system lasted less than 1 min. In preparation for the induction of photothrombosis, the CCA was free floated in saline by retracting the underlying longus coli and rectus capitii muscle laterally.

Common carotid artery thrombosis (Fig 1)

The photosensitizing dye rose bengal (15 mg/ml in 0.9 % saline) was injected intravenously to a body dose of 20 mg/kg over 2 min simultaneously with the start of the irradiation period. The right CCA was irradiated for 10 minutes with the focused beam of a tunable argon laser (Inova 70-4, Coherent, CA, U.S.A.) operated at 562 nm and focused rectangularly (~ 1.5 x 1mm) on the right CCA approximately 7 mm proximal to the CCA bifurcation. The laser was used at a power of 350 mW, corresponding to an average intensity of 23 W/cm^2 according to the efficiency criterion described previously (Watson et al, 1987). The blood flow was confirmed to

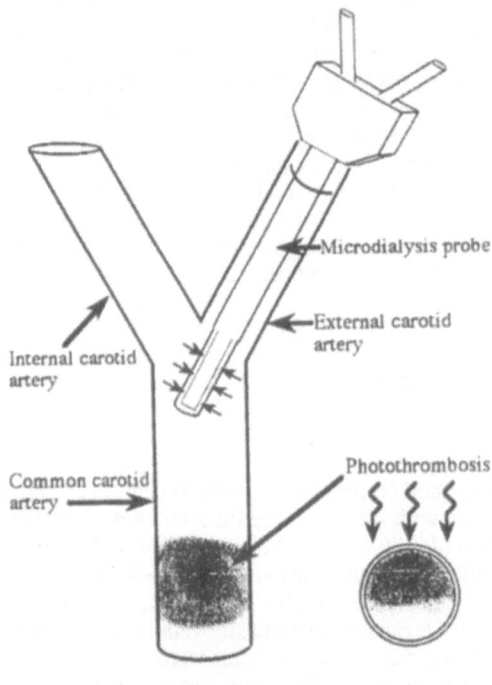

Fig 1

confirmed to move in an anterograde direction over the thrombosed CCA segment by visual inspection under the operating microscope. In a separate experiment, the effect of non-occlusive common carotid artery thrombosis on the temperature in CCA and rectum was studied. As can be seen in Fig 2, a very modest increase from 37.2 ± 0.1 at baseline to 37.7 ± 0.1 °C during laser irradiation was observed in the CCA immediately distal to the site of photothrombosis. In rectum, the temperature increased from 36.7 ± 0.1 to 36.8 ± 0.1 ° C during irradiation. Thus, thermal effects are not involved significantly in the production of an acute thrombus by rose bengal injection and 562 nm irradiation.

Plasma dialysis procedure (Fig 1)

The CMA/11 guide and microdialysis probe were inserted into the CCA as described above (see also Fig 1). The 0.24 mm diameter of the concentric microdialysis tip took up less than one third of the common carotid artery lumen (Ø 0.75-0.90 mm) and should therefore not have caused any significant obstruction of the blood flow in the right common and internal carotid arteries. Care was taken not to touch the artery wall with the microdialysis probe tip in order to prevent local endothelial damage. The microdialysis probe was perfused with Ringer's solution at a flow rate of 2.0 µl/min. Following half an hour of equilibration, baseline recordings were collected during the next 2 hours whereafter the thrombotic insult was initiated. Microdialysis samples were collected at 30-min intervals before, at 10 minutes during and after the first hour of CCAT and thereafter at 30-min intervals for an additional 2 hours. Samples were frozen and stored at -80 °C until biochemical measurements were undertaken.

Assessment of monoamine concentrations

Each collecting tube contained 10 or 30 µl of 0.1 M perchloric acid and 8 mM L-cysteine. This combination was shown to stabilize both the catecholamines and indoleamines at room temperature for at least 12 hours. Furthermore, the solution of PCA + L-cysteine resulted in over 90 % stability of all major monoamines, their precursors and metabolites when freezing the samples for up to 2 months and thereafter thawing them. This means a major improvement compared to other techniques including only PCA which only stabilize catecholamines or only L-cysteine which only stabilizes indoleamines. Concentrations of 5-hydroxytryptamine (5-HT = serotonin) were determined by reversed-phase isocratic liquid chromatography with electrochemical detection slightly modified compared to previous methods described elsewhere

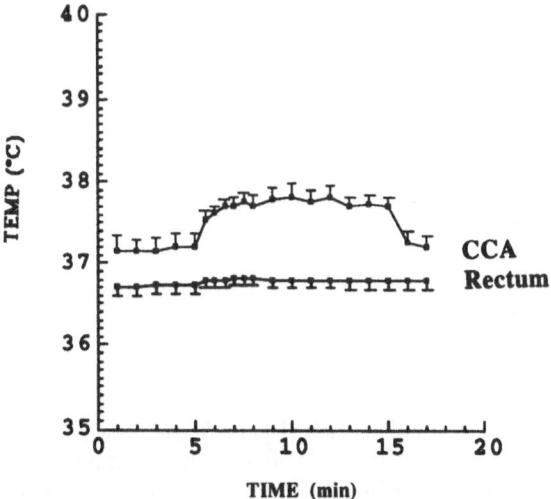

Fig 2. Effect of common carotid artery thrombosis (CCAT) on temperature as measured in the common carotid artery (CCA) immediately distal to the site of photothrombosis (at the same site as the microdialysis probe) and in rectum. Time 1-5 minutes represent baseline values, 5.5 - 15 minutes are temperature recordings during photothrombosis and 16 - 17 minutes represent postirradiation temperature. Values are mean ± S.E., n = 4.

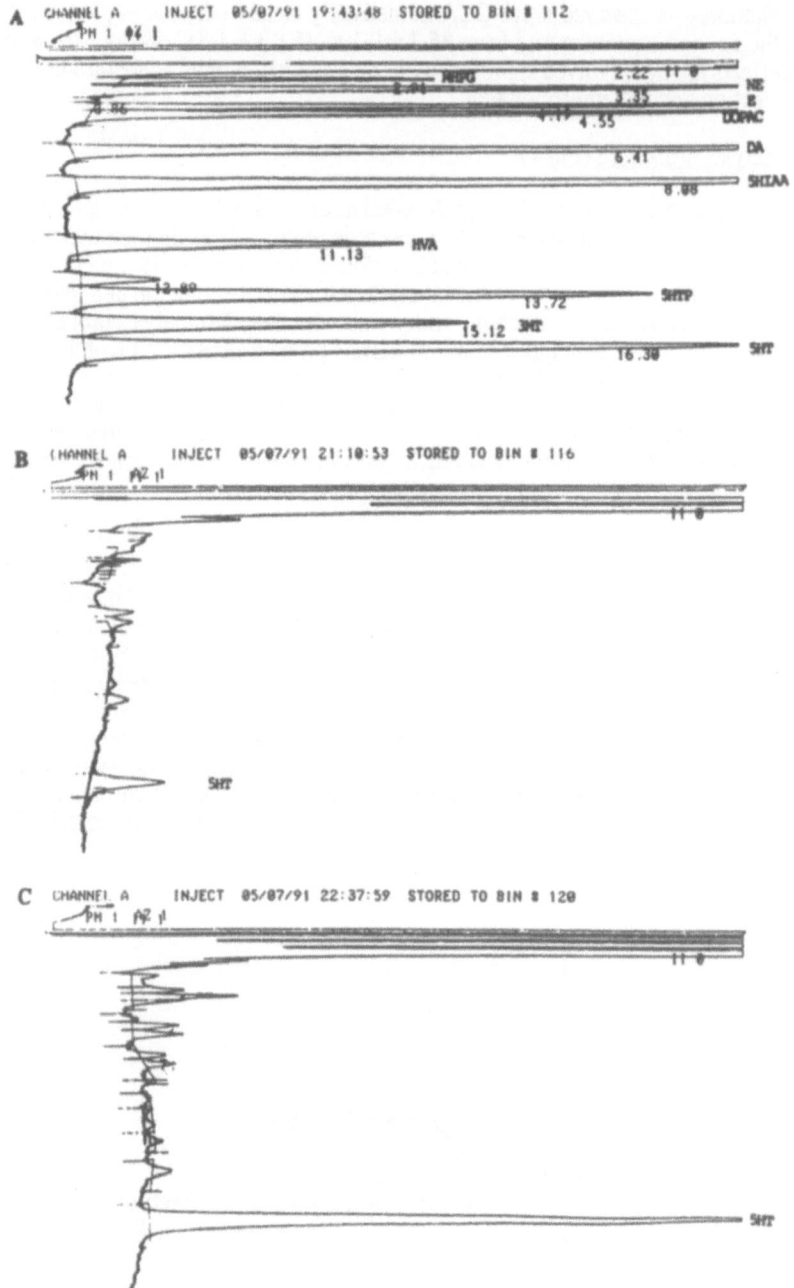

Fig 3. Chromatograms of a standard cocktail (1 pmol of each substance injected) (A); a plasma baseline sample (B); a plasma sample 10 min after common carotid artery thrombosis (C). Chromatographic conditions and sample preparation are described in text. Abbreviations: MHPG=3-methoxy-4-hydroxyphenylglycol; NE=norepinephrine; E=epinephrine; DOPAC=dihydroxyphenylacetic acid; DA=dopamine; 5HIAA=5-hydroxyindoleacetic acid; HVA=homovanillic acid; 5HTP=5-hydroxytryptophan; 3MT=3-methoxytyramine; 5HT= 5-hydroxytryptamine=serotonin.

(Wester et al, 1987b,c, 1990). The isocratic mobile phase, with a flow rate of 1.2 ml/min, consisted of a 100 mM citrate buffer including 0.3 mM Na_2EDTA, 0.334 mM octylsulphate; 6.0/94.0 % (vol/vol) acetonitrile/water at a pH of 2.35. The stationary phase consisted of a 150 x 4.6 mm stainless steel column packed with Nucleosil C-18, 5 μm from Macherey-Nagel & Co. KG. (Düren, Germany). The substances were detected electrochemically with a BAS 3 electrochemical detector (Bioanalytical Systems, West Lafayette, IN, U.S.A.) operated at a potential of +0.75 V versus Ag/AgCl reference electrode. The detector was coupled to an integrator, model SP4270 (Spectraphysics, San Jose, CA, U.S.A.). In Fig 3, representative chromatograms are shown of a standard cocktail (Fig 3A); a baseline plasma sample (Fig 3B) and a plasma sample 10 min after common carotid artery thrombosis (Fig 3C). The probe recovery of 5-HT as measured in vitro at 37°C was found to be 18%. 5-HT values presented in the Results section have however not been compensated for recovery.

Statistics

Analysis of variance (ANOVA) was performed with the system for statistics (SYSTAT; Wilkinson, 1989).

Physiological findings

Physiological findings recorded just before rose bengal injection (controls) and photochemical common carotid artery thrombosis (CCAT) were within normal ranges (Table 1). Continuous blood pressure monitoring before, during and after rose bengal with or without laser irradiation did not change significantly during the experiment but was somewhat low. This is most probably due to the fact that experiments were conducted with the rats placed on their backs during the entire experimental procedure.

TABLE 1. Physiologic Variables

	MAP (mm Hg)	pH	pO_2 (mm Hg)	pCO_2 (mm Hg)
Controls	94.8 ± 5.1	7.38 ± 0.07	144 ± 15.6	37.7 ± 4.71
CCAT	93.7 ± 8.0	7.42 ± 0.01	149 ± 8.7	39.8 ± 2.02

MAP=Mean arterial blood pressure, represents mean value from recordings every 10 minutes during the entire experiment. Values are mean ± S.D., n=3 for controls and n=5 for rats subjected to common carotid artery thrombosis (CCAT).

TABLE 2. Plasma microdialysis; 5-hydroxytryptamine concentrations

	Baseline	Rose bengal	Rose bengal + irradiation
Controls	9.7 ± 1.8	16.1 ± 2.5	-
CCAT	12.4 ± 4.4	-	108 ± 23*

Serotonin (5-hydroxytryptamine) values (expressed as nmoles/l perfusate) are given as mean ± S.E., n=3 for controls and n=5 for rats subjected to common carotid artery thrombosis (CCAT). * $p<0.001$ by ANOVA. Values of 5-HT concentrations have not been compensated for the 18% recovery of the microdialysis probe.

RESULTS

Plasma microdialysis (Table 2)

Stable baseline concentrations of extracellular plasma 5-HT were detected in 4 consecutive samples collected prior to CCAT (mean 12.4 ± 4.4 nmol/l of perfusate). During and immediatel' after the non-occlusive CCAT, 5-HT levels in plasma increased significantly to 108 ± 23.0 nmol/l ($p<0.001$, ANOVA), representing an almost 9-fold increase. The increased 5-HT-levels remained significantly elevated for 60 min but then gradually decreased; however, the serotonin concentrations did not reach baseline levels during the 3 hour post-irradiation period (data not shown). In control rats, the baseline of 5-HT was stable at 9.7 ± 1.8 nmol/l perfusate. During and after i.v. administration of rose bengal, a non-significant 5-HT increase was seen ($p>0.1$, ANOVA).

COMMENT

The microdialysis technique, as developed by Ungerstedt (1984), was in this study employed for intra-arterial recordings in the common carotid artery. This new application was found to be an indeed suitable and powerful approach that allows direct HPLC plasma serotonin measurements without any further extraction procedure.

These results indicate a substantial increase of plasma serotonin concentrations in downstream plasma in response to acute non-occlusive common carotid artery thrombosis (CCAT) which appears to be induced entirely by a photochemical and not a photothermal effect on the endothelium. The serotonin increase may be of major importance as a mediator of blood-brain barrier disruption and/ or abnormal cerebral blood flow and/or neuronal dysfunction in ischemic stroke and transient ischemic attacks (TIA's).

ACKNOWLEDGEMENTS

This study was supported by USPSS grants NS-05820, NS-27127 and NS-23244. P.W. was supported by a USPSS Fogarty International Research Fellowship (1 FO5 TW04558-01 ICP (AHR-5)), the Swedish Medical Research Council, the Swedish Heart and Lung Foundation, King Gustaf V´s Anniversary Foundation, 1987 Foundation for Stroke-research and the Swedish Society of Medicine.

REFERENCES

Dietrich W.D., Prado R., Watson B.D. Photochemically stimulated blood-borne factors induce blood-brain barrier alterations in rats. Stroke 19: 857-862, 1988.

Dietrich W.D., Busto R., Ginsberg M.D. Effect of the serotonin antogonist ketanserin on the hemodynamic and morphological consequences of thrombotic infarction. J. Cereb. Blood Flow Metab. 9: 812-820, 1989.

Dietrich W.D., Prado R., Watson B.D., Busto R., Ginsberg M.D. Hemodynamic consequences of common carotid artery thrombosis and thrombogenically activated blood in rats. J. Cerebral Blood Flow Metab. In press, 1991.

Jellinger K and Riederer P: Zentrale Neurotransmitter bei Zerebraler Ischämie und Hirninfarkt. Fortschr. Neurol. Psychiatr. 54: 91-123, 1983.

Shah A.B., Beamer N. Coull BM: Enhanced in vivo platelet activation in subtypes of ischemic stroke. Stroke, 16: 643-647, 1985

Ungerstedt U. Measurement of neurotransmitter release by intracranial dialysis, In: Measurement of neurotransmitter release in vivo, Marsen C.A. ed. John Wiley and Sons, New York, pp. 81-105, 1984.

Watson B.D., Dietrich W.D., Prado R., Ginsberg M.D. Argon laser-induced arterial photothrombosis. J. Neurosurg. 66: 748-754, 1987.

Welch K.M.A., Meyer J.S., Teraura T., Hashi K., Shinmaru S. Ischemic anoxia and cerebral serotonin level J. Neurol. Sci. 16: 85-92, 1972.

Welch K.M.A., Meyer J.S. Neurochemical alterations in cerebrospinal fluid in cerebral ischemia and stroke,

Wood J.H. ed. Neurobiology of cerebrospinal fluid. New York, Plenum press, pp 325-336, 1980.

Wester P., Puu G. Reiz S., Winblad B., Wester P.O. Increased monoamine metabolite concentrations and cholinesterase activities in cerebrospinal fluid of patients with acute stroke. Acta Neurol. Scand. 76: 473-479, 1987a.

Wester P., Gottfries J., Klintebäck F., Johansson K., Winblad B. Simultaneous determination of seventeen of the major monoamine neurotransmitters, precursors and metabolites I. Optimization of the mobile phase using factorial designs and a computer program to predict chromatograms. J. Chromatogr. 415: 261-274, 1987b.

Wester P., Gottfries J., Winblad B. Simultaneous determination of seventeen of the major monoamine neurotransmitters, precursors and metabolites II. Assessment of human brain and cerebrospinal concentrations. J. Chromatogr. 415: 275-288, 1987c.

Wester P., Bergström U., Gezelius C., Eriksson A., Hardy J., Winblad B. Ventricular cerebrospinal fluid monoamine transmitters and metabolite concentrations reflect human brain neurochemistry in autopsy cases. J. Neurochem. 54: 1148-1154, 1990.

Wilkinson, L. SYSTAT: The System for Statistics. Evanston, IL., SYSTAT, Inc., 1989.

Wurtman R.J., Zervas N.T. Monoamine neurotransmitters and the pathophysiology of stroke and central nervous system trauma. J. Neurosurg. 40: 34-36, 1974.

THE VULNERABILITY OF ADULT AND YOUNG BRAIN TO ISCHEMIA

Vesna Cvejic, Kazuo Kumami, Bogomir B. Mrsulja, and Maria Spatz

National Institutes of Health, National Institute of Neurological Disorders and Stroke, Stroke Branch. Bethesda, Maryland 20892 USA

Introduction

Adults are more susceptible than young animals to cerebral ischemia. 5-Hydroxytryptamine (5-HT), the monoaminergic inhibitory neurotransmitter, is implicated among a variety of factors (such as other transmitter systems and arachidonic acid cascade) to play an important role in the complex pathomechanism of ischemic brain injury of adults.[1-5] Accumulated data indicate a distinct association of ischemic 5-HT disturbance (synthesis, uptake, release, 5-HT_1- and 5-HT_2-receptors) with development and/or progression of cerebral edema. Particularly, a close correlation between the cerebrocortical increase of 5-HT turnover rate (= increased release) and the kinetic changes in 5-HT_2 (S_2- postsynaptic) binding sites was seen in cortical synaptosomes of adult gerbils subjected to bilateral ischemia (15 min) with recirculation (1 hr).[6,7] The cerebral changes in the level of 5-HT, its precursor tryptophan and its main metabolite, 5-hydroxyindoleacetic acid (5-HIAA), as well as norepinephrine (NE) unlike those of energy metabolites depended on the age of the animal and duration of ischemia.[8,9] The same ischemic insult which reduced the levels of 5-HT and NE in the adult cortex delayed or failed to induce 5-HT and NE changes in the young cortex. The present report reviews and summarizes our latest findings regarding the relationship between synaptosomal 5-HT release and the kinetic properties of 5-HT_2-(S_2-) postsynaptic receptors in the cerebral cortex of adult and young animals subjected to the same ischemic insult.

Materials and Methods

Groups of adult (3-4 months) and young (3 weeks), each containing 3-11 Mongolian gerbils, were subjected to ischemia induced by 15 min of bilateral carotid artery occlusion followed by 1 hr reflow under halothane (1%) anesthesia. Respective sham-operated animals served as controls. A synaptosomal fraction of the cerebral cortex obtained by the method of Booth and Clark[10] was used to determine (1) the levels of 5-HT, its precursor tryptophan and metabolite 5-HIAA, (2) uptake and release of [3H]-5-HT, and (3) to characterize 5-HT_2-(S_2-) postsynaptic receptor binding sites.

The concentrations of tryptophan, 5-HT, and 5-HIAA were assayed by high-performance liquid chromatography (Beckman, Berkeley, CA) with electro-chemical detection (Bioanalytic Systems, Lafayette, IN). The specific uptake and release of [3H]-5-hydroxytryptamine creatine sulfate [(3H)-5-HT], specific activity 26.3 Ci/mmol; New England Nuclear Company, Boston, MA] of synaptosomes was determined by a modification of the procedure of Kukar et

The Role of Neurotransmitters in Brain Injury, Edited by
M. Globus and W.D. Dietrich, Plenum Press, New York, 1992

al.[11] The uptake (pmol of ^{3}H-5-HT/mg protein/min) was calculated as the difference between values obtained at 37°C and values at 0°C. The difference in the amount of radioactivity (counts/mg protein/min) in each sample obtained by subtracting counts/mg protein/min of the reincubated sample from counts/mg protein/min of the once-incubated sample (first incubation) was used for calculating the percentage ^{3}H-5-HT release. Student's (two-sided) T test was used to evaluate the statistical significance of the obtained different values.

S_2-postsynaptic receptors were detected in synaptosomes by binding sites using [^{3}H]ketanserin (specific activity 90-96 Ci/mmol, New England Nuclear Company, Boston, MA) the potent 5-HT antagonist which binds selectively to S_2-postsynaptic sites. The specific binding was determined in assays containing methysergide [the only substance among the known serotonin antagonists

Table 1. Cortical synaptosomal metabolites

	Treatment	Duration	5-HT	5-HIAA	Try
		minutes		(pmol/mg protein)	
ADULT	Sham	15	31.9 ± 1.7	5.7 ± 0.2	90.8 ± 3.8
	Ischemia	15	11.1 ± 0.9*	13.2 ± 0.5*	96.1 ± 6.7
	Sham	75	26.6 ± 1.4	4.4 ± 0.3	72.6 ± 4.8
	Ischemia & reflow	15 60	19.0 ± 0.7*	6.2 ± 0.3*	306.2 ± 11.8*
YOUNG	Sham	15	17.0 ± 0.9	4.2 ± 0.2	140.0 ± 7.7
	Ischemia	15	6.1 ± 0.7*	4.9 ± 0.3*	177.9 ± 9.6
	Sham	75	18.8 ± 0.8	3.4 ± 0.2	160.4 ± 4.4
	Ischemia & reflow	15 60	14.6 ± 1.5	4.3 ± 0.3*	366.3 ± 5.7*

The data represent the mean s ± S.E.M. of 5-11 experiments.
*p<0.05 compared with respective controls.

which does not interact with α_1-adrenergic and histamine (H_1) receptors] in excess of [^{3}H]ketanserin concentrations. Scatchard analysis[12] was used to evaluate the results. Whenever the biphasic shape of the curves suggested more than one binding site, the data were examined according to the curve-fitting model for estimating the equilibrium constant.[13] Protein content was determined by the method of Lowry et al.[14]

Results

Ischemia decreased 5-HT levels in both groups of animals while it significantly increased 5-HIAA only in adults (Table 1). After recirculation, the 5-HT and 5-HIAA concentrations normalized although the 5-HT and 5-HIAA content were still significantly altered in adults as compared to their respective controls. In addition, the accumulation of tryptophan was greater in synaptosomes from adult than young animals. Synaptosomal 5-HT uptake was higher in adult than young cortex.

The reduction in the synaptosomal 5-HT uptake (38%) was similar to that (40%) observed in young animals after ischemia only when compared with their respective controls (Fig. 1a, b). An increased spontaneous 5-HT release was observed in adults after ischemia (47%) and recirculation (41%) as compared to sham controls (100%). On the other hand, a reduction of spontaneous 5-HT release (47%) from the synaptosomes was seen in young cortex after ischemia only when compared to the respective sham controls (Fig. 1c, d).

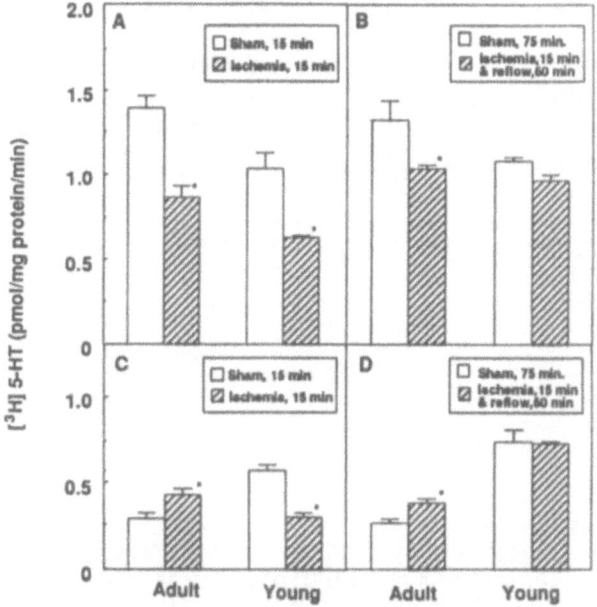

Fig. 1. The effect of ischemia and postischemia on [3H]-5-hydroxytryptamine synaptosomal uptake and release. A, B. Uptake; C, D. Release. The data represent the means ± S.E.M. of 4-8 experiments. * = p < 0.05 compared to sham controls.

The kinetic properties of synaptosomal binding sites for [3H]ketanserin were not significantly affected by ischemia without reflow in the young and adult cortex. The estimated values of K_D and B_{max} were similar to those observed in cortical synaptosomes isolated from the cerebral cortex of their respective controls. However, the synaptosomes obtained from adult in contrast to young synaptosomes of gerbils subjected to 1 hr recirculation after 15 min of ischemia showed changes in the kinetic properties of [3H]ketanserin binding sites (Table 2).

The evaluation of Scatchard plots and computerized nonlinear analysis of the data showed two binding sites for [3H]ketanserin; one with a similar affinity and another with a lower affinity (K_D) than that seen in synaptosomes of respective sham-operated gerbils. The estimated B_{max} value in each experimental group of adult and young animals were not significantly different from those observed in the synaptosomes obtained from their respective sham controls.

Table 2. Effect of cerebral ischemia and postischemia on the properties of [^{3}H] ketanserin binding sites in cerebral cortical synaptosomes

	Treatment	Duration	Kd	B$_{max}$
		minutes	*(nM)*	*(fmol/mg protein)*
ADULT	Sham	15	4.41 ± 0.54	477.5 ± 75.5
	Ischemia	15	5.48 ± 1.10	548.7 ± 177.2
	Sham	75	4.41 ± 0.54	477.5 ± 75.5
	Ischemia & reflow	15 } 60	2.90 ± 0.95 46.30 ± 16.70	187.0 ± 35.8 675.7 ± 194.5
YOUNG	Sham	15	3.00 ± 0.30	318.0 ± 9.6
	Ischemia	15	2.23 ± 0.39	158.9 ± 7.3
	Sham	75	1.80 ± 0.43	177.9 ± 10.7
	Ischemia & reflow	15 } 60	2.06 ± 0.12	264.0 ± 4.4

The data represent the means ± S.E.M. of 3-4 experiments.

Discussion

The data clearly indicate that cerebral ischemia (15 min) equally affected the synaptosomal content of 5-HT and tryptophan as well as the uptake of 5-HT in adult and young gerbils. However, the 5-HT uptake normalized in the young but not in the adult cortical synaptosomes of the gerbils subjected to the same period of ischemia and recirculation. The 5-HT uptake is known to be energy-dependent, requiring both oxygen and glucose. On the other hand, ischemia affects equally the energy-related metabolites in both the adult and young cerebral cortex.[8,9] The lack of correlation between the ischemic changes in monoamines and those of energy-related metabolites suggested that the ischemic response of monoamines is triggered by ischemia-induced processes unrelated to energy metabolism.[15] The persistent reduction in synaptosomal 5-HT uptake in postischemic adult but not young cortex strongly suggests that the adult structure is more sensitive than that of the young to the same ischemic insult. A similar conclusion can be drawn from the observed different 5-HT release and the changes in kinetic properties of [3H]ketanserin binding sites in adult as compared to young synaptosomes. The lingering decrease in 5-HT content and in the 5-HT uptake and the persistence of 5-HT release as well as the modulation of S$_2$-postsynaptic binding sites detected in adult but not young synaptosomes after ischemia and reflow strongly suggest a deficient recovery of the ischemically disturbed synaptosomal membranes in the adult cortex. This assumption agrees with our previous studies showing changes in 5-HT receptor binding sites (5-HT$_{1A}$, 5-HT$_{1B}$, and 5-HT$_2$), membrane fluidity and lipid peroxidation *in vitro* in the cortex even though a normalization of energy metabolites occurs during recirculation in the ischemia model.[7] The ischemically decreased 5-HT release which was not only seen in the young cortex but also in the hippocampus and striatum,[16] suggests that the synaptosomal release system in the young brain is "protected" by nature since the storage capacity for monoamines is comparatively smaller in young than adult brain.[17]

In conclusion, the results of this study indicate a close correlation between the synaptosomal 5-HT release (= presynaptic function) and properties of 5-HT$_2$-postsynaptic receptors in ischemia of adult and young gerbils. Only the increased release of 5-HT from synaptosomes modulated the profile of 5-HT$_2$-postsynaptic receptors. These findings also suggest that the observed greater susceptibility of adult gerbils to brain ischemia than that of the young animals is linked to their distinctive neuronal function.

References

1. Lavyne, M., Moskovitz, M.A., Larin, F., Zervas, N., and Wurtman, R. Brain ^{3}H-catecholamine metabolism in experimental cerebral ischemia. Neurology 25:483 (1975).
2. Kogure, K., Scheinberg, P., Matsumoto, A., Busto, R., and Reinmuth, O. Catecholamines and experimental ischemia. Arch. Neurol. 32:21 (1975).
3. Welch, K.M.A., Chabi, E., Dlodson, R.F., Wang, T.-P.F., Nell, J., and Bergin, B. The role of biogenic amines in the progression of cerebral ischemia and edema: Modification by p-chlorphenylalanine, methysergide and pentoxyfilline. In: "Dynamics of Brain Edema", H. Pappius and W. Feindel, eds.,Springer Verlag, Berlin, Heidelberg, New York (1976).
4. Mrsulja, B.B., Djuricic, B.M., Cvejic, V., Mrsulja, B.J., Spatz, M., and Klatzo, I. Biochemistry of experimental ischemic brain edema. Adv. Neurol. 28:217 (1980).
5. Matsumoto, M., Kimura, K., Fujisawa, A., Matsuyama, T., Fukanaga, R., Yoneda, S., Wada, H., and Hiroshi, A. Differential effect of cerebral ischemia on monoamine content of discrete brain regions of the Mongolian gerbil: *Meriones unguiculatus.* J. Neurochem. 42:647 (1984).
6. Wroblewska, B., Ueki, T., Mrsulja, B.B., Djuricic, B.M., and Spatz, M. Cerebro-cortical modulation of S$_2$-receptors and turnover rate of 5-hydroxytryptamine in ischemia. Neurochem. Res. 12:21, 1987.
7. Villacara, A., Kumami, K., Yamamoto, T., Mrsulja, B.B., and Spatz, M. Ischemic modification of cerebrocortical membranes: 5-HT receptors, fluidity and inducible *in vitro* lipid peroxidation. J. Neurochem. 53:595 (1989).
8. Spatz, M., Ueki, Y., Djuricic, B.M., and Mrsulja, B.B. 5-Hydroxytryptamine in cerebral cortex of adult and young ischemic gerbils, in: "Cerebral Vascular Disease 5," J.S. Meyer, J.S., H. Lechner, M. Reivich, and E.O. Otto, eds., Experimental Medicine, Amsterdam, 1985.
9. Kumami, K., Mrsulja, B.B., Ueki, Y., Djuricic, D., and Spatz, M. Effect of ischemia of noradrenergic and energy-related metabolites in cerebral cortex of young and adult gerbils. Metab. Brain Dis. 3:273 (1989).
10. Both, R.F. and Clark, J.B. A rapid method for the preparation of relatively pure metabolically competent synaptosomes from rat brain. Biochem. J. 176:365 (1978).
11. Kuhar, M.J., Roth, L.H., and Aghajanian, A. Synaptosomes from forebrains of rats with midbrain raphe lesions; Selective reduction of serotonin uptake. J. Pharmacol. Exp. Ther. 181:36 (1972)
12. Scatchard, G. The attraction of proteins for small molecules and ions. Ann. N.Y. Acad. Sci. 5:660 (1949).
13. Knott, G.D. Lab-a mathematical modeling tool. Comp. Prog. Biomed. 10:271.(1979).
14. Lowry, O.H., Rosebrough, N.J., Farr, A.L., and Randall, J.R. Protein measurement with folin phenol reagent. J. Biol. Chem. 193:265 (1951).
15. Mrsulja, B.B., Djuricic, B.M., Ueki, Y., Lust, W.D., and Spatz, M. Cerebral ischemia: Changes in monoamines are independent of energy metabolism. Neurochem. Res. 14:1 (1989).

16. Cvejic, V., Kumami, K., and Spatz, M. Effect of cerebral ischemia on synaptosomal uptake and release of ^{3}H-5-hydroxytryptamine in adult and young gerbils. Metab. Brain Dis. 5:1 (1990).

17. Karki, N., Kuntzman, R., and Brodie, B.B. Storage, synthesis and metabolism of monoamines in the developing brain. J. Neurochem. 9:53.

Chapter 4

The Role of Steroids, Adenosine, GABA, and Acetylcholine in Brain Injury

THE ROLE OF GLUCOCORTICOIDS IN ISCHEMIC CELL DEATH

James N. Davis, Gerald D. Miller,
and Joanne K. Morse

Neurology Research Laboratory
DVA Medical Center and
Departments of Medicine (neurology)
Pharmacology and Neurobiology

Duke University
Durham, N.C.

GLUCOCORTICOIDS AND ISCHEMIC CELL DEATH

Selective neuronal vulnerability

When the circulation to the brain is briefly interrupted
and then restored, only certain neurons are damaged. This
selective neuronal vulnerability after transient ischemia is
found in many brain regions, but is perhaps most dramatic in
the hippocampus. There, both CA_1 and CA_4 neurons are damaged,
but CA_1 neurons die only after a definite delay. This delay
has been the focus of intense research in the past decade with
the hope that a better understanding of this type of damage
will lead to effective therapies which can be administered to
patients after periods of transient ischemia.

This chapter focuses on the influence of adrenal glucocorticoid hormones on delayed ischemic CA_1 cell death in the
hippocampus. Glucocorticoids exacerbate cell death while removing glucocorticoids attenuates damage. We propose that
this delayed ischemic cell death is a form of apoptosis and
speculate about possible mechanisms underlying apoptotic cell
death which might be regulated by glucocorticoids.

While glucocorticoids are not usually thought of as
neurotransmitters, they have obvious and important roles in
brain function. In fact, plasma glucocorticoids increase
during virtually any form of "stress' including ischemia.

Glucocorticoids in the gerbil

Figure 1 illustrates plasma cortisol levels in gerbils.
To our knowledge, glucocorticoid levels have not been measured
previously in gerbils. Cortisol is the predominant glucocor-

The Role of Neurotransmitters in Brain Injury, Edited by W.D. Dietrich
and Y.T. Globus, Plenum Press, New York, 1992

ticoid in this species as in man, while corticosterone levels
are about one third of the cortisol levels under all
conditions (1). Corticosterone is the predominant glucocorticoid in rats. Plasma cortisol levels more than triple after
carotid surgery and stay elevated for at least 4 hours
afterwards. Plasma cortisol levels are increased in gerbils
just from handling. Anesthesia with halothane does not
prevent this increase in levels. Finally, even sham carotid
occlusion produces a long-lasting increase in plasma cortisol.

Glucocorticoids and the hippocampus

The hippocampus contains a relatively high concentration
of binding sites for glucocorticoid steroid hormones. In
fact, the hippocampus has been proposed as the brain
"receptor" for glucocorticoids (2). It is established that
the brain appears to monitor levels of the glucocorticoids and
use that information to regulate the release of CRF, cortisol
releasing factor, in the hypothalamus which in turn regulates
the release of ACTH from the pituitary and, ultimately, the
release of glucocorticoids from the adrenal cortex. During
periods of stress, plasma glucocorticoids are elevated until
the brain senses the increased levels and turns down production by the pituitary hypothalmic axis.

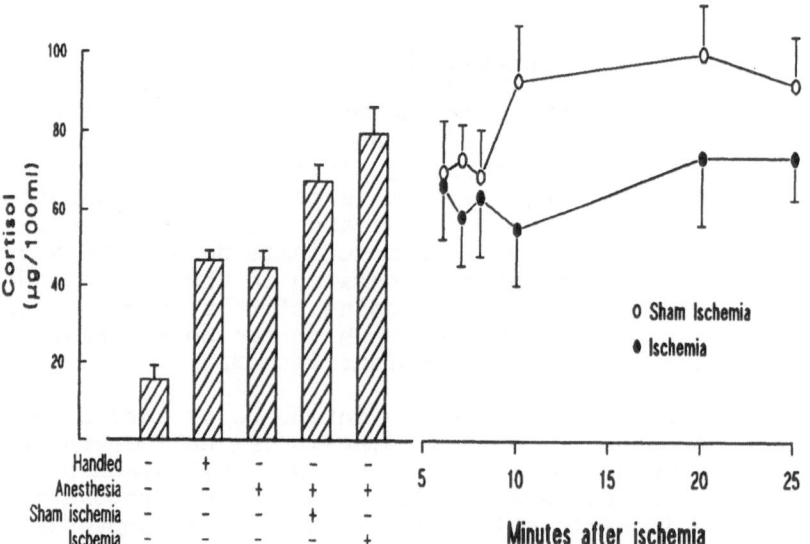

Figure 1. Plasma cortisol during ischemia, sham
         ischemia, anesthesia and handling. The bar
         graph to the left demonstrates levels of
         cortisol in a variety of conditions.
         Control animals were removed from their
         cages and decapitated within two minutes in
         an adjacent room. The right side inset of
         the accompanying figure demonstrates post-
         ischemic plasma cortisol levels in gerbils
         subjected to either ischemia or sham
         ischemia. Part of the data from this figure
         has been published (1).

Sapolsky and McEwen (3) proposed a hypothesis linking glucocorticoids with hippocampal damage from ischemia, excitotoxins, and neurotoxins. Furthermore, their theory relates the effects of aging to alterations in the brain's response to stress. They argue that as hippocampal neurons are lost with age, the ability of the brain to respond to stressful elevations of plasma glucocorticoids declines. Thus, in their theory, aged animals are subjected to prolonged elevations in plasma glucocorticoids because with fewer hippocampal neurons they cannot shut off stress-induced increases in these hormones as effectively as younger animals. This prolonged exposure to glucocorticoids in older animals then serves to cause further loss of hippocampal neurons. This important theory of aging induced loss of hippocampal neurons is based on the observation that elevations in plasma glucocorticoids render hippocampal cells susceptible to damage.

Glucocorticoids and ischemic cell death

We became initially interested in this phenomenon when Sapolsky and Pulsinelli (50) demonstrated that the immediate administration of adrenal glucocorticoids after a transient carotid occlusion in the four vessel occlusion model exacerbated ischemic neuronal damage while immediate adrenalectomy appeared to offer partial protection. Our first experiments aimed at replicating this finding in the gerbil.

Figure 2 demonstrates our results. The experiment was carried out exactly as Sapolsky and Pulsinelli had carried out their experiments in the rat and demonstrates that, in the gerbil, administration of a mixture of glucocorticoids exacerbates hippocampal damage while immediate adrenalectomy offers

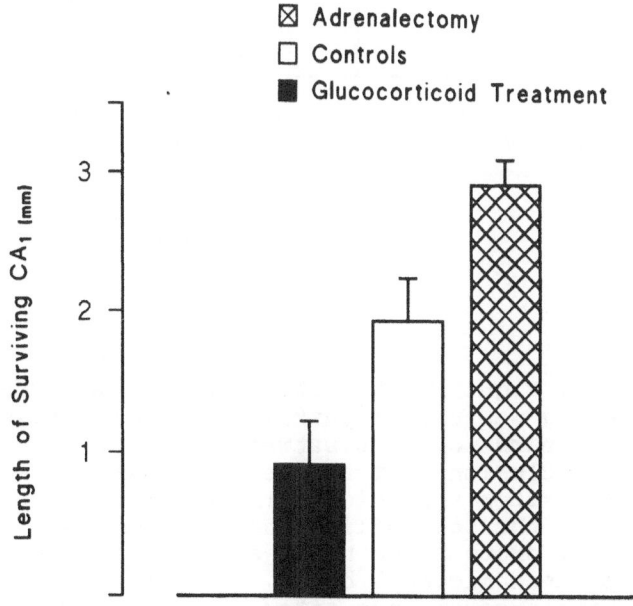

Figure 2. Effect of treatment with glucocorticoids or adrenalectomy on CA1 damage (modified from (1)).

apparent protection. Gerbils were exposed to 5 minutes of bilateral carotid occlusion and then either had immediate
adrenalectomy or were begun on twice daily injections of a
mixture of cortisol and corticosterone in sesame oil.
Controls had sham adrenalectomies and received sesame oil
vehicle injections

Since adrenalectomy appeared to partially protect the
hippocampus when performed immediately after carotid occlusion, we wondered how long after ischemia the adrenalectomy
could be performed and still be effective. Thus, in a second
experiment, adrenalectomy was performed 24 hours after bilateral carotid occlusion.

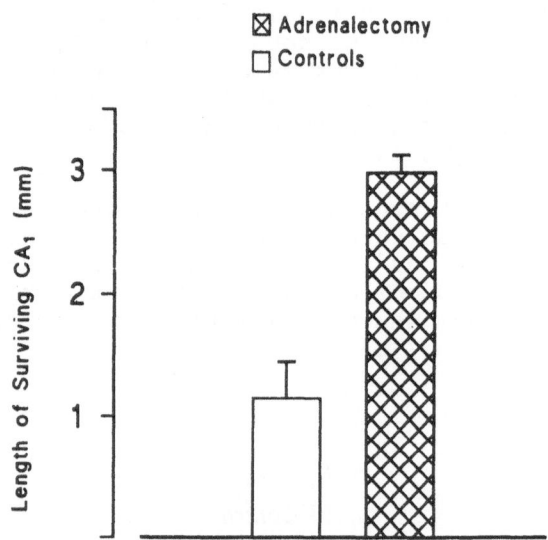

Figure 3. Effect of adrenalectomy carried out 24
        hours after carotid occlusion on CA1 damage
        (modified from (1)).

Once again, significant protection was observed (Figure
3). This finding was rather puzzling since ischemic cell
death in CA1 appears to begin between eight and twelve hours
after carotid occlusion so that the adrenalectomy was able to
exert an effect after ischemic cell death had begun. This observation raised the possibility that delayed adrenalectomy
did not actually protect the brain as much as perhaps delay
the process of ischemic cell death. To test this hypothesis,
we studied animals seven days after a carotid occlusion who
were subjected to adrenalectomy 24 hours after the carotid
occlusion. The results were striking. Adrenalectomy had no

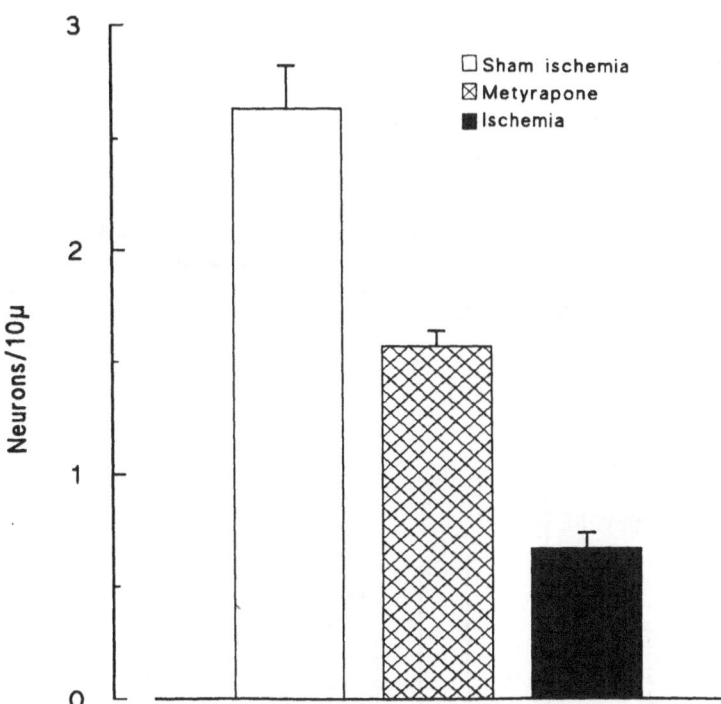

Figure 4. Effect of metyrapone treatment on
ischemic CA_1 damage when the animals are
sacrificed 72 hours after carotid occlusion.
The bars are means ± standard errors of the
means of CA_1 neuronal counts made from each
hemisphere and expressed per 10 μ of CA_1
length.

effect on CA_1 when damaged was assessed 7 days after a carotid
occlusion. Thus, adrenalectomy 24 hours after ischemia was
associated with less severe damage at 72 hours, but not at 7
days. Delayed adrenalectomy appears to slow the process of
ischemic cell death without actually protecting the neurons.
The mechanism behind this effect of delayed adrenalectomy is
not yet understood. It could represent a glucocorticoid
effect on one or more of the intracellular processes involved
in cell death. Alternatively, adrenalectomy could change the
histological appearance of cell death by delaying the removal
of dead CA_1 pyramidal cells by astrocytes and other glia.

Drugs which block glucocorticoid actions

    Since adrenalectomy represents a rather severe surgical
procedure, it has a number of systemic effects including the
removal of mineralocorticoid hormones. In order to be certain
that the effects of adrenalectomy on CA_1 pyramidal cell death
were mediated by glucocorticoids, we next investigated the
effects of the drug, metyrapone, on CA_1 cell death.

    Metyrapone inhibits 11-ß hydroxylation of steroids in the
adrenal cortex, blocking stress-related release of glucocorticoids. We found that the administration of metyrapone
(200 mg/kg twice daily) immediately after bilateral carotid

occlusion was similar to adrenalectomy in apparently protecting CA_1 neurons (Figure 4). These data support the hypothesis that the effect of adrenalectomy on ischemic cell damage is mediated by glucocorticoid hormones.

That hypothesis is further confirmed by a preliminary experiment using the glucocorticoid receptor blocker, RU 38 486 (Figure 5). When this compound was administered intraperitoneally, beginning immediately after 5 minutes of bilateral carotid occlusion, there is a mild but statistically significant blockade of CA_1 ischemic neuronal death. It should be

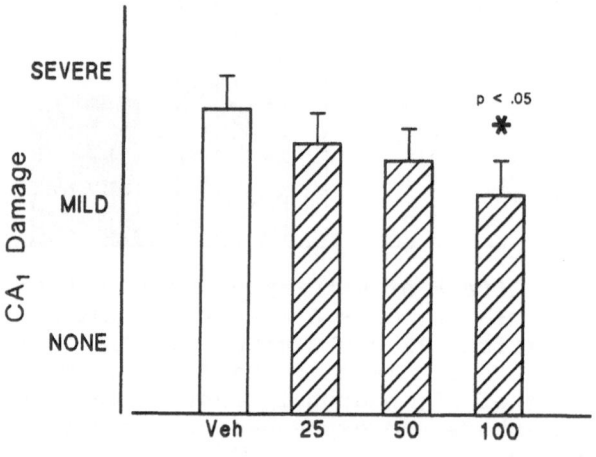

Figure 5. The effect of various doses of RU 38 486 given daily beginning immediately after ischemia on CA_1 cell death in gerbils sacrificed 72 hours later. Damage was rated with a qualitative 3 level scale and statistical significances calculated with a chi-square statistic.

noted that this is a preliminary experiment in which the drug was administered twice daily. It is likely that the action of the drug in blocking glucocorticoid receptors was short-lived lasting only a few hours. With continuous administration, a more significant effect will probably be apparent.

Glucocorticoids and brain temperature

In the course of these experiments, we observed that adrenalectomy and metyrapone administration were associated with drops in body and brain temperature. Figure 6 demonstrates the brain temperature in gerbils after a single dose of metyrapone. Temperatures measured in freely moving gerbils using microprobes inserted into the striatum declined by almost 2°C within 30 minutes of an injection of metyrapone (4).

Gerbils subjected to adrenalectomy either immediately or 24 hours after bilateral carotid occlusion whether sacrificed 72 hours or 7 days later all demonstrated statistically significant reductions in body temperature. These findings raise the possibility that the effects of glucocorticoids on ischemic neuronal damage might be mediated by changes in body temperature.

Post ischemic elevations in brain temperature have been described in gerbils after bilateral carotid occlusions (5). We have also observed the same elevation in brain temperature in our animals. It seems likely that post ischemic hyperthermia is due to the post ischemic surge in plasma glucocorticoid hormones.

Possible mechanisms for glucocorticoid action

These data taken together show the regulatory effect of glucocorticoids on ischemic neuronal death. One possible mechanism for the glucocorticoid effect is an alteration in glucose utilization. Sapolsky (6) has shown that providing alternate fuels such as mannose or beta-hydroxybutyrate to animals partially protects them from the damaging effects of glucocorticoids and various metabolic insults such as kainic acid or 3-acetylpyridine administration. These findings along with classical studies showing that glucocorticoids impair glucose utilization may be the mechanism rendering pyramidal cells vulnerable to insults. On the other hand, chronic and acute administration of glucocorticoids is associated with elevations in plasma glucose (7). Plasma glucose elevations during ischemia can worsen cellular damage presumably by increasing the formation of lactic acid (8).

Speculating about the mechanism of glucocorticoid exacerbation of ischemic damage is complicated by the many effects

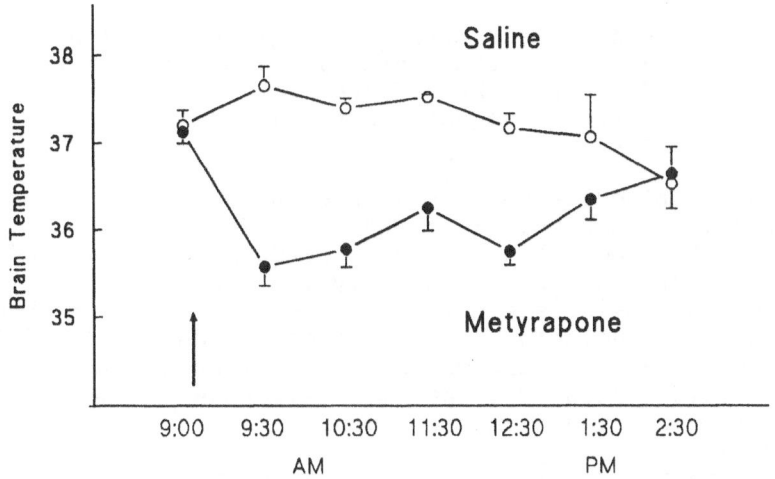

Figure 6. Effect of 100 mg/kg of metyrapone on
striatal temperature in the gerbil. Each
point is the mean of 3 or 4 determinations.
The vertical bars denote one standard error
of the mean.

of glucocorticoids on the adrenal-pituitary axis, on protein synthesis, and on metabolism. In fact adrenalectomy may be associated with disappearance of hippocampal granule cells (9-11). Recent studies of glucocorticoid effects on CA1 after-hyperpolarization have reported opposing effects for type I (mineralocorticoid) (12) receptor and type II (glucocorticoid) (13) receptor activation in the hippocampus. Finally, the importance of glucocorticoids in thermogenesis and body temperature homeostasis have long been recognized (14). We continue to explore the possibility that changes in brain temperature induced by glucocorticoids may be responsible for part of the regulatory effect of these hormones on ischemic cell death.

IS ISCHEMIC CELL DEATH PROGRAMMED?

Apoptosis

The process of ischemic cell death is not totally understood. While excitatory amino acids and increases in intracellular Ca^{++} are important in triggering the process, the mechanisms ultimately leading to cell death are not well defined. CA1 pyramidal cell death does not occur immediately, but after a delay of at least several hours in all species studied to date. Pathologists generally separate cell death into two broad classifications, necrosis or apoptosis (15). Apoptosis has become synonymous with the term "programmed cell death". It appears to be the process underlying cell death during development, during organ remodeling, after hormone stimulation of some tissues, during certain immune mediated killing, after radiation exposure, among other physiological signals (15,16). In the adult, apoptosis accounts for cell death in the intestinal crypts, the uterine lining, when cells are attacked by natural killer lymphocytes, when macrophages are exposed to the fungal toxin, gliotoxin, when lymphocytes or thymocytes are exposed to glucocorticoids, and when neurons are deprived of trophic factors.

Yet it is not clear that all programmed cell death is the same. For example, neurons that die naturally during development could differ from those dying from lack of target trophic support. At present the details of different programmed cell deaths are not known.

The distinction between apoptosis and necrosis was originally made on morphological grounds. Necrosis is characterized by mitochondrial swelling, breakdown of plasma and nuclear membranes, and cellular lysis. One of the earliest signs of apoptosis is a curious clumping of nuclear chromatin and often the nucleolus (17). The cell membranes then form vesicles containing portions of the cell organelles and cytoplasm. This vesicular compartmentalization of the cell is eventually ingested by macrophages.

Apoptosis is further characterized by the early appearance of DNA fragmentation. DNA from apoptitic cells is digested at the site of the "linker" DNA joining nucleosomes into pieces of approximately 176 bp length or a multiple of 176 bp. When this fragmented DNA is subjected to electrophoresis in an agarose gel, a characteristic "step ladder" appearance is found (18). This step ladder appearance has

174

been identified in many of the examples of apoptosis and has lead to the hypothesis that expression of a nuclease is the early step triggering apoptosis. The possibility that genetic expression of this nuclease is the primary step in programmed cell death has been suggested from experiments with gluco-corticoid-stimulated lymphocyte death (19), but not all obser-vations are consistent with this proposal (20). Ca^{++} influx seems to be an important early event leading to activation of the putative nuclease and has been found in several cells during apoptosis (21). This has led to the hypothesis that an early event in apoptosis is the activation of a Ca^{++} sensitive nuclease (19). However, a recent report shows that some forms of apoptosis are not associated with DNA fragmentation (22). This may well be the case in ischemic cell death since clump-ing of chromatin is not seen in CA_1 pyramidal cells after transient ischemia (see below).

The synthesis of specific proteins, which some have termed "suicide proteins", may be important in some forms of programmed cell death. Protein synthesis inhibitors prevent apoptitic cell death in chick neurons during development (23), in rat sensory and sympathetic neurons when trophic factors are removed (24), and in leukemia cells treated with gluco-corticoids (25). However protein synthesis inhibition elicits apoptosis in macrophages (26) and human leukemia cells (27). Thus the role of suicide proteins in apoptosis is not well understood.

There are a number of laboratories attempting to isolate and characterize the genes responsible for programmed cell death. Mutations of C. elegans have been identified in which certain neurons do not undergo normal developmental cell death (28). These mutations have been termed "cell death" mutations and the genes responsible for them have been identi-fied (29,30). Two of these genes, ced3 and ced4 appear to code for primary steps in cell death in these animals (31). In other systems, the use of molecular biological tools are being applied to the identification and characterization of programmed cell death. It seems likely that within a few years probes and sequences will be available for same of these genes allowing a better understanding of the molecular mecha-nisms underlying apoptosis.

Hypothetical sequence of events

A theoretical model of the events that occur in a hippocampal pyramidal cell after ischemia is depicted in Figure 7.

In this model glutamate release during ischemia triggers a massive influx of Ca^{++}. Glucocorticoid release during stress raises plasma glucocorticoid levels and glucocorticoid receptors are activated. They shed a heat shock protein 90 (HSP-90), bind glucocorticoid, and enter the cell nucleus where they bind to specific glucocorticoid recognition ele-ments (GRE) of DNA. Of the genes promoted by receptor binding to GRE's may be a Ca^{++}-sensitive nuclease that degrades DNA forming the characteristic step ladder appearance on gel electrophoresis. Other Ca^{++}-dependent proteins could also be regulated by glucocorticoids. Any of these expressed proteins could be the critical step in glucocorticoid modulation of ischemic cell death.

Morphological observations of ischemic cell death

The morphology of CA1 pyramidal cell death has been extensively characterized both in gerbil (32) and in rat (33). From these studies it is clear that chromatin clumping characteristic of apoptosis in other cells is not an early event in ischemia. On the other hand, the morphology of delayed cell death in CA1 pyramidal cells is not characteristic of necrosis either. The somatic mitochondria appear normal until the very end of the cell's existence. There is no breakdown of plasma or nuclear membranes even when the cells are no longer functional.

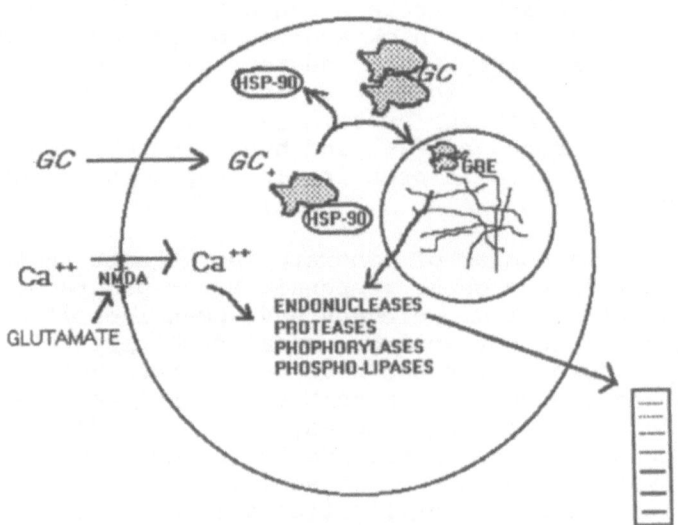

Figure 7. Drawing of a hypothetical model demonstrating interactions of glucocorticoids and ischemic cell death. The rectangle on the left signifies the "stepladder" seen on agarose gel electrophoresis of apoptotic cells undergoing DNA fragmentation.

Protein synthesis in ischemic cell death

The synthesis of new proteins during ischemic cell death has been characterized and new proteins are formed. Overall protein synthesis is decreased after ischemia (34-36), but there does appear to be real increases in synthesis of some proteins (37,38). Perhaps the best characterized of these is a protein or proteins from the heat shock family of approximately 70 kD in size (HSP-70) (39,40). In addition recent experiments in several laboratories have shown that protein synthesis inhibition will delay or prevent CA1 neuronal cell death (41,42) (Pulsinelli, personal communication).

176

DNA fragmentation

In another in vitro experiment, hippocampal neurons in culture were damaged by kainic acid in the presence of glucocorticoids and did not show a "step ladder" appearance when their DNA was studied by electrophoresis (43). However these experiments were carried out in mixed cultures where the amount of cell death was small (around 30% by LDH measurements) and where DNA in the supernatant was examined. Unless a variety of times were examined, DNA fragmentation could have occurred as an early event and been undetected. In fact, DNA fragmentation could have occurred in the nucleus of the cells and not been detected in the supernatant of the cultures.

Evidence for the hypothesis

There is considerable rationale for considering the hypothesis that ischemia triggers programmed cell death in CA_1 pyramidal cells. CA_1 pyramidal cells do not die immediately but only after several hours. It is not clear when the cell death process starts. The fact that glucocorticoids might regulate the _rate_ of cell death fits well with the idea that a glucocorticoid responsive step is critical in causing delayed CA_1 death. Protein synthesis inhibition (41), NGF (44) bFGF (Finkelstein, personal communication), or aFGF (45) (MacMillan and Davis, unpublished observations) infusion before the onset of ischemia protects CA_1 pyramidal cells from delayed neuronal death consistent with the hypothesis. Finally the leading hypothesis at the present time to explain delayed neuronal death, overexcitation of pyramidal cells after ischemia (46), has been very difficult to prove. There is little evidence of excessive CA_1 firing during the immediate post-ischemic period (47-49).

REFERENCES

1. J. K. Morse and J. N. Davis, Regulation of ischemic hippocampal damage by adrenal steroids in the gerbil: Adrenalectomy alters the rate of pyramidal cell death, Exp. Neurol., 110:86 (1990).
2. B. S. McEwen, E. R. De Kloet and W. Rostene, Adrenal steroid receptors and actions in the nervous system, Physiol. Rev., 66:1121 (1986).
3. R. M. Sapolsky, L. C. Krey and B. S. McEwen, The neuroendocrinology of stress and aging: The glucocorticoid cascade hypothesis, Endocrine Reviews, 7:284 (1986).
4. J. K. Morse and J. N. Davis, Chemical adrenalectomy decreases CNS temperature and protects hippocampal cells following ischemia, The Pharmacologist, 31:183 (1989).(Abstract)
5. T. Kuroiwa, P. Bonnekoh and K. A. Hossmann, Prevention of postischemic Hyperthermia Prevents Ischemic Injury of CA_1 neurons in gerbils, J. Cereb. Blood Flow Metab., 10:550 (1990).
6. R. M. Sapolsky, Glucocorticoid toxicity in the hippocampus: reversal by supplementation with brain fuels, J. Neurosci., 6:2240 (1986).
7. T. Koide, T. W. Wieloch and B. K. Siesjo, Chronic dexamethasone pretreatment aggravates ischemic neuronal necrosis, J. Cereb. Blood Flow Metab., 6:395 (1986).

8. B. K. Siesjo, M. Smith and D. S. Warner, Acidosis and ischemic brain damage, in: "Cerebrovascular Diseases," W. J. Powers and M. E. Raichle, eds., Raven Press, New York (1987).

9. R. S. Sloviter, G. Valiquette, G. M. Abrams, E. C. Ronk, A. L. Sollas, L. A. Paul and S. Neubort, Selective loss of hippocampal granule cells in the mature rat brain after adrenalectomy, Science, 243:535 (1989).

10. J. Maehlen and A. Torvik, Necrosis of granule cells of hippocampus in adrenocortical failure, Acta Neuropath., 80:85 (1990).

11. E. Gould, C. S. Woolley, D. L. Miller, G. M. Begany, R. E. Brinton and B. S. McEwen, The stress non-responsive period permits naturally occurring cell death in the developing dentate gyrus, Soc. Neurosci. Abstr., 16:328 (1990).(Abstract)

12. M. Joels and E. R. De Kloet, Mineralocorticoid receptor-mediated changes in membrane properties of rat CA1 pyramidal neurons in Vitro, Proc. Natl. Acad. Sci. U. S. A., 87:4495 (1990).

13. M. Joels and E. R. De Kloet, Effects of glucocorticoids and norepinephrine on the excitability in the hippocampus, Science, 245:1502 (1989).

14. C. J. Gordon and J. E. Heath, Integration and central processing in temperature regulation, Ann. Rev. Physiol., 48:595 (1986).

15. A. H. Wyllie, J. F. R. Kerr and A. R. Currie, Cell death: The significance of apoptosis, Int. Rev. Cytol., 68:251 (1980).

16. N. I. Walker, B. V. Harmon, G. C. Gobe and J. F. R. Kerr, Patterns of cell death, Meth. Achiev. exp. Pathol, 13:18 (1988).

17. A. H. Wyllie, G. J. Beattie and A. D. Hargreaves, Chromatin changes in apoptosis, Histochem. J., 13:681 (1981).

18. A. H. Wyllie and R. G. Morris, Hormone-induced cell death: Purification and properties of thymocytes undergoing apoptosis after glucocorticoid treatment, Adv. Neurol., 109:78 (1982).

19. M. M. Compton and J. A. Cidlowski, Identification of a glucocorticoid-induced nuclease in thymocytes, J. Biol. Chem., 262:8288 (1987).

20. G. D. Baxter, P. J. Smith and M. F. Lavin, Molecular changes associated with induction of cell death in a human T-cell leukaemia line: putative nucleases identified as histones, Biochem. Biophys. Res. Commun., 162:30 (1989).

21. D. J. McConkey, S. Orrenius and M. Jondal, Cellular signalling in programmed cell death (apoptosis), Immunology Today, 11:120 (1990).

22. R. A. Lockshin, A. Alles, A. Kodaman and Z. F. Zakeri, Programmed cell death and apoptosis: Early DNA degredation does not appear to be prominent in either embyonic cell death or metamorphosis in insects, FASEB J., 5:A371 (1991).(Abstract)

23. R. W. Oppenheim, D. Prevette, M. Tytell and S. Homma, Naturally occuring and induced neuronal death in the chick embryo in vivo requires protein and RNA synthesis: Evidence for the role of cell death genes, Develop. Biol., 138:104 (1990).

24. D. P. Martin, R. E. Schmidt, P. E. DiStefano, O. H. Lowry, G. C. Joyce and E. M. Johnson, Inhibitors of protein

synthesis and RNA synthesis prevent neuronal death caused by nerve growth factor deprivation, J. Cell Biol., 106:829 (1988).

25. U. Galili, R. Leizerowitz, J. Moreb, H. Gamliel, D. Gurfel and A. Polliack, Metabolic and ultrastructural aspects of the in vitro lysis of chronic lymphocytic leukemia cells by glucocorticoids, Cancer Res., 42:1433 (1982).

26. P. Waring, DNA fragmentation induced in macrophages by gliotoxin does not require protein synthesis and is preceded by raised inositol triphosphate levels, J. Biol. Chem., 265:14476 (1990).

27. S. J. Martin, S. V. Lennon, A. M. Bonham and T. G. Cotter, Induction of apoptosis (programmed cell death) in human leurkemic HL-60 cells by inhibition of RNA or protein synthesis, J. Immunol., 145:1859 (1990).

28. P. W. Sternberg, Genetic control of cell type and pattern formation in Caenorhabditis elegans, in: "Genetic regulatory hierarchies in development," T. R. F. Wright, ed., Academic Press, New York (1990).

29. H. M. Ellis and H. R. Horvitz, Genetic control of programmed cell death in the nematode C. elegans, Cell, 44:817 (1986).

30. H. R. Horvitz, P. W. Sternberg, I. S. Greenwald, W. Fixsen and H. M. Ellis, Mutations that affect neural cell lineages and cell fates during the development of the nematode Caenorhabditis elegans, C. S. H. Symp. Quant. Biol., 48:453 (1983).

31. J. Yuan and H. R. Horvitz, The caenorhabditis elegans genes ced-3 and ced-4 act cell autonomously to cause programmed cell death, Develop. Biol., 138:33 (1990).

32. B. J. Crain, D. A. Evenson, K. Polsky and J. V. Nadler, Electron microscopic study of the gerbil dentate gyrus after transient forebrain ischemia, Acta Neuropath., 79:409 (1990).

33. K. Magnusson, J. Deshpande, T. Linden, H. Kalimo and T. W. Wieloch, Dense deposits in rat hippocampal CA1 neurons during post ischemic delayed neuronal death, Soc. Neurosci. Abstr., 15:363 (1989).(Abstract)

34. M. Yoshidomi, T. Hayashi, K. Abe and K. Kogure, Effects of a new calcium channel blocker, KB-2796, on protein synthesis of the CA1 pyramidal cell and delayed neuronal death following transient forebrain ischemia, J. Neurochem., 53:1589 (1989).

35. W. Bodsch, K. Takahashi, A. Barbier, B. G. Ophoff and K. A. Hossmann, Cerebral protein synthesis and ischemia, Prog. Brain. Res., 63:197 (1985).

36. R. Thilmann, Y. Xie, P. Kleihues and M. Kiessling, Persistent inhibition of protein synthesis precedes delayed neuronal death in postischemic gerbil hippocampus, Acta. Neuropathol. (Berl)., 71:88 (1986).

37. M. Kiessling, G. A. Dienel, M. Jacewicz and W. A. Pulsinelli, Protein synthesis in postischemic rat brain: a two-dimensional electrophoretic analysis, J. Cereb. Blood. Flow. Metab., 6:642 (1986).

38. M. Jacewicz, M. Kiessling and W. A. Pulsinelli, Selective gene expression in focal cerebral ischemia, J. Cereb. Blood. Flow. Metab., 6:263 (1986).

39. G. A. Dienel, M. Kiessling, M. Jacewicz and W. A. Pulsinelli, Synthesis of heat shock proteins in rat brain cortex after transient ischemia, J. Cereb. Blood. Flow. Metab., 6:505 (1986).

40. T. S. Nowak, J. Ikeda and T. Nakajima, 70-kDa heat shock protein and c-fos gene expression after transient ischemia, <u>Stroke</u>, 21 (suppl. 3):107 (1990).

41. T. Shigeno, Y. Yasundo, G. Kato, K. Kusaka, T. Minia, K. Takakura, D. I. Graham and S. Furukawa, Reduction of delayed neuronal death by inhibition of protein synthesis, <u>Neurosci. Letters</u>, 120:117 (1990).

42. K. Goto, A. Ishige, K.Sekiguchi, S. Iizuka, A. Sugimoto, M. Yuzurihara, M. Aburada, E. Hosoya and K. Kogure, Effects of cycloheximide on delayed neuronal death in rat hippocampus. <u>Brain Res.</u>, 534:299-302 (1990).

43. J. N. Masters, C. E. Finch and R. M. Sapolsky, Glucocorticoid endangerment of hippocampal neurons does not involve deoxyribonucleic acid cleavage, <u>Endocrinology</u>, 124:3083 (1989).

44. A. M. Buchan, L. Williams and B. Bruederlin, Nerve growth factor: Pretreatment ameliorates ischemic hippocampal neuronal injury, <u>Stroke</u>, 21:177 (1990).(Abstract)

45. Y. Oomura, K. Sasaki, T. Muto, K. Suzuki, K. Hanai, I. Tooyama, H. Kimura and N. Yanaihara, Physiological actions of fibroblast growth factor (FGF) in central nervous system, <u>Soc. Neurosci. Abstr.</u>, 16:516 (1990).(Abstract)

46. R. Suzuki, T. Yamaguchi, Y. Inaba and H. G. Wagner, Microphysiology of selectively vulnerable neurons, <u>Prog. Brain Res.</u>, 63:59 (1985).

47. D. R. Armstrong, K. H. Neill, B. J. Crain and J. V. Nadler, Absence of electrographic seizures after transient forebrain ischemia in the Mongolian gerbil, <u>Brain Res.</u>, 476:174 (1989).

48. A. Mitani, H. Imon and K. Kataoka, High frequency discharges of gerbil hippocampal CA1 neurons shortly after ischemia, <u>Brain Research Bulletin</u>, 23:569 (1989).

49. G. Buzsaki, T. F. Freund, F. Bayardo and P. Somogyi, Ischemia-induced changes in the electrical activity of the hippocampus, <u>Exp. Brain Res.</u>, 78:268 (1989).

50. R. M. Sapolsky and W. A. Pulsinelli, Glucocorticoids potentiate ischemic injuries to neurons: therapeutic implications, <u>Science</u>, 229:1397-1400 (1985).

PREVENTION OF HYPOXIC-ISCHEMIC DAMAGE IN NEONATAL RAT BY

GLUCOCORTICOIDS

U.I. Tuor

Div. of Neonatology, Hospital for Sick Children and University of
Toronto, 555 University Ave, Toronto, M5G 1X8

Introduction

In the adult, glucocorticoids have been employed for the treatment of a
range of CNS disorders. Beneficial effects of glucocorticoid therapy have been
observed in a few studies, usually related to the improvement of cerebral edema
and recently for the treatment of spinal cord trauma (5,7,17). As a therapy for
either global or focal cerebral ischemia, the administration of glucocorticoids has
usually been ineffective and in some cases detrimental (9,12,14,18,23).

In the neonate, conflicting information is available concerning the influence
of glucocorticoids on ischemic brain damage. In 7 day old rats, a single very large
dose of dexamethasone administered immediately prior to a hypoxic-ischemic insult
was harmful in that it increased mortality with no effect on pathology (2).
Dexamethasone (20 mg/kg) administered 2 hours before an asphyxic insult in 5 day
old rats appeared to be beneficial in that there was a smaller decrease in brain ATP
in dexamethasone treated rats than controls (1). Similarly, in a piglet model of
brain damage due to pneumothorax, a large dose of dexamethasone was effective in
reducing brain edema when given 4 hr before the pneumothorax (25). Thus, most
available evidence in the adult discourages the use of glucocorticoids for the
treatment of cerebral ischemia whereas studies in the neonate are inconclusive.

Surprisingly, recent results of experiments in neonatal rats demonstrate that
the synthetic glucocorticoid, dexamethasone, prevents hypoxic-ischemic brain
damage (3). As discussed in further detail below, several known glucocorticoid
effects may influence the outcome of an episode of hypoxia-ischemia.

Glucocorticoids and hypoxic ischemic damage

The model of hypoxia-ischemia which we used to study the influence of
glucocorticoids on ischemic brain damage is a well established perinatal model of
unilateral incomplete reversible ischemia (13,21). Briefly, the right carotid artery
was ligated in seven day old rats which were then allowed to recover for 3 hours.
They were then placed in a chamber in an incubator at 37°C and subjected to three
hours of hypoxia (8% O_2, balance N_2). This results in cerebral pathology ipsilateral
to the side of ligation.

Unexpectedly, the administration of dexamethasone (0.5 mg/kg/d) for three days prior to the hypoxia-ischemia prevented the gross brain damage associated with the insult (3). Cerebral infarction ipsilateral to the carotid ligation occurred in the majority (31/39) of the vehicle treated animals but was absent in all dexamethasone treated animals. Paraffin embedded sections stained with cresyl violet, hematoxylin and eosin were examined and quantitative analysis confirmed that there was marked pathological damage in the cortex, striatum, thalamus and hippocampus ipsilateral to the carotid ligation which did not occur in dexamethasone pretreated animals.

Another surprising result was that dexamethasone was effective at rather low doses. Cerebral infarction was prevented in animals receiving a dose of dexamethasone as low as 0.01 mg/kg/d (Fig 1). This dose is 10-100 times less than the doses which have previously been administered to test the therapeutic efficacy of glucocorticoids for the treatment of adult cerebral ischemia, edema and trauma (1,2,5,7,9,12,14,17,18,23,25).

Gross hypoxic-ischemic brain damage could also be prevented with a single dose of dexamethasone but this effect was dependent on the time interval between injection and hypoxia-ischemia (Fig 1). Dexamethasone (0.1 mg/kg) injected 24 hours prior to the insult prevented cerebral infarction whereas dexamethasone administered either immediately or 3 hours before the insult had no such effect indicating a latency of action of dexamethasone of several hours.

Somewhat disappointing was the result that *post*-treatment was ineffective (3). In animals treated with dexamethasone (.1 or .5mg/kg/day) for 3 days after the hypoxia-ischemia, the incidence of gross unilateral cerebral infarction in dexamethasone and vehicle treated animals was similar.

Actions of Glucocorticoids Investigated

<u>Alterations in cerebral perfusion</u>

The pathogenesis of cerebral necrosis in the present model of hypoxia-ischemia is related to a reduction in local cerebral blood flow ipsilateral to the

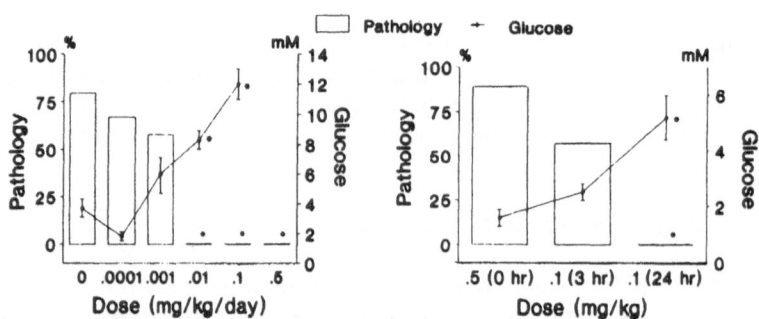

Fig. 1   The incidence (%/group) of pathology and blood glucose at the end of hypoxia are dependent on the dose of dexamethasone. [Mean±SEM]
Left: 'Chronic' treatment (ie. 48, 24 and 0 hrs pre-hypoxia).
Right: Single doses at various times pre-hypoxia.
[* p <0.01 compared to vehicle (0 mg/kg)] (Data from (3), with permission)

carotid occlusion (24,31). Thus, cerebral pathology might have been prevented by an improvement in the severity of the ischemia. Improvements can occur with an elevation in blood pressure; and, indeed, a side effect in human infants of dexamethasone therapy is hypertension and bradycardia (19,20). Although the 7 day old rat is too small to measure arterial blood pressure, heart rate was measured during hypoxia-ischemia and was found not to differ between the dexamethasone and vehicle treated groups (3). This suggests a lack of effect of dexamethasone on the systemic circulation.

Glucocorticoids have also been reported to affect cerebrovascular resistance and blood flow (10,33). We have performed preliminary measurements of cerebral blood flow in vehicle and dexamethasone treated animals (.1mg/kg 24 hrs pre-hypoxia) using quantitative autoradiography (15,29). Initial results demonstrate that cerebral blood flow is similar in these two groups (Fig. 2). Thus, improvements in cerebral perfusion do not appear to explain the neuroprotective effects of dexamethasone treatment.

Alterations in thermoregulation.

Changes in body temperature during or post ischemia are important determinants of brain damage since even modest reductions in body temperature can provide protection (8). Although dexamethasone might be expected to increase body temperature, we confirmed that thermoregulation is similar in dexamethasone and vehicle treated animals. Axillary temperature was equivalent in dexamethasone and vehicle treated groups (Fig 2).

Alterations in Blood glucose

It is clear that blood glucose levels influence the extent of brain damage resulting from a hypoxic-ischemic insult. In adult animals, the majority of studies indicate that hyperglycemia exacerbates pathology (16). In contrast, in infant animals, several studies suggest that hyperglycemia may be beneficial. In newborn rodents, anoxia or hypoxia-ischemia results in better survival of animals pretreated with glucose (26,30). In addition, brain injury produced by an episode of hypoxia-

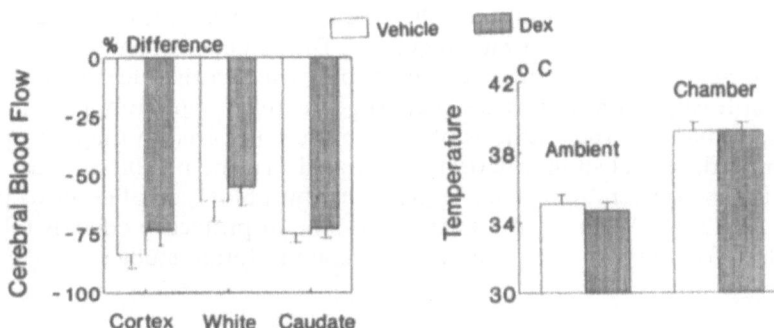

Fig. 2. Left: The reduction in cerebral blood flow on the ligated side (% of contralateral) is similar in animals treated with dexamethasone (DEX) or vehicle.
Right: Axillary temperature is equivalent in vehicle and DEX treated animals at room temperature or in the hypoxia chamber.

ischemia in neonatal rats is not worsened and may be improved by glucose treatment (11,32). Hyperglycemia has also been shown to improve the recovery of cerebral metabolism after a period of asphyxia in the newborn lamb (22) and to help maintain electroencephalographic activity during ischemia in fetal sheep (6).

In the infant rats treated with different doses of dexamethasone, pathology was correlated to the level of blood glucose at the end of the 3 hours of hypoxia (Fig 1). In particular, for animals in which dexamethasone was protective, the blood glucose level within the systemic circulation post hypoxia was elevated when compared with vehicle injected animals. This suggested that a relative hyperglycemia in dexamethasone treated rats could be the mechanism involved in the neuroprotective effect of dexamethasone.

To investigate whether an elevation in blood glucose throughout hypoxia-ischemia could improve outcome, we recently injected animals with 10% dextrose during hypoxia-ischemia, thereby elevating their blood glucose levels. Preliminary results indicate that there is a similar frequency of damage in saline or dextrose treated animals. Thus, relative elevations in blood glucose during hypoxia-ischemia also do not appear to explain the neuroprotective effects observed in dexamethasone treated animals.

Other Potential Mechanisms of Action

As in human infants, dexamethasone administration produces a reduction in somatic growth in a dose dependent manner (3). Growth retardation may influence hypoxic-ischemic encephalopathy although reports of its effects have been variable. For example, in postnatal growth retarded infant rats the damage caused by an episode of hypoxia-ischemia was reduced (28) whereas in intrauterine growth retarded newborn piglets damage was increased (4). One way in which growth retardation might be neuroprotective would be if brain growth and metabolism, in addition to somatic growth, are reduced. A decrease in local metabolic requirement could reduce the susceptibility of the brain to a hypoxic-ischemic insult. Dexamethasone's effects on local cerebral metabolism will be investigated in future experiments.

Irrespective of its overall physiological effect, dexamethasone likely acts on receptors within the cell. In general, glucocorticoids act by binding to glucocorticoid receptors within the cell. The hormone receptor complex is then considered to interact with specific regulating DNA elements resulting in either positive or negative control of gene expression. The fact that dexamethasone requires at least 3 hrs to induce its effect suggests that a minimum period of time is required, perhaps for the synthesis of some protein involved in the neuroprotective effect. Indeed, several studies have demonstrated that the beneficial effect of dexamethasone on cerebral edema is prevented by a blockade of de novo protein synthesis (25,27). Whether, dexamethasone's neuroprotective effect is related to the synthesis of a new protein will be determined in future studies.

Summary

We have demonstrated that pre-treatment of neonatal rats with dexamethasone prevents brain damage caused by a concurrent episode of cerebral hypoxia-ischemia. Infant rats treated prior to the insult with dexamethasone (doses of 0.01 to 0.5 mg/kg/day) had no infarction whereas 60-80% of animals treated with vehicle or low doses of dexamethasone ($\leq$.001 mg/kg/day) had an infarction. The neuro-protective effect is not immediate since a single dose of dexamethasone either immediately or 3 hours before the hypoxia was not effective. In contrast to pre-treatment, administration of dexamethasone following the insult was not effective.

Although the exact mechanism of action by which dexamethasone induces it neuroprotective effect is unknown, the importance of several effects of gluco-corticoid therapy have been investigated. Results indicate that several gluco-corticoid mediated changes in physiology which are also known to influence ischemic outcome do not appear to be involved in the effects of glucocorticoid therapy. Heart rate, local cerebral blood flow and body temperature are similar in dexamethasone and vehicle treated animals. Furthermore, although blood glucose is increased in dexamethasone treated animals, preliminary results indicate that such elevations in glucose do not prevent pathology. Glucocorticoid effects which have yet to be investigated include a potential reduction in metabolic requirements and a de novo synthesis of a new protein.

References

1) Altman DI, Young RSK, Yagel SK. Effect of dexamethasone in hypoxic-ischemic brain injury in the neonatal rat. Biol Neonate 46 (1989) 149-156.
2) Adlard BPF, SW De Souza. Influence of asphyxia and of dexamethasone on ATP concentrations in the immature rat brain. Biol Neonate 24 (1974)82-8.
3) Barks, JDE, Post, M, Tuor, UI. Dexamethasone prevents hypoxic-ischemic brain damage in the neonatal rat. Pediatr Res. 29 (1991) 558-563
4) Bauer, R, Zwiener, U, Buchenau, W, Hoyer, D, Witte, H, Lampe, V, Burgold, K, Zieger, M. Restricted cardiovascular and cerebral performance of intra-uterine growth retarded newborn piglets during severe hypoxia. Biomed Biochim Acta 48 (1989) 697-705.
5) Bracken MB, Shepard MJ, Collins WF, Holford TR, Young W, Baskin DS, Eisenberg HM, Flamm E, Leo-Summers L, Maroon J, Marshall LF, Perot PL, Piepmeier J, Sonntag VKH, Wagner FC, Wilberger JE, Winn HR. A randomized controlled trial of methylprednisolone or naloxone in the treatment of acute spinal cord injury. (1990) New Engl J Med 322: 1405-1411.
6) Chao, CR, Hohimer, AR, Bissonnette, JM. The effect of elevated blood glucose on the elctroencephalogram and cerebral metabolism during short-term brain ischemia in fetal sheep. Am J Obstet Gynecol 161 (1989) 221-228.
7) Fenske A, Fischer M, Regli F, Hase U. The response of focal ischemic cerebral edema to dexamethasone. J Neurol 220 (1979) 199-209.
8) Ginsberg, MD Local metabolic responses to cerebral ischemia. Cerebrovasc Brain Metab Rev 2 (1990) 58-93.
9) Goldstein, M, Barnett, HJM, Orgogozo, JM, Sartorius, N, Symon, L, Vereschagin, NV. Recommendations on stroke prevention, diagnosis and therapy: report of the WHO task force on stroke and other cerebrovascular disorders. Stroke 20 (1989) 1407-1431.
10) Hall ED, Braughler, JM. Glucocorticoid mechanisms in acute spinal cord injury: a review and therapeutic rationale. Surg Neurol 18 (1982) 320-327.
(11) Hattori, H, Wasterlain CG. Posthypoxic glucose supplement reduces hypoxic-ischemic brain damage in the neonatal rat. Ann Neurol 28(1990)122-8.
12) Jastremski, M, Sutton-Tyrell, K, Vaagenes, P, Abramson, N, Heiselman, D, Safar, P, (The brain resuscitation clinical trial I study group). Glucocorticoid treatment does not improve neurologic recovery following cardiac arrest. JAMA 262 (1989) 3427-3430.

13) Johnston, MV. Neurotransmitter alterations in a model of perinatal hypoxic-ischemic brain injury. Ann Neurol 13 (1983) 511-518.

14) Koide, T, Wieloch, T W, Siesjo, B K. Chronic dexamethasone pretreatment aggravates ischemic neuronal necrosis. J. Cereb. Blood Flow Metab. 6 (1986) 395-404.

15) Lyons, DT, Vasta, F, Vannuci, RC. Autoradiographic determination of regional cerebral blood flow in the immature rat. Pediatr Res 21 (1987) 471-476.

16) Marie, C, Bralet J. Blood glucose level and morphological brain damage following cerebral ischemia. Cerebrovasc Brain Metab Rev 3 (1991) 29-38.

17) Norris, JW, Hachinski, VC. High dose steroid therapy in cerebral infarction. Br Med J 292 (1986) 21-23.

18) Nakagawa H, Groothuis DR, Owens ES, Fenstermacher JD, Patlak CS, Blasberg RG. Dexamethasone effects on [125 I] albumin distribution in experimental RG-2 gliomas and adjacent brain. J Cereb Blood Flow Metab (1987) 7: 687-701.

19) Ohlsson A, Heyman E. Dexamethasone-induced bradycardia. Lancet II (1988) 1074.

20) Puntis, J W L, Morgan, M E I, Durbin, G M. Dexamethasone-induced bradycardia. The Lancet II (1988) 1372.

21) Rice JE, Vannucci RC, Brierley JB. The influence of immaturity on hypoxic-ischemic brain damage in the rat. Ann Neurol 9 (1981) 131-141.

22) Rosenberg, AA, Murdaugh, E. The effect of blood glucose concentration on postasphyxia cerebral hemodynamics in newborn lambs. Pediatr Res 27 (1990) 454-459.

23) Sapolsky, RM, Pulsinelli, WA. Glucocorticoids potentiate ischemic injury to neurons: therapeutic implications. Science 229 (1985) 1397-1400.

24) Silverstein, F, Buchanan, K, Johnston, M V. Pathogenesis of hypoxic brain injury in a perinatal rodent model. Neurosci Lett 49 (1984) 271-277.

25) Temesvari P, Joo F, Koltai M, Eck E, Adam G, Siklos L, Boda D. Cerebroprotective effect of dexamethasone by increasing the tolerance to hypoxia and preventing brain oedema in newborn piglets with experimental pneumothorax. Neurosci Lett 49 (1984) 87-92.

26) Thurston, JH, Hauhart, RE, Jones, EM. Anoxia in mice: reduced glucose in brain with normal or elevated glucose in plasma and increased survival after glucose treatment. Pediatr Res 8 (1974) 238-243.

27) Tosaki, A, M Koltai, F Joo, G Adam, P Szerdahelyi, I Lepran, I Takats, L Szekeres. Actinomycin d suppresses the protective effect of dexamethasone in rats affected by global cerebral ischemia. Stroke 16 (1985) 501-505.

28) Trescher, WH, Lehman, RA, Vannucci, RC. The influence of growth retardation on perinatal hypoxic-ischemic brain damage. Early Human Dev 21 (1990) 165-173.

29) Tuor, UI. Local cerebral blood flow in the newborn rabbit: an autoradiographic study of changes during development. Pediatr Res 29 (1991) 517-523.

30) Vannucci, RC, Vannucci, SJ. Cerebral carbohydrate metabolism during hypoglycemia and anoxia in newborn rats. Ann Neurol 4 (1978) 73-79.

31) Vannucci, RC, Lyons, DT, Vasta, F. Regional cerebral blood flow during hypoxia-ischemia in immature rats. Stroke 19 (1988) 245-250.

32) Voorhies TM, Rawlinson D, Vannucci RC. Glucose and perinatal hypoxic-ischemic brain damage in the rat. Neurology 36 (1986) 1115-1118.

33) Young, W, Flamm, ES Effect of high dose corticosteroid therapy on blood flow, evoked potentials and extracellular calcium in experimental spinal injury. J. Neurosurg 57 (1982) 667-673.

EFFECT OF CYCLOHEXYLADENOSINE ON ISCHEMIA-INDUCED INCREASES OF
HIPPOCAMPAL GLUTAMATE AND GLYCINE

S.L. Cantor, M.H. Zornow, P.J. Kelly, A.J. Baker, M.S. Scheller,
T.L. Yaksh, L.P. Miller[*], and H.M. Shapiro

Neuroanesthesia Research Laboratory, 0829 UCSD, La Jolla, CA 92093
and [*]Gensia Pharmaceuticals, San Diego, CA 92121

INTRODUCTION

Hippocampal cells are known to be particularly vulnerable to ischemic insult, suffering a delayed neuronal damage which usually takes 24 hours to develop[1]. The excitatory amino acid (EAA) hypothesis is one theory which attempts to explain this pattern of injury. It proposes that ischemia causes release of high levels of extracellular EAA's, such as glutamate and aspartate, which then act on postsynaptic NMDA and non-NMDA receptors to trigger an influx of Na+, Cl-, and especially Ca++, which leads to cell swelling and activation of a host of secondary destructive mechanisms[1,2,3]. In support of this hypothesis, ischemia has been shown to result in several-fold increases in levels of amino acids in the hippocampus, including glutamate (GLU), aspartate (ASP)[4,5] and glycine (GLY)[6]. One potential means of preventing this delayed neuronal damage would be to inhibit this peri-ischemic rise in EAA's. Adenosine A1 agonists are a class of compounds which have been shown to inhibit EAA release in the hippocampus *in vitro* in response to both potassium[7,8] and electrical stimulation[9,10]. We investigated whether a selective A1 agonist, cyclohexyladenosine (CHA) could prevent the rise of EAA's provoked by ischemia in the rabbit hippocampus.

METHODS

30 New Zealand white rabbits were anesthetized with halothane (1-2%) and mechanically ventilated. Esophageal temperature was servocontrolled to 38°C. Monitoring included continuous arterial and central venous pressures, heart rate, end tidal C02 concentrations and the electroencephalogram. Frequent determinations of arterial blood gases, hematocrit and glucose were performed. Rabbits were randomly assigned to receive 0 (n=5), 0.1 (n=8), 1.0 (n=6), 10 (n=6), or 100 (n=5) µM CHA dissolved in artificial CSF (vehicle) and administered via a microdialysis probe placed sterotactically into the right (ipsilateral) dorsal hippocampus. A microdialysis probe was also placed into the left (contralateral) hippocampus of each animal which was perfused with vehicle alone. Ten minutes of transient global cerebral ischemia (TGCI) was induced by neck tourniquet inflation and deliberate hypotension. Microdialysis samples were collected at frequent intervals during ischemia and reperfusion and then analyzed for their concentration of GLU and GLY by HPLC. The GLU or GLY Release Index was defined as the fractional increase in amino acid concentration x time during ischemia and reperfusion. Data were analyzed by regression analysis where the amino acid release index was plotted as a function of the log(dose CHA + 0.001).

The Role of Neurotransmitters in Brain Injury, Edited by
M. Globus and W.D. Dietrich, Plenum Press, New York, 1992

RESULTS

All values are presented as mean ± SEM. Ten minutes of ischemia caused levels of both GLU and GLY to increase several-fold over pre-ischemic levels. Figure #1 shows a typical profile of the fractional change in GLU release as a function of time for a vehicle treated group. With the onset of ischemia, GLU levels increased to a peak of 3.28 ± 0.55 times baseline in the vehicle treated probes and returned to baseline during reperfusion. Figure #2 is a similar graph for a vehicle treated glycine group. Glycine levels increased to a peak of 5.41 ± 0.91 fold over baseline and remained elevated for the duration of the study.

There was no significant difference in resting release between the left and right sides for either glutamate or glycine. (See Table #1). Also shown in the table are the GLU and GLY Release Indices of the control group for both the left and right sides, as well as the

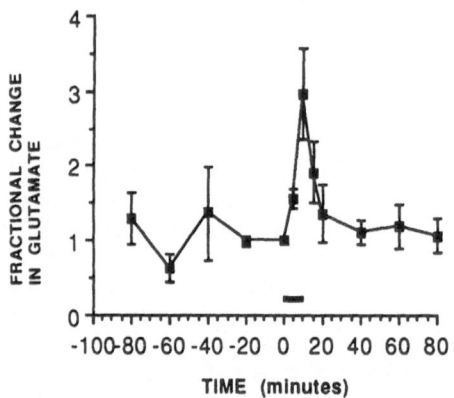

Figure 1. Fractional increase in glutamate vs. time for a vehicle treated group. All values are mean ± SEM. Solid bar indicates 10 minutes of global ischemia.

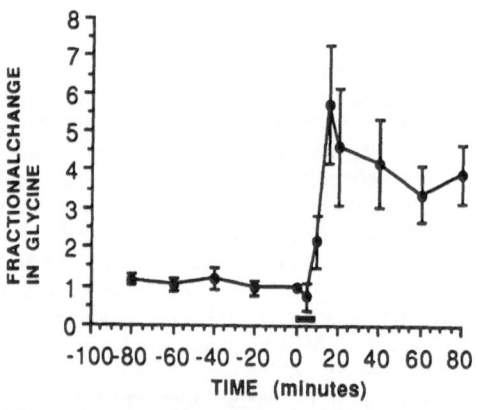

Figure 2. Fractional increase in glycine vs. time for a vehicle treated group. All values are mean ± SEM. Solid bar indicates 10 minutes of global ischemia.

percent change from control in each of the CHA dosage groups. The left sides of each dosage group received vehicle alone, whereas the right sides received the varying concentrations of CHA. When the GLU Release Index is plotted as a function of the log dose of CHA, regression analysis reveals a tendency toward a dose-dependent reduction in the GLU Release Index (p=0.054) on the CHA-treated (right) side, whereas the left (vehicle-treated) side showned no such tendency. Regression of the GLY Release Index of the right, or CHA-treated sides on CHA dose yielded a significant dose-dependent reduction (p=0.03) whereas the contralateral side showed no such attenuation (p=0.49).

Table 1. All values are mean ± SEM. Resting release is the amino acid concentration measured at least 70 minutes after microdialysis probe placement. Left refers to vehicle-treated probes; right refers to CHA treated-probes, except in the case of the right control group which received vehicle only. *= p<0.05.

DOSE DEPENDENT INHIBITION BY CHA OF RELEASE OF GLU AND GLY FROM HIPPOCAMPUS EVOKED BY GLOBAL CEREBRAL ISCHEMIA

AMINO ACID	RESTING RELEASE (μM)	CONTROL AMINO ACID RELEASE INDEX	AMINO ACID RELEASE INDEX AS % OF CONTROL IN CHA TREATED GROUPS				REGRESSION SLOPE	p VALUE
			0.1 uM	1.0 uM	10 uM	100 uM		
GLUTAMATE								
LEFT	7.5±1.2	83.5±20.2	215%	129%	127%	263%	+16.3	0.24
RIGHT	11.7±2.0	145.8±24.9	100%	84%	61%	76%	-11.1	0.054
GLYCINE								
LEFT	5.6±0.7	322.1±122.6	97%	94%	57%	87%	+18.2	0.49
RIGHT	6.5±0.9	296.7±65.9	70%	58%	66%	54%	-24.5	0.03*

DISCUSSION

Many have advocated the use of adenosine agonists as a therapy for ischemic brain injury [2,11,12,13]. Several studies have shown cerebral protection by adenosine analogs in animals following transient global cerebral ischemia[14,15,16]. Reduction in EAA release following ischemia is the presumed mechanism of protection advocated in these studies, since adenosine agonists have clearly been shown to decrease the release of glutamate and aspartate *in vitro* in response to such stimuli as potassium and field potentials, and A1 receptors are known to be abundant in the hippocampus. However, the ability of adenosine agonists to inhibit ischemically-induced rises in EAA's has not been reported.

Our *in vivo* results show that CHA, a highly selective A1 agonist, does cause a dose-dependent reduction in the fractional increase in glycine levels following ischemia, and a tendency toward a reduction in glutamate levels. These reductions however, were mild, and it is not clear if they can account for the neuroprotection seen in other animal models.

REFERENCES

1.   Buchan, A.M., Do NMDA antagonists protect against cerebral ischemia: Are clinical trials warranted? Cerebrovascular and Brain Metabolism Reviews, 1990. 2(1): p. 1-26.

2.   Choi, D.W., Methods for antagonizing glutamate neurotoxicity. Cerebrovascular and Brain Metabolism Reviews, 1990. 2(2): p. 105-147.

3.   Rothman, S.M., Thurston, J.H., and Hauhart, R.E., Delayed neurotoxicity of excitatory amino acids in vitro. Neuroscience, 1987. 22(2): p. 471-480.

4.   Hagberg, H., Lehmann, A., Sandberg, M., B.Nystrom, Jacobson, I., and Hamberger, A., Ischemia-induced shifts of inhibitory and excitatory amino acids from intra-to extracellular compartments. Journal of Cerebral Blood Flow and Metabolism, 1985. 5(3): p. 413-419.

5.   Benveniste, H., Drejer, J., Schourboe, A., and Diemer, N., Elevation of the extracellular concentrations of glutamate and aspartate in rat hippocampus during transient cerebral ischemia monitored by intracerebral microdialysis. Journal of Neurochemistry, 1984. 43: p. 1369-1374.

6.   Baker, A.J., Zornow, M.H., Grafe, M.R., Scheller, M.S., Skilling, S.R., Smullin, D.H., and Larson, A.A., Hypothermia prevents ischemia-induced increases in hippocampal glycine concentrations in rabbits. Stroke, 1991. 22: p. 666-673.

7.   Dolphin, A.C. and Archer, E.R., An adenosine agonist inhibits and a cyclic AMP analogue enhances the release of glutamate but not GABA from slices of the rat dentate gyrus. Neuroscience Letters, 1983. 43: p. 49-54.

8.   Burke, S.P. and Nadler, J.V., Regulation of glutamate and aspartate release from slices of the hippocampal CA1 area: effects of adenosine and baclofen. Journal of Neurochemistry, 1988. 51(5): p. 1544-1551.

9.   Corradetti, R., Conte, G.L., F.Moroni, Passani, M.B., and Pepeu, G., Adenosine decreases aspartate and glutamate release from rat hippocampal slices. European Journal of Pharmacology, 1984. 104: p. 19-26.

10.   Fastbom, J. and Fredholm, B.B., Inhibition of 3[H]glutamate release from rat hippocampal slices by L-phenylisopropyladenosine. Acta Physiologica Scandanavia, 1985. 125: p. 121-123.

11.   Marangos, P.J., Adenosinergic approaches to stroke therapeutics. Medical Hypotheses, 1990. 32: p. 45-49.

12.   Marangos, P.J., Lubitz, D.v., Daval, J.L., and Deckert, J., Adenosine: its relevance to the treatment of brain ischemia and trauma. Progress in Clinical and Biological Research, 1990. 361: p. 331-349.

13.   Dragunow, M. and Faull, R.L.M., The neuroprotective effects of adenosine. Trends in Pharmacological Science, 1988. 7: p. 194-196.

14.   Evans, M.C., Swan, J.H., and Meldrum, B.S., An adenosine analog, 2-chloroadenosine, protects against long term development of ischaemic cell loss in the rat hippocampus. Neuroscience letters, 1987. 83: p. 287-292.

15.   Daval, J., Lubitz, D.K.J.E.v., Deckert, J., Redmond, D.J., and Marangos, P.J., Protective effect of cyclohexyladenosine on adenosine A1-receptors, guanine nucleotide and forskolin binding sites following transient brain ischemia: a quantitative autoradiographic study. Brain Research, 1989. 491: p. 212-226.

16.   Lubitz, D.K.E.J.v., J.M. Dambrosia, and Redmond, D.J., Protective effect of cyclohexyl adenosine in treatment of cerebral ischemia in gerbils. Neuroscience, 1989. 30(2): p. 451-462.

ADENOSINE RECEPTOR BLOCKADE AUGMENTS INTERSTITIAL EXCITATORY AMINO ACIDS

DURING CEREBRAL ISCHEMIA

Veronica M. Sciotti, Francis M. Roche,
Margaret C. Grabb, David G.L. Van Wylen

Departments of Surgery and Physiology
School of Medicine and Biomedical Sciences
State University of New York at Buffalo
Buffalo, New York

The role of excitatory amino acid neurotransmitters in the genesis of ischemic neuronal damage is of current interest. As shown by microdialysis techniques, interstitial fluid (ISF) levels of the excitatory amino acid neurotransmitters glutamate and aspartate are markedly elevated during cerebral ischemia (Benveniste et al., 1984; Hagberg et al., 1985; Hillered et al., 1989; Shimada et al., 1990). This data, in conjunction with extensive documentation of the toxic effects of glutamate on neurons in vitro (Rothman and Olney, 1986), has led to the theory that the ischemia-induced elevation of ISF excitatory amino acids overstimulate post-synaptic glutamate receptors, leading to a cascade of cellular events which ultimately cause cell death.

The neuromodulatory effects of adenosine are well-known. Primarily through in vitro work, adenosine has been shown to induce a receptor-mediated inhibition of neurotransmitter release from presynaptic terminals. Since adenosine also accumulates in the ISF during cerebral ischemia (Van Wylen et al., 1986; Hillered et al., 1989; Sciotti et al., 1990), adenosine-mediated inhibition of neurotransmitter release may be important in the attenuation of excitatory amino acid accumulation during cerebral ischemia. In vitro, adenosine and adenosine agonists have been shown to inhibit neuronal activity and decrease hypoxic neuronal damage by reducing neurotransmitter release from presynaptic terminals (Dolphin and Archer, 1983; Burke and Nadler, 1988; Terrian et al., 1989). If endogenous adenosine inhibits the release of excitatory amino acids during cerebral ischemia, then adenosine receptor blockade should be associated with enhanced accumulation of ISF glutamate and aspartate during cerebral ischemia. We tested this hypothesis using the microdialysis technique to deliver 8(p-sulphophenyl)theophylline (SPT), an adenosine receptor antagonist, locally to the brain during cerebral ischemia and reperfusion. In addition, microdialysis was used to simultaneously measure changes in cerebral blood flow (CBF) and ISF levels of glutamate, aspartate, adenosine and adenosine metabolites.

The Role of Neurotransmitters in Brain Injury, Edited by
M. Globus and W.D. Dietrich, Plenum Press, New York, 1992

METHODS

Animal Preparation

All experiments were performed on 300-500g adult male Wistar rats. Animals were anesthetized initially with sodium thiamylal (50 mg/kg), after which a tracheostomy was performed and the animals mechanically ventilated with a mixture of oxygen, nitrogen and 1.5-2.0% halothane. Core body temperature was maintained between 37°C and 38°C. The left femoral artery was cannulated for the measurement of arterial blood pressure and the withdrawal of arterial blood samples. The right femoral artery was cannulated for hemorrhage during cerebral ischemia. Snare ligatures were placed bilaterally around each carotid artery. Microdialysis probes were implanted bilaterally in the caudate nuclei and perfused at 0.5 μl/min with artificial cerebrospinal fluid (CSF). A description of the brain microdialysis technique in conjunction with local CBF measurements by hydrogen clearance has been published previously (Van Wylen et al., 1988 and 1989).

Protocol

A 90 minute period was allowed for recovery from microdialysis probe implantation. Thereafter, dialysate samples were collected at 20 minute intervals with the exception of ischemia and early reperfusion, during which time 10 minute samples were collected. Following a baseline measurement of CBF and a collection of dialysate with both microdialysis probes being perfused with artificial CSF, the microdialysis probe on one side of the brain was perfused with artificial CSF containing 10^{-3} M SPT, an adenosine receptor antagonist, for the remainder of the protocol. The contralateral side continued to be perfused with artificial CSF alone. CBF measurements were made and one dialysate sample was collected during unilateral SPT infusion prior to cerebral ischemia. Immediately prior to the induction of ischemia, hydrogen was delivered to the animal. Reversible cerebral ischemia was induced by bilateral carotid occlusion and arterial exsanguination to a blood pressure of 50 mmHg. Hydrogen administration was then discontinued and a single CBF measurement was made during ischemia. Following 20 minutes of ischemia, blood pressure was restored by reinfusion of the shed blood and the carotid snares were removed. Dialysate samples were collected and CBF determinations were made during a 60 minute reperfusion period following ischemia.

Dialysate from six animals was used for the analysis of glutamate and aspartate. Another six rats were subjected to the same protocol for the determination of dialysate adenosine and adenosine metabolites. In an additional four rats, dialysate samples were used for the determination of amino acids as well as adenosine, inosine and hypoxanthine.

Analytical Procedures

The concentrations of glutamate and aspartate in the dialysate were detected fluorometrically using high performance liquid chromatography (HPLC) after precolumn derivatization with o-Phthalaldehyde. Determination of dialysate adenosine, inosine, and hypoxanthine concentrations was performed with reverse phase HPLC and ultraviolet detection at a wavelength of 254 nm.

Statistical Analysis

Mean values and standard errors of the means were calculated for all the data. Differences between mean values were determined using one of two

Table 1. Dialysate Levels and Cerebral Blood Flow Measurements

CONTROL SIDE

	Glutamate	Aspartate	Adenosine	Inosine	Hypoxanthine	CBF
Base	3.0+0.4	0.9+0.3	0.6+0.1	1.2+0.2	3.7+0.5	85+6
Base	2.6+0.4	0.9+0.2	0.5+0.1	0.9+0.1	2.8+0.3	83+8
10' ISC	5.8+1.0	2.6+0.7*	3.7+1.1	1.5+0.2	5.3+0.8	13+1*
20' ISC	10.8+2.7*	2.9+0.9*	5.7+1.5*	3.6+0.6	12.4+2.2	##
10' REP	7.4+1.2*	2.6+0.5*	6.0+2.5*	8.0+2.1*	20.2+3.9*	104+13*
30' REP	3.3+0.5	1.0+0.2	1.7+0.6	8.2+1.2*	33.1+6.8*	65+7
50' REP	3.3+0.5	0.8+0.2	0.5+0.1	2.6+0.6	14.4+4.2*	44+5*
70' REP	3.5+0.8	0.9+0.2	0.6+0.1	0.9+0.2	9.4+3.2	46+6*

SPT-TREATED SIDE

	Glutamate	Aspartate	Adenosine	Inosine	Hypoxanthine	CBF
Base	4.2+0.6	1.0+0.2	0.5+0.1	1.1+0.2	4.1+1.2	99+12
Pre-ISC	3.9+0.6#	0.9+0.2	0.5+0.1	1.0+0.1	2.8+0.4	89+10
10' ISC	14.6+2.7*#	3.8+1.2	5.9+1.4*#	2.6+0.5#	7.7+1.7	12+1
20' ISC	43.7+7.0*#	5.8+1.0*#	13.0+2.0*#	5.5+1.1	20.9+3.8*	##
10' REP	18.3+2.7*#	7.7+3.7*	11.1+3.2*#	12.2+2.6*	32.8+7.6*#	120+15*
30' REP	5.4+0.8#	1.2+0.3	3.4+1.3	12.2+3.2*	38.2+8.3*	75+13
50' REP	5.1+0.8#	1.0+0.3	0.6+0.2	2.8+0.7	20.3+4.6*	41+5*
70' REP	5.5+1.0	0.9+0.3	0.6+0.2	1.6+0.5	11.2+2.9	41+5*

Mean $\pm$ SEM dialysate levels (μM) and cerebral blood flow (CBF; ml/min/100g) measurements before (Base), during (ISC), and after (REP) 20 min of ischemia. On the SPT-treated side, 10^{-3} M 8(p-sulphophenyl)theophylline (SPT) in artificial CSF was begun 20 min prior to ischemia (Pre-ISC). *$p<0.05$ vs Base, #$p<0.05$ vs Control Side, ## = CBF measurements unobtainable.

types of statistical analysis: 1) paired t-test for comparisons of dialysate concentrations from the untreated side of the brain with those collected at the same time from the treated side; 2) analysis of variance followed by Dunnett's test for comparisons of dialysate concentrations for a given side to baseline. A p < 0.05 was accepted as indication of a statistically significant difference.

RESULTS

The local infusion of SPT did not affect arterial blood pressure, PO_2, PCO_2, or pH. Arterial PO_2 (138+8 mmHg), PCO_2 (40+1 mmHg), pH (7.4+0.01), and heart rate (338+9 b/min) did not change significantly during the course of the protocol. However, there was a tendency for mean arterial blood pressure to decline over time, with pressures being significantly lower during the last reperfusion period (79+3 mmHg) when compared to baseline (95+3 mmHg).

Dialysate levels of measured purine metabolites and amino acids are presented in Table 1. The addition of 10^{-3}M SPT to the dialysis probe prior to induction of ischemia did not significantly change dialysate levels of purine metabolites or amino acids. On the untreated side of the brain, ischemia resulted in an elevation of dialysate adenosine, inosine, hypoxanthine, glutamate and aspartate. Dialysate glutamate, aspartate, and adenosine increased approximately 4-fold, 3-fold, and 10-fold, respectively, during ischemia and then returned toward basal levels during reperfusion. Although dialysate levels of inosine and hypoxanthine increased during ischemia, the highest levels were observed in the reperfusion period.

A similar pattern of purine metabolites and amino acids during ischemia and reperfusion was observed on the side of locally infused SPT. However, dialysate levels of the measured purine metabolites and amino acids were significantly higher in the presence of SPT. Glutamate levels on the side of locally infused SPT were increased 11-fold over basal levels and were 4-fold greater than dialysate levels observed during the same ischemic time period on the untreated side of the brain. Dialysate aspartate was increased 6-fold over base and was 2-fold greater on the side of locally infused SPT during ischemia when compared to the untreated side. Significant increases in dialysate adenosine and adenosine metabolites above the untreated side were also observed during ischemia and reperfusion. Adenosine levels observed during ischemia on the side of locally infused SPT were approximately 2-fold greater than levels on the untreated side.

Local CBF responses to ischemia in the absence and presence of SPT are also depicted in Table 1. Although a 12% decrease in CBF was observed when SPT was added to the CSF perfusing the microdialysis probe, this difference was not significantly different when compared to baseline blood flow levels. CBF levels of 12-13 ml/min/100g were measured during 20 minutes of cerebral ischemia. An increase in CBF was observed on both sides of the brain upon reperfusion, with the hyperemic response to ischemia tending to be greater on the side of locally infused SPT. On both the treated and untreated side of the brain, a significant decline in postischemic CBF was observed during the final 20 minutes of reperfusion.

DISCUSSION

Adenosine has been proposed to modulate excitatory amino acid release. Few studies, however, have explored the relationship between endogenous adenosine and the ischemia-evoked release of amino acid neurotransmitters in vivo. Using the microdialysis technique, we have evaluated excitatory amino acid release from the caudate nucleus of rats before, during, and after cerebral ischemia in the absence and presence of adenosine receptor blockade. The major finding of this study is that ISF levels of the excitatory amino acids glutamate and aspartate were enhanced during ischemia and reperfusion in the presence of SPT. In addition, SPT administration resulted in increased ischemia-induced release of adenosine and adenosine metabolites when compared to the untreated side of the brain.

The excitotoxic hypothesis, as described by Rothman and Olney (1986), has been implicated in the neuronal damage resulting from ischemia, hypoxia, and certain neurodegenerative disease states. According to the excitotoxic hypothesis, elevated extracellular levels of excitatory amino acids overstimulate post-synaptic neuronal receptors, including the well-characterized N-methyl-D-aspartate (NMDA) receptors. NMDA receptor activation opens ion channels, leading to increases in intracellular calcium and, consequently, to mitochondrial damage, activation of enzymes involved in free radical formation and in the degradation of cytoskeletal proteins. This concept is consistent with the observation that areas of the brain most vulnerable to ischemia-induced damage possess the highest density of NMDA- or other glutamate-sensitive receptors.

Excitotoxicity resulting from ischemia-induced elevations of excitatory amino acid neurotransmitters could be minimized by: 1) post-synaptic antagonism of glutamate receptors, or; 2) pre-synaptic attenuation of excitatory amino acid release. The protective effect of blocking the action of glutamate at post-synaptic ion channels or receptor sites with such compounds as MK-801 and 2-amino-7-phosphonoheptanoate (APH) is widely accepted (Simon et al., 1984; Gill et al., 1988; Swan and Meldrum, 1990).

However, there is a relatively sparse amount of information concerning neuroprotection provided in the intact brain as a result of pre-synaptic inhibition of neurotransmitter release (von Lubitz et al., 1988; Bielenberg, 1989). In addition, while adenosine has been shown, in vitro, to modulate neurotransmitter release from brain (Dolphin and Archer, 1983; Burke and Nadler, 1988; Terrian et al. 1989), this phenomenon, to date, has not yet been demonstrated in vivo.

Our data provide evidence that endogenous adenosine is involved in the modulation of excitatory amino acid release during cerebral ischemia. Local administration of SPT markedly potentiated the ischemia-induced accumulation of glutamate and aspartate. Our interpretation of this response is that SPT, by blocking presynaptic adenosine A_1 receptors, interferes with the adenosine-mediated attenuation of neurotransmitter release that normally occurs during cerebral ischemia. These results are in agreement with the data of Prestwich et al. (1987), who observed increases in glutamate release from cultured cerebellar granule cells exposed to the adenosine receptor antagonist 8-phenyltheophylline. This response was attenuated by the adenosine A_1 agonist R(-)-phenylisopropyladenosine. Similar effects on glutamate release have been demonstrated in vitro by Dolphin and Archer (1983) using the adenosine agonist and antagonist 2-chloroadenosine and theophylline, respectively.

We also observed a greater increase in ISF purine metabolites during ischemia and early reperfusion on the side of locally infused SPT. This greater increase in adenosine and adenosine metabolites in the presence of adenosine receptor blockade is likely due to an increased metabolic activity of cells in the presence of enhanced excitatory neurotransmitters. Hoehn and White (1990) demonstrated that glutamate causes increased adenosine release from slices of rat parietal cortex through both NMDA and non-NMDA receptors. These investigators also demonstrated that the glutamate-evoked adenosine release was not decreased by inhibition of ecto-5'-nucleotidase, indicating that adenosine was not derived from the extracellular degradation of adenine nucleotides.

Adenosine is proposed to be involved in CBF regulation (Van Wylen et al., 1991). However, despite the fact that SPT is a non-selective adenosine receptor antagonist and, therefore, should have blocked adenosine A_2 receptors, we observed no significant difference in CBF between the two sides of the brain in response to local infusion of the receptor blocker. This was evident even during early reperfusion, when the ischemia-induced increase in ISF adenosine would be expected to contribute to reactive hyperemia. There are two possible explanations for this observation. First, ISF levels of glutamate and aspartate achieved with SPT infusion could be high enough to directly dilate cerebral resistance vessels. Busija and Leffler (1989) have demonstrated a dilator effect of topically applied amino acids on piglet pial arterioles. Alternatively, high levels of excitatory amino acids may increase local brain activity and thus, via adenosine-independent mechanisms, result in enhanced CBF as a consequence of tissue oxygen and energy supply/demand imbalances. In the latter scenario, SPT may be attenuating a potentially greater hyperemic response to increased metabolic activity. Van Wylen et al. (1988) have reported a 236% increase in CBF induced by local infusion of the glutamate agonist kainic acid. This hyperemic response to kainic acid was attenuated but not eliminated in the presence of SPT.

In summary, ISF purine metabolites and excitatory amino acids are potentiated during cerebral ischemia in the presence of adenosine receptor blockade. These data support a role for adenosine in the inhibition of excitatory amino acid release during cerebral ischemia and suggest that adenosine may have a cerebroprotective action.

REFERENCES

Benveniste H, Drejer J, Schousboe A, Diemer N (1984) Elevation of extracellular concentrations of glutamate and aspartate in rat hippocampus during transient cerebral ischemia monitored by intracerebral microdialysis. J Neurochem 43:1369-1374.

Bielenberg GW (1989) R-PIA protects cortical tissue in permanent MCA occlusionin the rat. J Cereb Blood Flow Metab 9:(suppl 1)S645.

Burke SP, Nadler JV (1988) Regulation of glutamate and aspartate release from slices of the hippocampal CA1 area: effects of adenosine and baclofen. J Neurochem 51:1541-1551.

Busija DW, Leffler CW (1989) Dilator effects of amino acid neurotransmitters on piglet pial arterioles. Am J Physiol 257:H1200-1203.

Dolphin AC, Archer ER (1983) An adenosine agonist inhibits and a cyclic AMP analogue enhances the release of glutamate but not GABA from slices of rat dentate gyrus. Neurosci Lett 43:49-54.

Gill R, Foster AC, Woodruff GN (1988) MK-801 is neuroprotective in gerbils when administered during the postischemic period. Neuroscience 25:847-855.

Hagberg H, Lehmann A, Sandberg M, Nystrom B, Jacobson I, Hamberger A (1985) Ischemia-induced shift of inhibitory and excitatory amino acids from intra-to extracellular compartments. J Cereb Blood Flow Metab 5:413-419.

Hillered L, Hallstrom A., Segersvard S, Persson L, Ungerstedt U (1989) Dynamics of extracellular metabolites in the striatum after middle cerebral artery occlusion in the rat monitored by intracerebral microdialysis. J Cereb Blood Flow Metab 9:607-616.

Hoehn K, White TD (1990) Role of excitiatory amino acid receptors in K^+-and glutamate-evoked release of endogenous adenosine from rat cortical slices. J Neurochem 54:256-265.

Prestwich SA, Forda SR, Dolphin AC (1987) Adenosine antagonists increase spontaneous and evoked transmitter release from neuronal cells in culture. Brain Research 405:130-139.

Rothman SM, Olney JW (1986) Glutamate and the pathophysiology of hypoxic-ischemic brain damage. Ann Neurol 19:105-111.

Sciotti V, Litchmore T, Van Wylen DGL (1990) Interstitial fluid (ISF) purine metabolite levels during and after cerebral ischemia (ISC). FASEB J 4:A1095.

Shimada N, Graf R, Rosner G, and Heiss WD (1990) Differences in ischemia-induced accumulation of amino acids in the cat cortex. Stroke 21:1445-1451.

Simon RP, Griffiths T, Swan JH, Meldrum BS (1984) Blockade of N-methyl-D-aspartate receptors may protect against ischemic damage in the brain. Science 226:850-852.

Swan JH, Meldrum BS (1990) Protection by NMDA antagonists against selective cell loss following transient ischaemia. J Cereb Blood Flow Metab 10:343-351.

Terrian DM, Hernandez PG, Rea MA, Peters RI (1989) ATP release, adenosine formation, and modulation of dynorphin and glutamic acid release by adenosine analogues in rat hippocampal mossy fiber synaptosomes. J Neurochem 53:1390-1399.

Van Wylen DGL, Park TS, Rubio R, Berne RM (1986) Increases in cerebral interstitial fluid adenosine concentration during hypoxia, local potassium infusion, and ischemia. J Cereb Blood Flow Metab 6:522-528.

Van Wylen DGL, Park TS, Rubio R, Berne RM (1988) Cerebral blood flow and interstitial fluid adenosine during hemorrhagic hypotension. Am J Physiol 255:H1211-H1218.

Van Wylen DGL, Moffe A (1988) Attenuation of kainic acid (KA) induced increases in cerebral blood flow (CBF) with adenosine (ADO) receptor blockade. Physiologist 31:A219.

Van Wylen DGL, Park TS, Rubio R, Berne RM (1989) The effect of local infusion of adenosine and adenosine analogues on local cerebral blood flow. J Cereb Blood Flow Metab 9:556-562.

Van Wylen DGL, Sciotti VM, Winn HR (1991) Adenosine and the regulation of cerebral blood flow. In: Adenosine and Adenine Nucleotides as Regulators of Cellular Function (Phillis JW, ed) Florida, CRC Press, pp 191-202.

von Lubitz DKJE, Dambrosia JM, Kempski O, Redmond DJ (1988) Cyclohexyl adenosine protects against neuronal death following ischemia in the CA1 region of gerbil hippocampus. Stroke 19:1133-1139.

METABOLIC DEPRESSION AS A POSSIBLE MECHANISM OF NEURONAL

PROTECTION BY ADENOSINE

P. Méric, E. Pinard, J.L. Corrèze, S. Roussel,
P. Roucher, E. Le Peillet°, D. Lekieffre°,
B. Tiffon[+], J. Mispelter[+], M. Plotkine°,
R. Boulu°, J.M. Lhoste [+], J. Seylaz

C.N.R.S. U.A. 641
° Université Paris V
[+] INSERM U219, Orsay, France

INTRODUCTION

It has been suggested that endogenous adenosine protects neurones from the damage induced by such pathological conditions as ischemia, hypoxia and epilepsy (1). Adenosine has many effects on the central nervous system, several of which could be involved in neuroprotection. Adenosine can improve the cerebral blood supply (2), limit the energy needs of cerebral tissue by preserving ionic homeostasis and hyperpolarizing neurons (3), or block specific events such as release of excitatory amino acids (4) or free radical production (5) that trigger neuronal death. We have investigated the possible roles of adenosine in two experimental pathological conditions that elicit localized neuronal damage in the rat: long lasting seizures induced by kainic acid (KA) and transient forebrain ischemia. Two strategies were used : blocking adenosine receptors with theophylline in KA-induced seizures, and enhancing the effects of adenosine with R-phenyl-isopropyl adenosine (R-PIA) in forebrain ischemia.

I. KAINATE-INDUCED SEIZURES

Kainate-induced seizures were investigated using the following protocol: 2 groups of unanesthetized spontaneously breathing rats were given kainic acid (9 mg/kg i.p.). One group was given kainic acid alone, the other was given theophylline (15 mg/kg i.p) 15 min before injection of kainic acid.

The following parameters were monitored before and after drug injections :
- Blood flow and tissue PO_2 in the cortex and hippocampus by *in vivo* mass spectrometry (6).
- Energy metabolism (ATP, phosphocreatine, inorganic phosphate) and lactate of the whole brain by alternate ^{31}P and ^{1}H *in vivo* NMR spectroscopy using a surface coil (12 x 8 mm) in the bore of the 4.7 T horizontal magnet of a Bruker spectrometer.

The effects of KA alone and theophylline + KA were measured for two hours after KA injection. Diazepam (2 mg/Kg, i-m) was then given to stop the seizures. Forty eight hours after KA injection, the rats were anesthetized and perfused transcardiacally with FAM (1/10 formaldehyde, 1/10 acetic acid, 8/10 methanol). The brain was then removed and processed for classical histology with cresyl violet staining.

The Role of Neurotransmitters in Brain Injury, Edited by
M. Globus and W.D. Dietrich, Plenum Press, New York, 1992

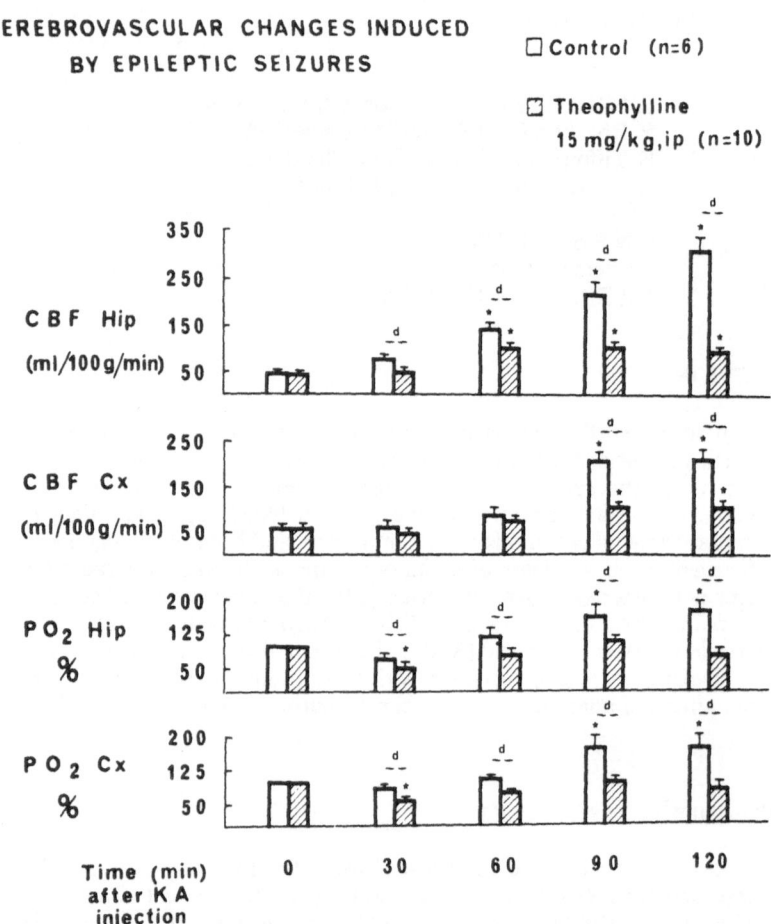

Figure 1. Cerebral blood flow (CBF) and tissue PO$_2$ in the cortex and the hippocampus, monitored by *in vivo* mass spectrometry, in 2 groups of unanesthetized spontaneously breathing rats. One (6 rats) was given kainic acid (9 mg/kg i.p., KA), and the other (10 rats) was given theophylline (15 mg/kg i.p.), 15 min before kainic acid injection.

The rats given KA alone maintained a basal cerebral energy homeostasis during the seizures (7), indicating that there was no energy failure. Cerebral blood flow increased, inducing brain tissue hyperoxia (8, 9), but it was associated with mild lactate accumulation which remained at a plateau thoughout the seizures (7).

The rats given theophylline prior to KA showed significantly smaller increases in cerebral blood flow and tissue PO_2 associated with seizures (10) (figure 1).

The lactate accumulation during seizures was clearly higher in theophylline-treated rats than in control rats given KA only (figure 2).

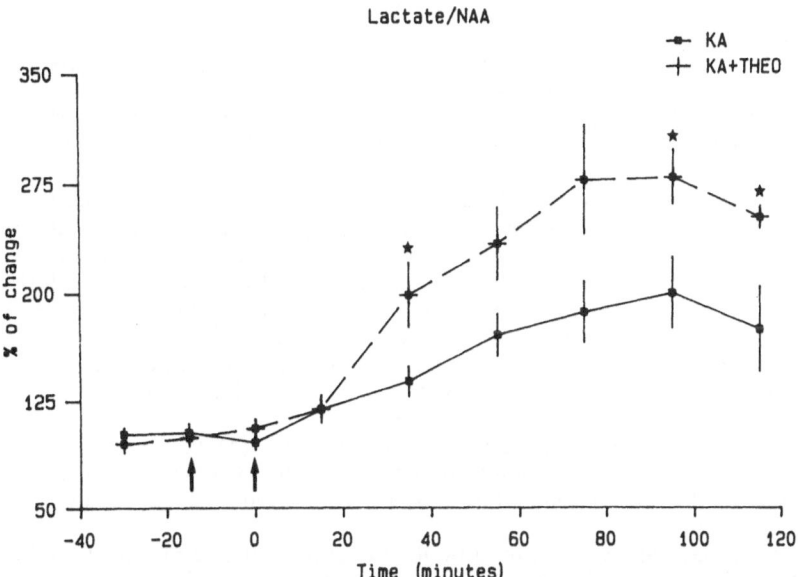

Figure 2 . Lactate accumulation in the whole brain, estimated by proton NMR spectroscopy, in 2 groups of unanesthetized spontaneously breathing rats. One (8 rats) was given kainic acid (9 mg/kg i.p., KA), and the other (13 rats) was given 15 mg/kg i.p. of theophylline (THEO), 15 min before kainic acid injection.

However, the cerebral levels of ATP, phosphocreatine and inorganic phosphate as well as the intracellular pH were not significantly altered by theophylline given prior to KA. Histological examination showed that theophylline treatment worsened the neuronal damage resulting from KA-induced seizures. Rats given KA alone showed typical neuronal damage, particularly in the CA3 layer of the hippocampus, while the thephylline-treated rats showed additional hippocampal damage at the CA1 and CA2 layers.

These effects of theophylline on the cerebrovascular and metabolic consequences of KA-induced seizures suggest that adenosine receptors are involved in hippocampal neuroprotection. They may act by inducing vasodilatation and/or limiting anaerobic glycolysis in cerebral tissue. The blockade of adenosine receptors by theophylline has been shown to reduce cerebral blood flow and to enhance glucose comsumption (11) under control conditions without inducing any brain damage. Such effects can be tentatively attributed to the blockade of respectively A_2 (vascular) and A_1 (neuronal) receptors.

II. TRANSIENT FOREBRAIN ISCHEMIA

The putative neuroprotective role of adenosine was investigated using the four vessel occlusion model (12) of reversible forebrain ischemia in the rat. Transient ischemia (30 min) was induced in two groups of paralyzed, ventilated, anesthetized rats (alpha-chloralose, 30mg/kg s.c.). One group was given the adenosine agonist R-phenyl-isopropyl-adenosine (R-PIA 20 μg/kg i.p.) before ischemia and hourly 10 μg/kg for up to 5 hours after ischemia. The other group was given saline at the same times as the treated group was given R-PIA. The cerebrovascular and metabolic effects of ischemia and R-PIA were estimated from:
- Blood flow and tissue PO_2 in the caudate nucleus and hippocampus, measured by *in vivo* mass spectrometry (6).
- Energy metabolism, measured by alternate ^{31}P and ^{1}H *in vivo* NMR spectroscopy in the whole brain using the same surface coil and magnet as in seizure experiments.

A separate group of freely moving rats underwent a similar 30 min-forebrain ischemia with and without R-PIA treatment. Their recovery was assessed using a neurological score index (13). Classical histology was performed on these rats 7 days after the induction of ischemia.

The control rats showed higly significant changes in all the variables monitored by NMR spectroscopy during ischemia. The levels of ATP and phosphocreatine (PCr) decreased throughout the ischemic period, while inorganic phosphate (Pi) and lactate (LA) accumulated and the intracellular pH dropped. All these variables returned to their basal values within about 30 minutes after ischemia.

R-PIA treatment had a statistically significant effect on the changes in all the energy metabolism parameters. The decreases in ATP and PCr were slower and smaller than in control rats. The accumulations of inorganic phosphate and lactate were also smaller in R-PIA-treated rats than in control rats. In addition, there was less intracellular acidosis at the end of ischemia in the R-PIA treated rats (pH = 6.40) than in the control rats (pH = 6.30). All the energy metabolism variables returned to their basal levels significantly faster in R-PIA-treated rats when carotid clamping was discontinued, . This can be summarized as a significant reduction in ischemia-induced energy failure in R-PIA-treated rats, though the dose of agonist used was lower than in previous studies.

The data for hippocampus and caudate nucleus blood flow (5) indicate that the doses of R-PIA used had no effect on cerebral blood flow and tissue PO_2 before ischemia, but that the post-ischemic hyperperfusion and hyperoxia were delayed by about twenty minutes in R-PIA-treated rats.

Neurological recovery was improved by R-PIA, as indicated by the significantly higher scores of the R-PIA-treated rats from 3 hours to 7 days after ischemia. However, the gradings of the histological neuronal damage were not significantly modified by R-PIA in the hippocampus and the striatum, and indicated only minimal protection in the neocortex.

The smaller decreases in ATP and PCr and the smaller increases in inorganic phosphate and lactate suggest that R-PIA treatment improved neurological recovery by shortening the duration and amplitude of the energy failure. But, this global protection of cerebral energy metabolism by R-PIA had little effect on local neuronal damage.

DISCUSSION

The multidisciplinary studies reported in the present paper describe two different pharmacological strategies designed to further elucidate the mechanisms involved in adenosine neuroprotection. Two experimental pathologies were investigated whose common feature is their deleterious consequences in the hippocampus (though at different layers) attributed to a massive release of glutamate triggering an excessive calcium entry into cells (14). However, the two pathologies differ in their cerebrovascular and metabolic consequences, since KA-induced seizures are accompanied by an increase in cerebral blood flow and tissue PO_2 without energy failure, whereas transient forebrain ischemia leads to an arrest of blood circulation with energy failure, these phenomena being reversible.

All taken together, our results suggest that adenosine is involved in the control of cerebral circulation and energy metabolism, and that the activation of its receptors leads to an improvement in neurological outcome and neuroprotection, although to a small extent. However while the theophylline dose used has been shown to block A_1 and A_2 receptors

(15), the total amount of R-PIA administered was very low, in order to avoid its hypotensive effect. A much higher dose of R-PIA results in significant neuroprotection (16).

Theophylline enhances the glutamate release *in vivo* (17) while adenosine blocks glutamate release *in vitro* (4). These mechanisms do not appear to be involved in the histological consequences of KA-induced seizures and transient forebrain ischemia in our experimental conditions. Our results suggest that adenosine acts mainly by inducing cerebral vasodilatation and limiting glycolysis during KA-induced seizures, and by inducing metabolic depression in ischemia. This latter phenomenon is not related to the inhibition of glutamate release, since a broad spectrum excitatory amino acid antagonist, kynurenic acid, does not modify the energy failure provoked by ischemia (18).

REFERENCES

1. Dragunow, M. and Faull, R.L., 1988, Neuroprotective effects of adenosine, Trends Pharmacol. Sci., 9: 193-194.

2. Puiroud, S., Pinard, E. and Seylaz, J., 1988, Dynamic cerebral and systemic circulatory effects of adenosine, theophylline and dipyridamole, Brain Research, 453: 287-298.

3. Phillis, J.W. and Wu, P.H., 1981, The role of adenosine and its nucleotides in central synaptic transmission, Prog. Neurobiol., 16: 187-239.

4. Coradetti, R., LoConte, G., Moroni, F., Passani, M.B. and Pepeu, G., 1984, Adenosine decreases aspartate and glutamate release from rat hippocampal slices, Eur. J. Pharmacol., 104: 19-26.

5. Cronstein, B.N., Rosenstein, E.D., Kramer, S.B., Weissman, G. and Hirschhorn, R., 1985, Adenosine : a physiologic modulator of superoxide anion generation by human neutrophils. Adenosine acts via an A_2 receptor of human neutrophils, J. Immunol., 135: 1366-1371

6. Barrère, B., Péres, M., Méric, P., Gillet, B., Beloeil, J.C. and Seylaz, J., 1989, Effects of kainate-induced seizures on cerebral metabolism: a combined ^{1}H and ^{31}P study in unanesthetized rats. J. Cereb. Blood Flow Metab, 9 (suppl 1): S233.

7. Barrère, B., Péres, M., Méric, P., Gillet, B., Beloeil, J.C., Seylaz, J., 1989, Effects of kainate-induced seizures on cerebral metabolism: a combined ^{1}H and ^{31}P study in unanesthetized rats. J. Cereb. Blood Flow Metab, 9 (suppl 1): S233.

8. Pinard, E., Tremblay, E., Ben Ari, Y. and Seylaz, J., 1984, Blood flow compensates oxygen demand in the vulnerable CA_3 region of the hippocampus during kainate-induced seizures., Neuroscience, 13: 1039-1049.

9. Pinard, E., Rigaud, A.S., Riche, D., Naquet, R. and Seylaz, J., 1987, Continuous determination of the cerebrovascular changes induced by bicuculline and kainic acid in unanaesthetized spontaneously breathing rats. Neuroscience, 23, 943-952.

10. Pinard, E., Riche, D., Puiroud, S. and Seylaz, J., 1990, Theophylline reduces cerebral hyperaemia and enhances brain damage induced by seizures, Brain Research, 511: 303-309.

11. Grome, J.J. and Stefanovitch, V., 1986, Differential effects of methylxanthines on local cerebral blood flow and glucose utilization in the conscious rat, Arch. Pharmacol., 333: 172-177.

12. Pulsinelli, W.A. and Brierley, J.B., 1979, A new model of bilateral hemispheric ischemia in the unanesthetized rat. Stroke, 10: 267-272.

13. Bures, J., Buresova, O. and Huston, J., 1976, "Techniques and basic experiments for the study of brain and behavior", Elsevier, Amsterdam.

14. Siesjö, B.K. and Bengtsson, F., 1989, Calcium fluxes, calcium antagonists, and calcium related pathology in brain ischemia, hypoglycemia, and spreading depression : a unifying hypothesis, J. Cereb. Blood Flow Metab., 9: 127-140.

15. Morii, S., Ngai, A.C., Ko, K.R. and Winn, R.H., 1987, Role of adenosine in regulation of cerebral blood flow : effects of theophylline during normoxia and hypoxia, Am. J. Physiol., 253: H165-H175.

16. Block, G.A. and Pulsinelli, W.A., 1987, The adenosine agonist, R-phenyl isopropyl adenosine, attenuates ischemic neuronal damage, J. Cereb. Blood Flow Metab., 7 (suppl. 1) S258.

17. Lekieffre, D., Callebert, J., Plotkine, M. and Boulu, R.G., 1991, Theophylline enhancement of ischemia-induced glutamate release does not influence the neurological and histological outcome in the 4-vessel model in rats, J. Cereb. Blood Flow Metabol., 11 (suppl. 2) S764.

18. Roucher, P., Méric, P., Correze, J.L., Mispelter, J., Tiffon, B., Lhoste, J.M. and Seylaz, J., 1991, Metabolic effects of kynurenate during reversible forebrain ischemia studied by in vivo ^{31}P-nuclear magnetic resonance spectroscopy. Brain Research, 550: 54-60.

DYNAMICS OF ISCHEMIC INJURY FOLLOWING GLOBAL CEREBRAL

ISCHEMIA IN A RAT CARDIAC ARREST MODEL

Kensuke Kawai, Liliana Nitecka, Christl Ruetzler,
Julia Lohr, Nobuhito Saito, Ferenc Joo, Günter Mies,
Thaddeus S. Nowak Jr. and Igor Klatzo

Stroke Branch, Clinical Neurosciences Program,
Division of Intramural Research, NINDS, Bethesda, MD

It has been recognized increasingly that in the pathophysiology of cerebral ischemia the final outcome of resulting injury is determined by events taking place after the ischemic insult. Among potential factors and mechanisms involved in the development of postischemic injury, the role of neurotransmitters has drawn considerable attention. Investigations in this area have been focused primarily on selective, delayed injury of CA1 pyramidal neurons of the hippocampus, which was shown to be associated with increased extracellular release of glutamate and aspartate (Benveniste et al., 1984) and with intraneuronal influx of calcium. It was also demonstrated that the CA1 injury can be prevented by eliminating the glutamatergic input to the CA1 sector (Wieloch et al., 1985; Johansen et al., 1986). Otherwise, it is apparent from recent studies that, beside glutamate and aspartate release in the hippocampus, a variety of neurotransmitters can be activated in various locations, and this may be associated with different patterns of ischemic injury recognizable in various experimental models of cerebral ischemia.

Global cerebral ischemia, an important clinical condition in view of the frequency of cardiac arrests, has some distinct pathophysiological features, which may be related to damage of other body organs (Safar, 1989). The pattern of injury in global ischemia is characterized by the main impact of injury being shifted from the CA1 sector of the hippocampus to the nucleus reticularis thalami (NRT), which is composed of exclusively GABAergic neurons (Blomqvist and Wieloch, 1985; Ross and Duhaime, 1989). As an experimental model for global cerebral ischemia associated with cardiac arrest, a simple procedure allowing compression of the main cardiac vessels without thoracotomy, and resulting in instantaneous cessation of effective cerebral circulation, was introduced by Korpatchev (1982). It has been used in the present study to elucidate a possible involvement of neuroexcitatory mechanisms in the development of neuronal injury following global cerebral ischemia.

MATERIALS AND METHODS

Global cerebral ischemia was produced in Sprague-Dawley rats anesthetized with 1.5% halothane in nitrous oxide/oxygen by inserting a bent wire into the mediastinum and placing it under the bundle of the major cardiac blood vessels. Simultaneous lifting of the device and application of outside pressure resulted in complete interruption of the circulation. The physiological parameters, which included the blood pressure (BP), cerebral blood flow assessed by laser Doppler flowmetry, EEG and EKG, were monitored during and after cessation of the blood circulation. Resuscitation was started 5 to 6 minutes after applying compression, and included artificial respiration and cardiac massage. Successful resuscitation was associated with a vigorous rise in BP, occurring approximately 10 min after onset of the cerebral ischemia. The rats were sacrificed in groups at various time intervals starting at 15 min and ending at 7 days after resuscitation. Following transcardiac perfusion with 4% buffered paraformaldehyde, the brains were embedded in paraffin or cut on the vibratome for staining with cresyl violet or with specific antiparvalbumin antibodies (obtained from Prof. M. R. Celio, Universität Zürich), respectively. Four rats sacrificed after 30 min recirculation were used for electron microscopic (EM) observations limited to the NRT. As a marker for ischemic injury, ^{45}Ca was injected 6 hrs before sacrifice, and the autoradiograms were prepared from frozen sections.

RESULTS

The autoradiograms prepared from animals sacrificed 6 h after ischemia revealed dense radioactivity sharply outlining the NRT. At 24 h, the areas with increased ^{45}Ca uptake extended to the ventral thalamic nuclei, and only after 48 h was there noticeable abnormal accumulation of ^{45}Ca in the CA1 sector of the hippocampus.

The sections stained with cresyl violet revealed striking neuronal changes visible in animals sacrificed as early as 15 min after resuscitation. The changes consisted of clear peripheral zones compressing the remaining neuronal cytoplasm towards the nucleus. This change was most pronounced in the NRT neurons and was also noticeable in the nucleus dorsalis medialis thalami (NDMT) and the nucleus centralis amygdalae (NCA) (Fig.1). Similar clear cytoplasmic halos could be seen in numerous hippocampal interneurons. One hour after global ischemia, the clear halos at the periphery of the neurons appeared to be converted into clear vacuoles visible in the neurons in the described locations (Fig.2). Neuronal vacuolation was largely absent in animals sacrificed after 3 h, whereas the NRT and the ventral thalamic nuclei showed signs of injury in cresyl violet staining consisting of abnormally dark or pale cells, frequently with markedly increased staining of neuronal processes. At 24 h and after following days, the hippocampal interneurons, the NDMT and NCA neurons appeared to be fully recovered and showed no changes, whereas the NRT revealed noticeable neuronal loss, the surviving neurons showing dark

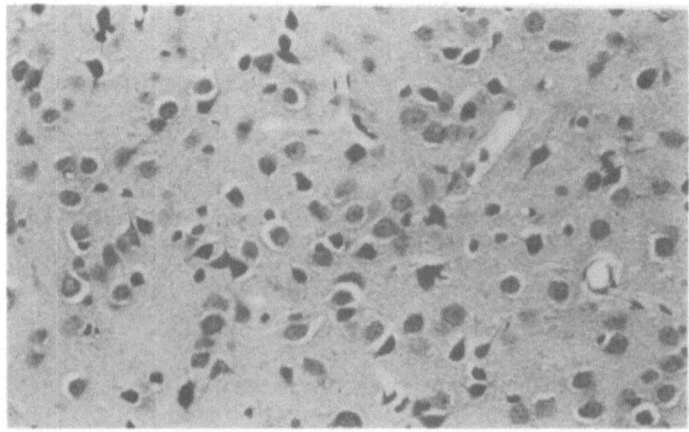

Fig.1. Neurons of the NCA with the clear cytoplasmic halos after 15 min recirculation. Cresyl violet staining; x125

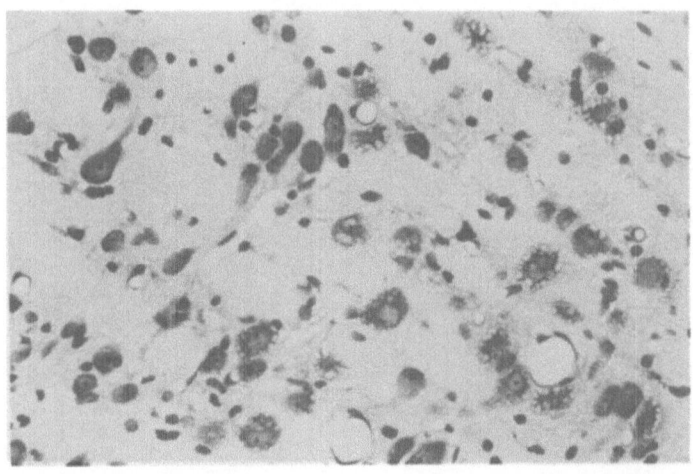

Fig.2. After one hour recirculation, neurons of the NRT show intense vacuolization of peripheral cytoplasm, whereas the adjacent ventral thalamic neurons in the left upper corner appear normal. Cresyl violet staining; x250

staining, shrinkage and becoming spindle-shaped. The neuronal damage of the pyramidal CA1 neurons became apparent 4 days following global ischemia. It became pronounced at 7 days with pyramidal neurons being reduced to a single-cell layer.

The neurons of the thalamic nuclei, striatum and the pars reticulata of the substantia nigra frequently revealed a chronic type of injury, associated with dark cresyl violet staining or loss of Nissl substance and increased stainability of neuronal processes. EM observations revealed predominantly peripheral membrane-bound vacuoles and mostly well-preserved mitochondria in the NRT neurons.

The immunostaining for parvalbumin (PV) showed a striking loss of PV immunoreactivity in NRT neurons in animals sacrificed after 3 h, particularly in the central portions of this nucleus (Fig.3B). The loss of PV immunoreactivity in the NRT appeared to be less marked during the next days. Some of the animals sacrificed after 6 h revealed, in addition to loss of staining in the NRT, a marked reduction of PV immunoreactivity in dendritic processes of the interneurons in the stratum pyramidale and stratum radiatum of the CA1 sector. A similar appearance in the hippocampus was observed in some of the animals sacrificed after 4 and 7 days.

DISCUSSION

One of the interesting findings in the present study pertains to very early changes in predominantly GABAergic neurons, noticeable as early as 15 min after global cerebral ischemia. These changes, characterized by abnormal influx of water into neuronal cytoplasm, appeared to be mostly reversible, with the exception of the NRT which showed after 7 days a noticeable neuronal loss and presence of neurons with a chronic type of injury. It can be assumed that the widespread involvement of GABAergic neurons may be associated with their dysfunction. This may lead to interruption of their inhibitory role resulting in hyperexcitability which in regions connected by transmitter circuitry may have an excitotoxic effect or significantly potentiate an existing ischemic damage in those areas. In this respect, although the prolonged survival of

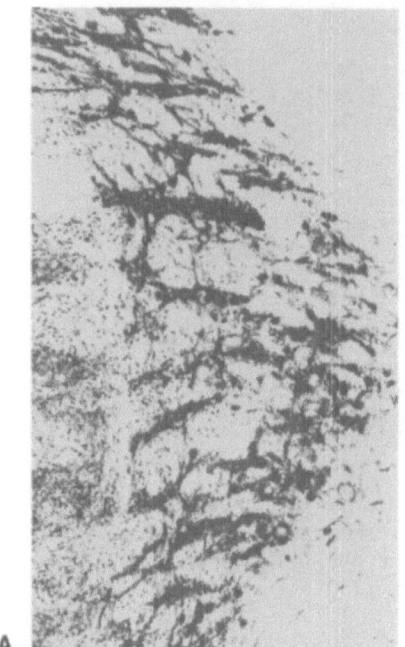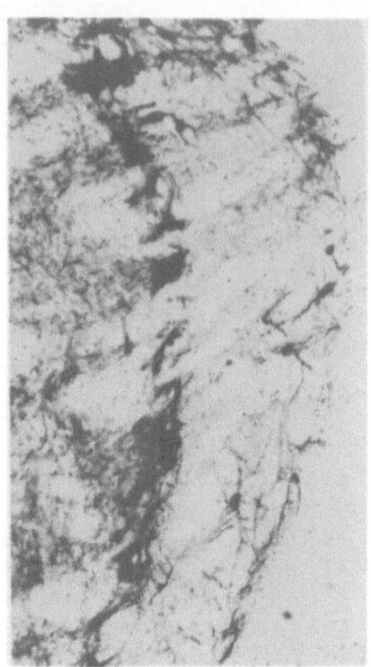

Fig.3. Immunostaining for parvalbumin in the NRT. A. Control rat. B. Animal sacrificed 3 hours after ischemia showing a pronounced reduction in staining of the GABAergic NRT neurons. x50

GABAergic interneurons and boutons was demonstrated by Nitsch et al. (1989), it remains a matter of conjecture how much a transitory dysfunction of interneurons, even of limited duration, may subsequently contribute to severe neuronal damage demonstrable in the CA1 pyramidal cells.

Concerning the NRT, our assumption that the dysfunction and damage of this nucleus may be related to a neuroexcitatory effect originating mainly from the cerebral cortex and thalamus is supported by recent neuropathological study of Ross and Graham (1991) on human cardiac arrest cases. This study revealed that in patients who suffered a cardiac arrest associated with severe and extensive damage to the cerebral cortex and thalamus, the NRT remained well preserved, whereas the less severe cases without marked cortical or thalamic injury, showed a conspicuous selective degeneration of NRT neurons. In our study, it was evident that damage to the cerebral cortex and thalamus was clearly of insufficient intensity to prevent a potential for glutamatergic excitation which, when relayed to the NRT, would eventually enhance ischemic injury in other interconnected regions.

SUMMARY

The global cerebral ischemia associated with cardiac arrest is characterized by early hydrophilic changes in predominantly GABAergic neurons. Although these changes appear to be largely reversible, it is assumed that they may be associated with temporary disinhibition, which may lead to enhancement of ischemic injury in interconnected regions.

REFERENCES

Benveniste H, Drejer J, Schousboe A, Diemer NH (1984) Elevation of the extracellular concentrations of glutamate and aspartate in rat hippocampus during transient cerebral ischemia monitored by intracerebral microdialysis. J Neurochem 43:1369-1374

Blomqvist P and Wieloch T (1985) Ischemic brain damage in rats following cardiac arrest using a long-term recovery model. J Cereb Blood Flow Metab 5:420-431

Johansen FF, Jorgensen MB and Diemer NH (1987) Ischemia induced delayed neuronal dealth in the CA-1 hippocampus is dependent on intact glutamatergic innervation. In: Excitatory Amino Acid Transmission (Hicks TP, Lodge D, McLennan, eds), Alan R. Liss, New York, pp245-248

Korpatchev WG, Lysenkov SP and Thiel LZ (1982) Modielirowanije Kliniczeskoj smerti i postreanimatioznoj bolezni u krys. Patol Fiziol Eksp Ter 3:78-80

Nitsch C, Goping G and Klatzo I (1989) Preservation of GABAergic perikarya and boutons after transient ischemia in the gerbil hippocampal CA1 field. Brain Res 495:243-252

Ross DT and Duhaime A-C (1989) Degeneration of neurons in the thalamic reticular nucleus following transient ischemia due to raised intracranial pressure: excitotoxic degeneration mediated via non-NMDA receptors? Brain Res 501:129-143

Ross DT and Graham DI (1991) Selective loss and selective sparing of neurons in the thalamic reticular nucleus following human cardiac arrest. J Cereb Blood Flow Metab 11(suppl.2):S855

Safar P (1986) Cerebral resuscitation after cardiac arrest: a review. Circulation 74(suppl IV):IV-138-IV-153

Wieloch T, Lindvall O, Blomqvist P and Gage FH (1985) Evidence for amelioration of ischaemic neuronal damage in the hippocampal formation by lesions of the perforant path. Neurol Res 7:24-26

MUSCARINIC RECEPTOR BLOCKADE REVEALS CHANGES IN CHOLINERGIC

FUNCTION IN TWO MODELS OF TRAUMATIC BRAIN INJURY (TBI)

C. Edward Dixon, Robert J. Hamm, Guy L. Clifton, and Ronald L. Hayes

Division of Neurosurgery
University of Texas Health Science Center at Houston
6431 Fannin St.
Houston, Texas 77030

Department of Psychology
VCU Box 2018
Virginia Commonwealth University
Richmond, Virginia 23294

INTRODUCTION

Along with other laboratories, we have provided data suggesting that abnormal agonist-receptor interactions produce sublethal excitotoxic changes in neural function that contribute to traumatic brain injury (TBI). Several lines of evidence indicate that activation of muscarinic cholinergic and/or NMDA glutamate receptors induces pathological processes ultimately producing the enduring neurological changes characteristic of TBI. This receptor activation is likely due to a massive release of neurotransmitters caused by a TBI-induced membrane depolarization. Membrane depolarization may be a phenomenon common to all forms of mechanical trauma whether focal or generalized and may contribute to long term changes in brain information flow (Hayes et al., 1989).

While excitoxicity contributes to the induction of deficits, the mechanisms contributing to their maintenance are less clear. It is known, however, that these mechanisms do not involve neuronal death, since mild and moderate traumatic brain injury can produce enduring pathological alterations in hippocampal neural function not accompanied by tissue damage detectable by light or electron microscopy. For example, we have reported (Lyeth et al., 1990) that mild and moderate TBI can produce prolonged (> 15 days) hippocampal dependent spatial memory deficits in rats, as assessed by an 8-arm radial maze. In that study, no qualitative or quantitative evidence of neuronal death was observed in rats showing persistent memory deficits. Additionally, no overt evidence of axonal injury was observed in any forebrain structure including major intrinsic and extrinsic connecting hippocampal pathways. Since this study detected no injury in neural substrates necessary for spatial memory in rats (Lyeth et al., 1990), we concluded that TBI is capable of producing prolonged spatial memory deficits in the absence of cell death or axonal injury. Furthermore, studies of long term potentiation (LTP), thought to be an electrophysiological correlate of memory, have provided evidence for sublethal changes in synaptic function potentially mediating these memory deficits. This research has shown that even low levels of TBI can suppress LTP in CA1 of the hippocampus (Miyazaki et al., 1988, in Press).

The Role of Neurotransmitters in Brain Injury, Edited by
M. Globus and W.D. Dietrich, Plenum Press, New York, 1992

Other lines of evidence confirm that TBI produces sublethal selective neuronal vulnerability, especially within the CA1 sector of the hippocampus. These data include observations of post-traumatic changes in excitability of CA1 neurons (DeWitt et al., 1989; Jenkins et al., 1989a,c, Miyazaki et al, 1988). Furthermore, mild levels of TBI not associated with cell death or decreases in CA1 blood flow (Chou et al., submitted) preferentially enhance the sensitivity of CA1 neurons to secondary ischemia (Jenkins et al. 1989a). This enhanced sensitivity can be dramatically attenuated by administration of specific muscarinic and NMDA receptor antagonists prior to TBI (Jenkins, 1988, 1989b). Even more extensive anatomical studies have recently confirmed that the levels of TBI employed in these studies were not accompanied by light or electron microscopic evidence of cell death or axotomy.

Spatial memory deficits are produced by cortical impact devices (White, et al., 1990), as well as by fluid percussion injury (Lyeth, et al., 1990). These same spatial memory tasks are also disrupted by muscarinic receptor blockers. For example, acquisition of the Morris water maze task can be impaired by pretrial injection of scopolamine prior to training (Buresova, Bolhuis, and Bures, 1986). Thus, the Morris water maze task is disrupted both by damage to hippocampal regions that are particularly vulnerable to TBI and by receptor blockade of the same neurotransmitter systems responsible for induction of TBI excitotoxic pathology.

With these observations in hand, we sought to determine whether mild and moderate TBI produce enduring disturbances in cholinergic receptor function, as indicated by an enhanced sensitivity of spatial memory performance to disruption by blockade of the muscarinic cholinergic receptor. We used the Morris water maze to examine the effects of a muscarinic antagonist (scopolamine, 1 mg/Kg) on spatial memory in injured and uninjured rats in two models of TBI. Although sufficient doses of scopolamine can disrupt spatial memory, this study employed a dose of scopolamine that did not adversely effect performance in uninjured rats. We report here that concussive, as well as more severe injury can produce increased sensitivity to disruption of cholinergically mediated memory function.

MATERIALS AND METHODS

Subjects

Forty male Sprague-Dawley rats (n=8 per group) weighing between 300 and 350 g were used. Animals were housed in individual cages with food and water continually available. The animal colony was maintained at a temperature of 20-22°C with a 0600(light)/1800(dark) cycle.

Apparatus

Injury Device: The controlled cortical impact injury device (see Dixon, et al., in print, for greater details) consists of a small (1.975 cm) bore, double acting, stroke-constrained, pneumatic cylinder with a 5.0 stroke. The cylinder is rigidly mounted in a vertical position on a crossbar, which can be precisely adjusted in the vertical axis. The lower rod end has an impactor tip attached (i.e., that part of the shaft that comes into contact with the exposed dura mater). The upper rod end was attached to the transducer core of a linear velocity displacement transducer (LVDT). The velocity of the impactor shaft can be controlled gas pressure. Impact velocity is directly measured by the LVDT (Shaevitz Model 500 HR) which produces an analog signal that is recorded by a PC-based data acquisition system (R.C. Electronics) for analysis of time/displacement parameters of the impactor.

Morris water maze: The water maze employs a 180 cm diameter and 60 cm pool. The pool was painted white and was filled with water to a depth of 28 cm. The water was made opaque by the addition of approximately 500 ml of nontoxic white latex paint. A platform 10 cm in diameter and 26 cm high (i.e. 2 cm below the water's surface) was used as the hidden goal platform. The pool was located in a 2.5 x 2.5 m room with numerous extra-maze ques (e.g. windows, pipes, bookcase) that remained constant throughout the experiment.

Experiment 1 - Controlled Cortical Impact

Production of cortical impact brain injury: All rats were initially anesthetized with 4% isoflurane with a 2:1 N_2O/O_2 mixture in a vented anesthesia chamber. Following endotracheal intubation, rats were mechanically ventilated with a 2% isoflurane mixture. Animals receiving a controlled cortical contusion were mounted in the injury device stereotaxic frame in a supine position secured by ear bars and incisor bar. The head was held in a horizontal plane with respect to the interaural line. A midline incision was made, the soft tissues reflected, and a 10 mm craniectomy was made centrally between bregma and lambda. Cortically impacted rats received impacts at 6 meters/sec, 2 - 2.5 mm deformation. Sham rats underwent identical surgical procedures but were not injured. Core body temperature was monitored continuously by a rectal thermistor probe and maintained at 37-38°C. After injury, the scalp was sutured closed, and the animal extubated.

Morris water maze: In order to avoid the motor deficits observed following injury, testing started on Day 11 after injury. Rats were given 4 trails per day for 5 consecutive days. For each daily block of 4 trials, rats were placed in the pool by hand facing the wall. Rats started a trial once from each of the 4 possible start location (north, east, south, west). The order of starting location was randomized. The goal platform was positioned 45 cm from the outside wall and was placed in either the northeast, southeast, southwest, or northwest quadrant of the maze. The location of the platform was held constant for each animal, but varied between animals. Rats were given a maximum of 120 sec to find the hidden platform. If the rat failed to find the platform after 120 sec, it was placed on the platform by the experimenter. All rats were allowed to remain on the platform for 30 sec before being placed in a heated incubator during a 4-min intertrial interval. Rats were tested as described on Days 11-15 and 30-34.

Scopolamine challenge: On Day 35 rats from both groups were injected (i.p.) with 1 mg/kg of scopolamine 15 prior to being tested in the maze for an additional 4 trials as describe above. The next day (day 36), after scopolamine had washed out, rats were retested for an additional 4 trials.

Results-Experiment 1

Figure 1 plots the latency to find the hidden platform on Days 11-15, 30-34, Day 35 (15 min pre-test administration) and Day 36 (1 day post-administration) The 1 mg/kg dose of scopolamine had no significant effect on the performance of the sham-injured animals. However, the same dose of scopolamine increased the latency to find the platform in the injured group. The change in goal latencies between Day 34 and Day 35 was significantly different for the sham-injured and injured groups, indicating that the injured animal's performance was more sensitive to disruption by the administration of a cholinergic antagonist.

Experiment 2 - Closed Head Injury

Production of closed head brain injury: The method of anesthesia was identical to the previous experiment. Rats were placed onto a soft foam block that was 3 cm thick. A midline incision was made, the soft tissues reflected to allow direct impact of the skull. Impact was centered between bregma and lambda, and 3 mm laterally right of sagittal suture. The velocity of the impact was 6 m/sec or 7 m/sec. The stroke of the impact was constrained to 2 mm to prevent skull fracture. Sham rats underwent identical surgical procedures but were not injured. Core body temperature was monitored continuously by a rectal thermistor probe and maintained at 37-38°C. After injury, the scalp was sutured closed, and the animal extubated.

Morris Water Maze: Since chronic motor deficits, as indexed by beam balance and beam walking tasks, were not observed following the injury magnitude employed in this study, testing started on Day 8 following injury. Rats underwent the same training protocol as Experiment 1.

Scopolamine challenge: On Day 13 rats from both groups were injected (i.p.) with 1 mg/kg of scopolamine 15 prior to being tested in the maze for an additional 4 trials as describe above.

Results-Experiment 2

Figure 2 plots the latency to find the hidden platform on Days 8-12 , Day 13 (15 min pre-test administration) and Day 14 (1 day post-administration) The 1 mg/kg dose of scopolamine had no-significant effects on the performance of the sham-injured animals. However, the same dose of scopolamine significantly increased the latency to find the platform in the injured group. The change in goal latencies between Day 12 and Day 13 was significantly different for the sham-injured and injured groups indicating that the injured animal's performance was more sensitive to disruption by the administration of a cholinergic antagonist.

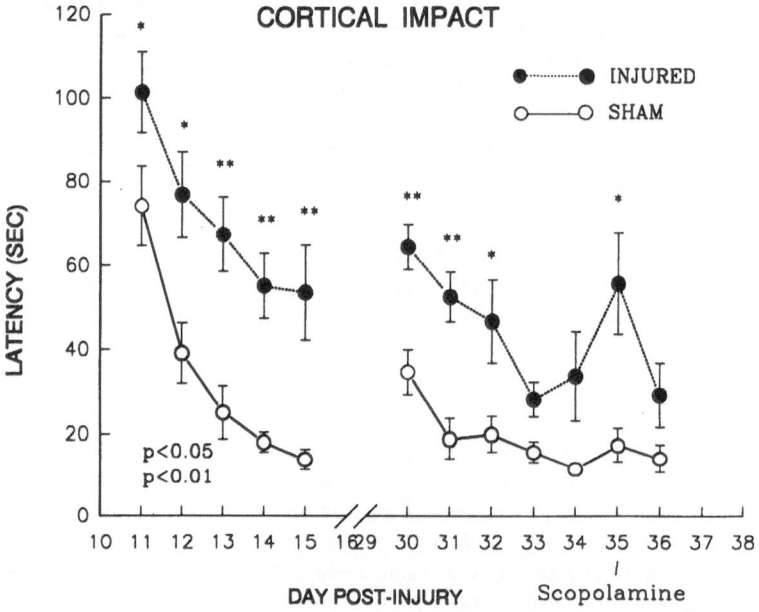

Figure 1. Latency to find the hidden platform is plotted against day post injury. Days 30-34 show acquisition of the spatial memory task. Day show differential performance after scopolamine administration. (*=p<0.05)

DISCUSSION

These data provide the first insights into features of receptor-coupled disturbances that could contribute to the maintenance of enduring cognitive deficits following TBI. Our observations suggest that changes in signal transduction by muscarinic cholinergic receptors have rendered this system more vulnerable to the effects of receptor blockade. At present, the nature of these changes are unknown. However, possible changes include alterations in: (1) the number of receptors, (2) the affinity of the receptors, (3) receptor coupling efficiency and/or (4) interactions with other neurotransmitter systems.

216

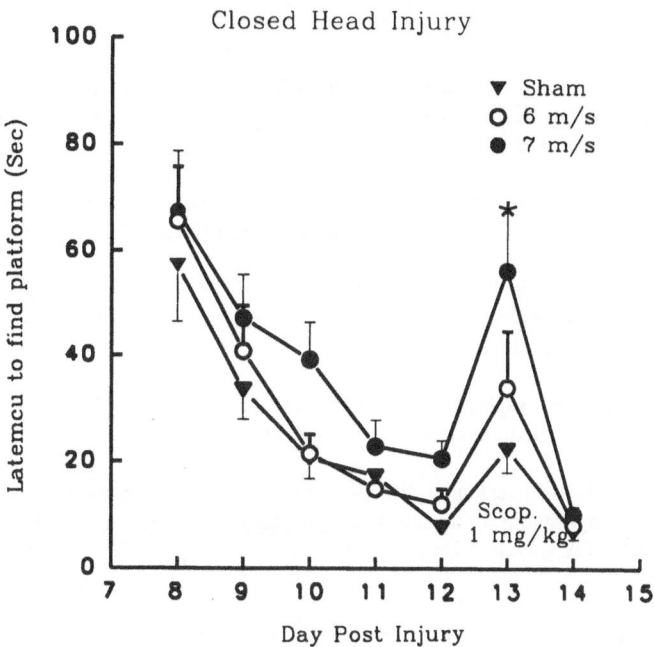

Figure 2. Latency to find the hidden platform is plotted against day post-injury. Days 8-12 show acquisition of the spatial memory task. Day 13 shows differential performance after scopolamine administration. (*=p<0.05)

It is unlikely that peripheral influences on autonomic functions (e.g. compromised vision produced by pupillary dilation) contributed to the behavioral effect of scopolamine in this study. Since scopolamine has no effect on spatial memory performance in uninjured control groups, prior injury to the CNS was necessary for scopolamine administration to elicit these deficits.

Behavioral changes following TBI include both transient and long-term changes. These long term changes are (1) "overt", that is, detected by routine neurological assessment procedures, or (2) "covert", that is, undetected by routine procedures in the absence of secondary challenges such as stress or drugs. Our experiments have shown in rats that very low (head impact acceleration) and moderate (cortical impact) levels of TBI, not associated with detectable histopathology, can produce overt disturbances in hippocampal cholinergic circuits mediating spatial memory that mimic components of human post-traumatic cognitive deficits. Covert disturbances persist in rats that have recovered normal spatial memory capacity and can only be revealed by pharmacological secondary challenge. Thus, secondary pharmacological challenges are useful, sensitive, non-invasive research tools to assess residual deficits. While the mechanisms contributing to covert deficits are not yet known, future studies will assess the contribution of excitotoxic mechanisms by determining the effects of pre-injury, or early post-injury, administration of receptor antagonists on covert deficits.

Acknowledgement

Supported by CDC R49/CCR303547, NIDRR H11B80029, NIH NS12587, and NS21458.

REFERENCES

Buresove, O., Bolhuis, J.J., and Bures, J., 1986, Differential effects of cholinergic blockade on performance of rats in the water tank navigation task and in a radial water maze. Behavioral Neuroscience, 100:476-482.

Chou, C. L., Jenkins, L.W., Hayes, R.L., Lyeth, B.G. and Povlishock, J.T., Submitted, Regional Cerebral Blood Flow Following Moderate Traumatic Brain Injury in the Rat. J Neurosurg.

DeWitt, D.S., Marlow, G.A., Jenkins, L.W., Butterworth, J.F., Prough, D.S., 1989, Effects of traumatic brain injury on hippocampal physiology in vitro and cortical DC potentials in vivo. J CBF & Metab (9)Supplement 1:S93.

Dixon, C.E., Lighthall, J.W., Yaghami, A.A., and Hayes, R.L., (In Press), A controlled cortical impact model of traumatic brain injury in the rat. J Neuroscience Methods.

Hayes, R.L., Lyeth, B.G., and Jenkins, L.W., 1989, Neurochemical mechanisms of mild and moderate head injury: Implications for treatment. In: Mild Head Injury, Levin, H.S., Eisenberg, H.M., and Benton, A.L. (Eds.), Oxford University Press, New York, pp 54-79.

Jenkins, L.W., Lyeth, B.G., LeWelt, W., Moszynski, K., DeWitt, D.S., Balster, R.L., Miller, L.P., Clifton, G.L., Young, H.F. and Hayes R.L., 1988, Combined pre-trauma scopolamine and phencyclidine attenuates post-traumatic increased sensitivity to delayed secondary ischemia. J. Neurotrauma 5(4):303-315.

Jenkins, L.W., Moszynski, K., Lyeth, B.G., Lewelt, W., DeWitt, D.S., Allen, A., Dixon, C.E., Povlishock, J.T., Majewski, T.J., Clifton, G.L., Young, H.F., Becker, D.P., and Hayes, R.L., 1989a, Increased vulnerability of the mildly traumatized brain to cerebral ischemia: The use of controlled secondary ischemia as a research tool to identify common or different mechanisms contributing to mechanical and ischemic brain injury. Brain Res. 477:211-224.

Jenkins, L.W., Lyeth, B.G., Lewelt, W., Moszynski, H.F., Young, H.F., Clifton, G.L., and Hayes, R.L., 1989b, Muscarinic and NMDA receptor blockade attenuates increased post-traumatic vulnerability to cerebral ischemia. Journal of Cerebral Blood Flow and Metabolism 9:S750.

Jenkins, L.W., Lyeth, B., DeWitt, D., Hamm, R., Phillips, L., Young, H., and Hayes, R., 1989c, Muscarinic and NMDA receptor blockade reduces post-ischemic EEG spike frequency following TBI and acute secondary ischemia. Society for Neuroscience Abstracts 15(2):1113.

Lyeth, B.G., Jenkins, L.W., Hamm, R.J., Dixon, C.E., Phillips, L.L., Clifton, G.L., Young, H.F., and Hayes, R.L., 1990, Prolonged memory impairment in the absence of hippocampal cell death following traumatic brain injury in the rat. Brain Research 526:249-258.

Makiyama Y., Jenkins L.W., Lyeth B.G., Phillips L.L., Dixon C.E., Hamm R.J., DeWitt D.S., Povlishock J.T., Clifton, G.L. and Hayes R.L., (In Review), An ultrastructural analysis of the CA1 sector of the excitotoxic rodent hippocampus following mild and moderate traumatic brain injury. Acta Neuropathol.

Miazaki, S., Newlon, P., Goldberg, S., Jenkins, L., Lyeth, B., Katayama, Y., and Hayes, R., 1988, Cerebral concussion suppresses hippocampal long-term potentiation (LTP) on rats. In: Intracranial Pressure VII, Hoff, J. and Betz, A. (Eds.), Springer-Verlag, New York, 1988, pp.651-653.

Miyazaki, S., Goldberg, S.J., Newlon, P.G., Katayama, Y., Lyeth, B.G., Jenkins, L.W., DeWitt, D.S. and Hayes, R.L., (In Press), Enduring Suppression of Hippocampal Long-Term Potentiation Following Traumatic Brain Injury In Rat. Brain Res.

White, D., Singha, A., Hamm, R.J., Dixon, C.E., Jenkins, L.W., Lyeth, B.G., and Hayes, R.L., 1990, Memory deficits following traumatic brain injury produced by controlled cortical contusion, Proceedings Soc. Neurosci., 16:780.

INVOLVEMENT OF THE CHOLINERGIC SYSTEM IN THE EFFECTS OF DM-9384
ON CARBON MONOXIDE (CO)-INDUCED ACUTE AND DELAYED AMNESIA

T. Nabeshima[1,2], M. Hiramatsu[2], T. Koide[2], S. Ishihara[2], A. Katoh[2], H. Ishimaru[2], H. Ichihashi[2], T. Shiotani[3] and T. Kameyama[2]

[1]Department of Hospital Pharmacy, Nagoya University School of Medicine, Nagoya 466, [2]Department of Chemical Pharmacology, Faculty of Pharmaceutical Sciences, Meijo University, Nagoya 468 and [3]Daiichi Pharmaceutical Co. Ltd., Tokyo 134, Japan

INTRODUCTION

Multi-infarct dementia (MID) may be caused from a deficiency in the supply of oxygen and glucose due to brain circulation insults. A transient ischemic attack is known to induce irreversible neuronal damage very slowly in the hippocampal CA1 subfield. "Delayed neuronal death" can also occur even after recovery from changes induced in biochemical and electrophysiologic parameters due to ischemic insult (Take et al., 1984b). There is wide evidence showing that the cholinergic neuronal system is involved in the formation of memory function (Beninger et al., 1989; Newhouse, 1990), and thus the lower degree of the cholinergic neuronal function may be one mechanism of memory dysfunction following ischemic insult since it has been proved that cholinergic neuronal function is lower in ischemic rats than in control rats (Take et al., 1984a).

Hypoxia can induce brain circulation failure, a deficiency in the supply of oxygen as well as ischemia. In fact, carbon monoxide (CO) has been shown to cause memory function to deteriorate in humans (Ando et al., 1987) and in mice (Nabeshima et al., 1991a). Such findings suggest that, like ischemia, CO-exposure can provide a good amnesic model to demonstrate memory deterioration and neuronal death.

N-(2,6-dimethylphenyl)-2-(2-oxo-1-pyrrolidinyl) acetamide (DM-9384), a pyrrolidone derivative, is a novel compound which has recently been proposed as treatment for cognitive disorders of senility. Several phase II clinical studies have indicated that the compound is effective in reducing various forms of cognitive disturbance (unpublished results). A preliminary report has shown that DM-9384 antagonizes hypobaric- or $NaNO_2$-induced hypoxia (Sakurai, et al., 1990) and ameliorates drug-induced impairment of passive avoidance learning such as in cases of amnesia induced by cycloheximide, bicuculline, scopolamine or chlordiazepoxide (Nabeshima, et al., 1990a, 1990b, 1991b).

We recently developed a CO-induced amnesia model in mice and demonstrated that animals exposed to CO exhibit dysfunction in the cholinergic neurons in the frontal cortex, striatum and hippocampus (Nabeshima et al., 1991a). In the present study, we investigated whether DM-9384 ameliorates CO-induced amnesia and morphological changes.

The Role of Neurotransmitters in Brain Injury, Edited by
M. Globus and W.D. Dietrich, Plenum Press, New York, 1992

METHODS

Animals

Eight-week-old male mice of the ddY strain (Nihon SLC, Shizuoka, Japan) were housed in transparent plastic cages, given food and tap water ad libitum and in a regulated environment (23 ± 1 °C, 50 ± 5 % humidity), with a 12-hr light/dark cycle (light on 8 a.m. - 8 p.m.).

Drugs

DM-9384 (N-(2,6-dimethylphenyl)-2-(2-oxo-1-pyrrolidinyl) acetamide) (Daiichi Pharmaceutical Co., Ltd., Tokyo) was synthesized at Daiichi Pharmaceutical Co., Ltd. Scopolamine hydrobromide was purchased from Tokyo Chemical Industry, Co., Ltd. and MK-801 was a gift from Dr. A.K. Cho (UCLA, California, U.S.A.). Drugs were dissolved in saline and mice received drugs i.p. or p.o. in a volume of 10 ml/kg body weight.

Experimental Schedules

Experiment 1: Training was carried out twice in 30-min intervals. DM-9384 was administered i.p. 15 min before the first training period. Mice were exposed to CO twice at 45-min intervals 30 min after the second training period. Retention testing was carried out 24 hr after the first CO-exposure.

Experiment 2: Training was carried out twice 7 days after the first CO-exposure. DM-9384 was administered i.p. 15 min before and retention testing was done 24 hr after the first training period. Scopolamine was administered i.p. immediately after the first training period.

Experiment 3: DM-9384 was administered i.p. 15 min before and again 24 hr after the first training period. Mice were exposed to CO 3 times. Retention testing was done 7 days after the first CO-exposure. Scopolamine was administered i.p. immediately after the first training period.

Experiment 4: Training was carried out twice in the same manner as Experiment 3. Mice were exposed to CO 3 times 24 hr after the first training period. MK-801 and DM-9384 were respectively administered i.p. 30 and 15 min before the first CO-exposure. Retention testing was done 7 days after the first CO-exposure. For histologic evaluation, mice were given DM-9384 (p.o.) 15 min before the CO-exposure and exposed to CO 24 hr after the training until they started gasping. Hematoxylin staining was done 1 day after the retention test.

Step-Down Type Passive Avoidance Task

The step-down type passive avoidance task was used, as previously described with some modifications (Kameyama et al., 1986).

Apparatus: A transparent acrylic rectangular cage (30 x 30 x 40 cm) with a grid floor and a semi-soundproof wooden outer box (35 x 35 x 90 cm) containing a 15 W illumination lamp were used. In the center of the grid floor, a wooden platform (4 x 4 x 4 cm) was set up. Electric shock (1 Hz, 500 msec, 35 V DC) was delivered to the grid floor by an isolated stimulator (Nihon Koden, Japan).

Training: Each mouse was placed on the wooden platform. When the mouse stepped down from the platform onto the grid floor on all fours, electric shock was delivered for 15 sec (first training). In order to minimize the variance of training, training was done twice at 2-hr intervals (except for experiment 1 which was done in 30-min intervals). In the second training, electric shock was delivered again for 15 sec when the mouse stepped down from the platform. Training was terminated if the mouse escaped from the grid floor back up onto the platform. Training was also considered terminated if the mouse did not step down onto the grid floor within 60 sec.

Retention test: Retention testing was carried out in a manner similar to that in the training except that electric shock was not delivered to the grid floor. Each mouse was placed on the platform, and step-down latency (SDL) was recorded. An upper cut-off time of 300 sec was set.

CO-Exposure

Each mouse was put into a transparent plastic vessel (radius of 3 cm, hight of 10 cm) with a pipe feeding into it, and was exposed to pure CO gas 3 times for 20, 25 and 25 sec at a rate of 10 cc/min. In the histology experiment, each mouse was exposed to CO once until it began gasping.

Histology

Mice were decapitated and brains were removed and frozen. Coronal sections 10 μm in thickness were cut by cryostat, stained with hematoxylin solution (Merck) and examined by a microcomputer imaging device (BRS2; Imaging Research Inc., Canada) to capture the image through a CCD camera (MS-4030; Siera Scientific, Canada) which was attached to a light microscopic device (BHT-322, Olympus, Japan).

Statistical Analysis

Data were expressed in terms of median, interquartile, the 10th and the 90th range. Significant differences were evaluated using the Mann-Whitney U test in two groups and using the Kruskal-Wallis test followed by the Bonferroni's test for multiple comparison.

Table I. Effect of DM-9384 on CO-induced amnesia

Treatment	Dose (mg/kg)	Step-down latency (sec)	
<Experiment 1>			
Control		143.5 (79.0-300.0)	
CO + DM-9384	0	27.0 (15.5-55.8)	**
	1	86.0 (36.5-184.3)	
	5	196.0 (42.8-255.8)	#
<Experiment 2>			
Control		280.0 (204.8-300.0)	
CO + DM-9384	0	35.5 (24.0-101.0)	**
	1	128.5 (39.0-223.0)	
	5	273.0 (157.0-300.0)	##
	10	211.0 (146.0-300.0)	##
<Experiment 3>			
Control		300.0 (266.0-300.0)	
CO + DM-9384	0	75.5 (28.0-198.5)	**
	1	119.0 (36.0-191.0)	
	5	275.0 (172.8-300.0)	##
	10	263.0 (175.0-300.0)	##
<Experiment 4>			
Control		290.0 (208.0-300.0)	
CO + DM-9384	0	38.5 (22.5-229.5)	**
	5	49.0 (16.5-175.5)	
	10	36.5 (21.0-239.0)	

Data were expressed in terms of median and interquartile ranges as indicated in parentheses. **$p<0.01$ vs. Control, #$p<0.05$, ##$p<0.01$ vs. CO alone (Bonferroni's test).

RESULTS

Effect of DM-9384 on the CO-induced acute amnesia <Experiment 1>

Experiment 1 demonstrated that memory deficiency occurred when mice were exposed to CO before memory was completely consolidated after training. The median for SDL in the retention test of the CO-exposed group was significantly shorter than that of the control group (p<0.01). This indicates that acute amnesia developed 24 hr after CO-exposure. DM-9384 (5 mg/kg) significantly ameliorated the CO-induced acute amnesia (p<0.05).

Effect of DM-9384 on delayed amnesia induced by pre-training CO-exposure <Experiment 2>

Experiment 2 demonstrated that memory acquisition was prevented when mice were exposed to CO at 7 days, but not at 1 or 2 days before training (Nabeshima et al., 1990c). The median for SDL in the retention test of the CO-exposed group was significantly shorter than that of the control group (p<0.01), indicating that amnesia was induced by exposing mice to CO at 7 days ahead of training (delayed amnesia). DM-9384 (1 - 5 mg/kg) prolonged the SDL in a dose-dependent manner in the CO-induced amnesia group. The effect of DM-9384 (5 and 10 mg/kg) was significant when compared with the CO-exposed group. To clarify whether the effect of DM-9384 was mediated via the cholinergic neuronal system, we attempted to block its action by using the muscarinic acetylcholine (ACh) receptor antagonist, scopolamine. Scopolamine blocked the anti-amnesic effect of DM-9384 (5 mg/kg) on delayed amnesia which had been induced by pre-training CO-exposure. There was a significant effect at a dose of 2 mg/kg of scopolamine, although scopolamine itself had no effect on the SDL in the control group.

Effect of DM-9384 on delayed amnesia induced by post-training CO-exposure <Experiment 3>

Experiment 3 demonstrated that delayed amnesia was induced 7 days after mice were exposed to CO 24 hr after training. It is already known that since memory consolidation was complete within 4 hrs after training, CO induced the dysfunction of memory retention and/or retrieval after memory was consolidated. DM-9384 at 5 and 10 mg/kg given 15 min before training ameliorated the CO-induced delayed amnesia. Scopolamine (1 mg/kg) antagonized the anti-amnesic effect of DM-9384 (5 mg/kg) on delayed amnesia which had been induced by post-training CO-exposure.

Effect of MK-801 and DM-9384 on delayed amnesia induced by post-training CO-exposure <Experiment 4>

Experiment 3 demonstrated that CO induced dysfunctions in memory retention and/or retrieval 7 days after exposure. It is well known that N-methyl-D-aspartate (NMDA) receptor antagonists such as MK-801 can block delayed neuronal death after ischemic insult. Furthermore, DM-9384 antagonized hypobaric- or $NaNO_2$-induced hypoxia (Sakurai, et al., 1990). Therefore, we investigated whether administrations of DM-9384 before CO-exposure could prevent delayed amnesia and morphological changes which would otherwise be induced by CO. In reflection of our previous results, we found that MK-801 (0.3 mg/kg) ameliorated CO-induced delayed amnesia. However, DM-9384 (5 and 10 mg/kg) could not prevent amnesia when it was administered just before CO-exposure, although pre-training administrations were effective as shown in Experiment 3. DM-9384 also had a tendency of protecting the cells of the CA1 subfield from swelling from CO-exposure.

Discussion

Learning and memory presumably consist of a series of steps in acquisition, consolidation, retention and retrieval. Nabeshima et al. (1990c) have reported that deficiencies in acquisition, consolidation and retention occur when CO-exposure is initiated before training, at the time of acquisition of memory and after memory consolidation.

222

Experiment 1 demonstrated that memory deficiency is induced when mice are exposed to CO after training before memory is completely consolidated (acute amnesia). Pre-training administrations of DM-9384 significantly ameliorated CO-induced acute amnesia and may have potentiated consolidation. The hypoxia-induced amnesic animal model may be similar to the memory deficit occurs in humans after gas poisoning. It is known that neuropsychiatric problems may develop insidiously over the days and weeks following recovery from CO intoxication in humans (Ginsberg, 1979). When CO-exposure is carried out before training, delayed amnesia develops more than 3 days after CO-exposure (Nabeshima et al., 1991a). Knowing this, we decided to conduct training 7 days after CO-exposure and to perform the retention test 1 day after the training in Experiment 2. Pre-training administrations of DM-9384 served to facilitate consolidation, retention and/or retrieval of the passive avoidance task in mice even after delayed amnesia had developed. Furthermore, when the retention test is carried out 1-2 days following CO-exposure conducted after memory consolidation, mice were able to maintain memory. However, delayed amnesia was observed 7 days after CO-exposure (Nabeshima et al., 1990c, Experiment 3). Pre-training administrations of DM-9384 also ameliorated this memory dysfunction as demonstrated in the retention test.

The possible participation of cholinergic mechanisms in learning and memory has been reported by many investigators (Beninger et al., 1989; Newhouse, 1990). The results of our present experiments indicate that DM-9384 ameliorates hypoxia-induced memory deficit. In our previous study, DM-9384 ameliorated drug-induced amnesia from dysfunction in the cholinergic neuronal system in mice, such as from cycloheximide- and chlordiazepoxide-induced amnesia (Nabeshima, et al., 1990a, 1990b, 1991b). Other reports have indicated that DM-9384 facilitates the acquisition of shuttle avoidance and discrimination learning in rats (Kawajiri et al., 1988), ameliorates scopolamine-induced memory impairment (Sakurai, et al., 1989), and stimulates ACh release by acting on the GABAergic neuronal system in the rat cortex (Watabe et àl., 1990). Kawajiri et al. (1990) have also shown that DM-9384 increases ACh release in the frontal cortex of rats with in vivo dialysis and that ACh in the dialysate is doubled when compared with control values with these effects lasting for approximately 1 hr. Therefore, it may be speculated that DM-9384 potentiates the cholinergic neuronal function by releasing ACh and that it may modify acquisition and/or consolidation of memory.

Neuronal death in the hippocampus is observed not only in the ischemic model, which is regarded as the experimental model for multi-infarct dementia, but also in the hypoxic model (Myers, 1970). The onset mechanism of amnesia as induced by CO-exposure has yet to be clarified; however, ischemia has been demonstrated as one source, nerve cells in the hippocampal CA1 region have been shown to process a high density of NMDA receptors after CO-exposure (Maragos, et al., 1988) and they die in a delayed manner (Nabeshima et al., 1991a). These observations suggest that an excitotoxic mechanism involving NMDA receptors may cause CO-induced delayed amnesia as well as ischemia-induced neuronal degeneration. MK-801 is already known to protect delayed neuronal degeneration induced by ischemia (Gill et al., 1987). It has also been shown to protect CO-induced neuronal death in the CA1 subfield of the hippocampus (Ishimaru et al., 1991). DM-9384 had a tendency to protect the cells of the CA1 subfield from swelling after CO-exposure. Sakurai et al. (1990) have also reported recently that DM-9384 has a protective effect against cerebral anoxia. From these findings we believe that DM-9384 also has a protective effect against ischemia, although its effect is weak.

In conclusion, we believe that DM-9384 may protect cerebral anoxia and improve the dysfunction of AChergic neuronal systems with beneficial effects on learning and memory.

References

Ando, S., Kametani, H., Osada, H., Iwamoto, M. and Kimura, N., 1987, Delayed memory dysfunction by transient hypoxia, and its prevention with forskolin, Brain Res., 405: 371-374.

Beninger, R.J., Wirsching, B.A., Jhamandas, K. and Boegman, R.J., 1989, Animal studies of brain acetylcholine and memory, Arch. Gerontol. Geriatr. Suppl. 1: 71-89.

Gill, R., Foster, A.C. and Woodruff, G.N., 1987, Systemic administration of MK-801 protects against ischemia-induced hippocampal neurodegeneration in the gerbil, J. Neurosci., 7: 3343-3349.

Ginsberg, M.D., 1979, Delayed neurological deterioration following hypoxia, Adv. Neurol., 26: 21-44.

Ishimaru, H., Katoh, A., Suzuki, H., Fukuta, T., Kameyama, T. and Nabeshima, T., 1991, Characterization of carbon monoxide (CO)-induced delayed neuronal death in mice, Japan. J. Pharmacol., Suppl. I, 55: 141.

Kameyama, T., Nabeshima, T. and Noda, Y., 1986, Cholinergic modulation of memory for step-down type passive avoidance task in mice, Res. Commun. Psychol. Psychiat. Behav., 11: 193-205.

Kawajiri, S., Sakurai, T., Ojima, H., Hatanaka, S., Yamasaki, T., Kojima, H. and Akashi, A., 1988, Effect of DM-9384, a new pyrrolidone derivative, on learning behavior and cerebral choline acetyltransferase activity in rats, Psychopharmacology 96 (Suppl.): 306.

Kawajiri, S., Taniguchi, K., Sakurai, T., Kojima, H. and Yamasaki, T., 1990, Effect of DM-9384 on extracellular acetylcholine in the rat frontal cortex measured with microdialysis, Eur. J. Pharmacol., 183: 928.

Maragos, W.F., Penney, J.B. and Young, A.B., 1988, Anatomic correlation of NMDA and [^{3}H]TCP-labeled receptors in rat brain, J. Neurosci., 8: 493-501.

Myers, R.E., 1970, Brain damage induced by umbilical cord compression at different gestational ages in monkey. in: "Medical Primatology", E.I. Goldsmith and J. MoorJankowski, eds., Karger, Basal, pp. 394-425.

Nabeshima, T., Katoh, A., Ishimaru, H., Yoneda, Y., Ogita, K., Murase, K., Ohtsuka H., Inari, K. Fukuta, T. and Kameyama, T., 1991a, Carbon monoxide-induced delayed amnesia, delayed neuronal death and change in acetylcholine concentration in mice, J. Pharmacol. Exp. Ther., 256: 378-384.

Nabeshima, T., Tohyama, K., Murase, K., Ishihara, S., Kameyama, T., Yamasaki, T., Hatanaka, S., Kojima, H., Sakurai, T., Takasu, Y. and Shiotani, T., 1991b, Effects of DM-9384, a cyclic derivative of GABA, on amnesia and decreases in GABA$_A$ and muscarinic receptors induced by cycloheximide, J. Pharmacol. Exp. Ther., 257: 271-275.

Nabeshima, T., Noda, Y., Tohyama, K., Itoh, J. and Kameyama, T., 1990a, Effects of DM-9384 in a model of amnesia based on animals with GABAergic neuronal dysfunctions, Eur. J. Pharmacol., 178: 143-149.

Nabeshima, T., Tohyama, K. and Kameyama, T., 1990b, Effects of DM-9384, a pyrrolidone derivative, on alcohol- and chlordiazepoxide-induced amnesia in mice, Pharmacol. Biochem. Behav., 36: 233-236.

Nabeshima, T., Yoshida, S., Morinaka, H., Kameyama, T., Thurkauf A., Rice, K.C., Jacobson, A.E., Monn, J.A. and Cho, A.K., 1990c, MK-801 ameliorates delayed amnesia, but potentiates acute amnesia induced by CO, Neurosci. Lett., 108: 321-327.

Newhouse, P.A., 1990, Cholinergic drug studies in dementia and depression, Adv. Exp. Med. Biol., 282: 65-76.

Sakurai, T., Hatanaka, S., Tanaka, S., Yamasaki, T., Kojima, H. and Akashi, A., 1990, Protective effect of DM-9384, a novel pyrrolidone derivative, against experimental cerebral anoxia, Japan. J. Phamacol., 54: 33-43.

Sakurai, T., Ojima, H., Yamasaki, T., Kojima, H. and Akashi, A., 1989, Effects of N-(2,6-dimethylphenyl)-2-(2-oxo-1-pyrrolidinyl) acetamide (DM-9384) on learning and memory in rats, Japan. J. Pharmacol., 50: 47-53.

Take, Y., Narumi, S., Nagai, Y., Kuriyama, E., Saji, Y. and Nagawa, Y., 1984a, Neurochemical study of the temporary cerebral ischemic rats produced by bilateral vertebral and carotid artery occlusion, Folia Pharmacol. Jpn., 84: 485-498.

Take, Y., Yamazaki, N., Fukuda, N., Saji, Y. and Nagawa, Y., 1984b, Physiological and biochemical study of temporary cerebral ischemic rats produced by bilateral vertebral and carotid artery occlusion, Folia Pharmacol. Jpn., 84: 471-483.

Watabe, S., Yamaguchi, H. and Ashida, S., 1990, Effects of DM-9384, a new cognitive enhancing agent, on cholinergic system in rat cortex, 20th Neurosci. abstr., p. 137.

Chapter 5

Intracellular Messengers and Brain Injury

CHANGES IN GENE EXPRESSION AFTER TRANSIENT ISCHEMIA

AS POTENTIAL MARKERS FOR EXCITOTOXIC PATHOLOGY

Thaddeus S. Nowak, Jr., Olive C. Osborne and Sadao Suga

Stroke Branch
National Institute of Neurological Disorders and Stroke
NIH, Bldg. 36, Rm. 4D04, Bethesda, MD 20892

INTRODUCTION

Changes in gene expression have come to be recognized as prominent features of neuronal and glial responses to various stimuli. Of these responses perhaps the best characterized is the induction of c-*fos* and related immediate early genes that encode proteins functioning as transcription factors that in turn regulate the expression of other genes (Morgan and Curran, 1991). Several signal transduction pathways can mediate c-*fos* induction in response to different inputs (Sheng and Greenberg, 1990). There is good evidence that glutamate agonists are directly or indirectly capable of inducing neuronal c-*fos* expression (Sonnenberg et al., 1989; Szekely et al., 1989). Glutamate receptors of the N-methyl-D-aspartate (NMDA) subtype appear to mediate c-*fos* induction after spreading depression (Herrera and Robertson, 1990), and such a mechanism is sufficient to account for the generalized increases in c-*fos* mRNA throughout the ipsilateral hemisphere following focal ischemia (Welsh et al., 1991). The signals mediating c-*fos* induction in hippocampus and other regions following transient global ischemia remain to be explicitly delineated, although NMDA antagonists have also been shown to reduce Fos protein expression in such models (Uemura et al., 1991).

Induction of heat shock / stress proteins is another prominent transcriptional response to ischemia (Nowak, 1990). The 70 kDa stress protein, hsp70, and its mRNA are expressed solely in neurons after moderate global ischemic insults (Vass et al., 1988; Nowak, 1991; Simon et al., 1991), although glial and vascular elements show a response under conditions of focal injury (Gonzalez et al., 1989; Ferriero et al., 1990; Sharp et al., 1991). In the case of the neuronal heat shock response a direct link with excitatory neurotransmission remains to be established, although kainic acid has been shown to induce hsp70 expression (Uney et al., 1988; Vass et al., 1989; Gonzalez et al., 1989). There is also evidence that regulatory elements upstream of a human hsp70 gene can mediate a cAMP-dependent response (Choi et al., 1991). Together with results demonstrating that hsp70 and c-*fos* mRNAs are induced with overlapping initial distributions after ischemia (Nowak et al., 1990), such observations support the suggestion that the postischemic stress response may also reflect the activation of stimulus-coupled signal transduction mechanisms in neurons. A notable feature of the postischemic stress response is the limited detection of immunoreactive hsp70 in vulnerable CA1 neurons that abundantly express hsp70 mRNA (Vass et al., 1988; Nowak 1991). In contrast, short 2 min ischemic insults, at the threshold for induction of both hsp70 and c-*fos* (Nowak and Osborne, 1991, Nowak et al., 1991), result in the appearance of hsp70 protein in CA1 neurons and produce tolerance to more severe ischemic stress (Kitagawa et al., 1990; Kirino et al., 1991). The present studies were designed to characterize expression of Fos- and Jun-related proteins encoded by immediate early genes after ischemia, and to examine the relationship between changes in transcription factor expression and the progression of neuronal injury and recovery.

METHODS

Ischemia was produced in Mongolian gerbils (female, 50-70 g) by bilateral common carotid artery occlusion for intervals of 2 min or 5 min under anesthesia with 2% halothane in 30% O_2, 70% N_2O. Animals were perfused at various recirculation times with 4% paraformaldehyde in 100 mM sodium phosphate (pH 7.4), and 50 μm vibratome sections were prepared and processed for immunocytochemistry essentially as described previously (Vass et al., 1988). Fos and Jun antibodies were affinity purified rabbit polyclonal IgG preparations obtained from Oncogene Science, Inc., Manhasset, NY, directed against the N-terminal domain of Fos and the DNA binding domain of Jun, respectively. The detection system consisted of biotinylated goat anti-rabbit second antibodies and streptavidin-conjugated peroxidase (Kirkegaard and Perry Laboratories, Gaithersburg, MD), using diaminobenzidine as substrate. For quantitative measurements of Fos immunoreactivity the number of positively stained neuronal nuclei was counted in standard fields obtained at comparable hippocampal levels and averaged for the various experimental groups.

RESULTS

Fos immunoreactivity was expressed in only a few scattered neurons in control gerbil brain but was strongly induced after transient ischemia. The main features of hippocampal Fos protein expression after ischemia are illustrated in Fig. 1, demonstrating its expected nuclear distribution and the absence of detectable expression in vulnerable CA1 neurons. While generally similar patterns of Fos staining were seen after ischemic insults of varied duration, the time course and magnitude of Fos expression in the several hippocampal cell populations were reproducibly dependent on the severity of the ischemic insult (Fig. 2). Dentate granule cells showed the most striking Fos induction, with a significant number of cells staining at 1 h and virtually every cell showing immunoreactivity at 2 h recirculation following either 5 min or 2 min ischemia. Fos expression was more sustained after the

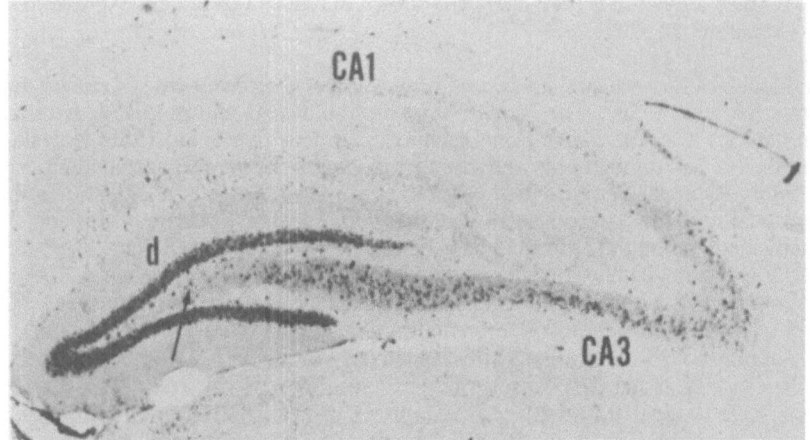

Fig. 1.   Fos immunoreactivity in gerbil hippocampus after transient ischemia. Representative vibratome section from an animal subjected to 5 min ischemia and 3 h reperfusion, and stained with a Fos antibody as described in the text. Positive nuclei are evident in dentate granule (d) and CA3 pyramidal fields, as well as in scattered hilar neurons (arrow), but not in CA1.

longer insult, with a higher proportion of cells remaining positive at 3 h. Neurons of CA3 and dentate hilus showed a relative delay in Fos expression and the time course of Fos staining was comparable in these cells after 2 min and 5 min insults, but the number of positive neurons was higher after the mild 2 min challenge. CA1 neurons remained unstained after either 2 min or 5 min ischemia.

Jun-like immunoreactivity was prominent in dentate granule cells and CA3 neurons of

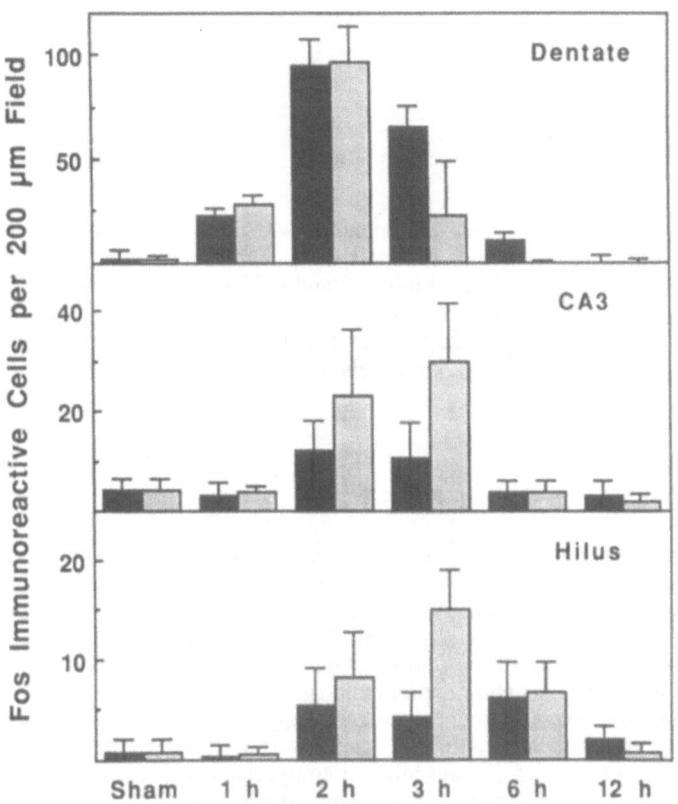

Fig. 2.   Time course of Fos immunoreactivity in gerbil hippocampus after 2 min and 5 min ischemia. The number of Fos-positive nuclei in a standard field was determined for dentate granule cells, CA3 pyramidal cells and hilar neurons at the indicated recirculation times after ischemia of 5 min (solid bars) or 2 min (stippled bars). Plotted values are means ± SD for data from 5-6 animals.

control gerbil brain, and these cell populations showed a further increase in staining intensity after ischemia, reaching a maximum at approximately 6 h (not shown). CA1 neurons showed minimal Jun immunoreactivity in control animals and no increase in Jun expression was seen after 5 min ischemia, but a striking increase was seen after 2 min ischemia (Fig. 3). As in the case of Fos, the proportion of Jun-positive neurons in dentate hilus was also greater after 2 min than after 5 min ischemia.

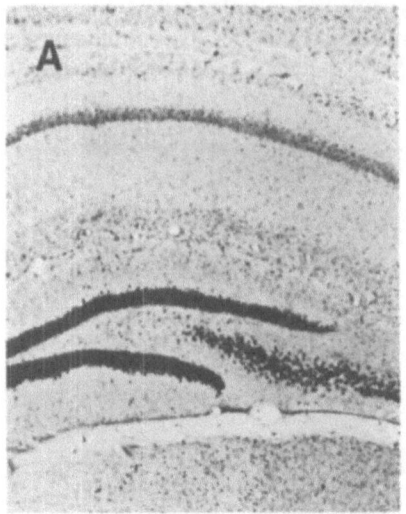
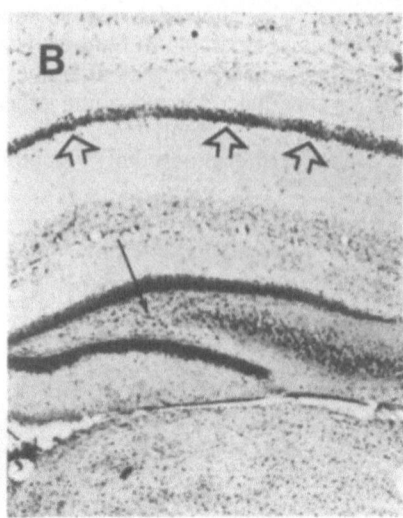

Fig. 3. Distribution of Jun-like immunoreactivity in gerbil hippocampus after transient ischemia. Animals were subjected to carotid occlusion for either 5 min (A) or 2 min (B) and perfused at 6 h recirculation. The shorter 2 min insult gave rise to striking increases in Jun staining of neurons in CA1 (large arrows) and dentate hilus (small arrow) that were not seen after 5 min ischemia.

DISCUSSION

The above results confirm recent studies of Fos protein localization in the gerbil ischemia model that showed minimal expression of Fos-like immunoreactivity in CA1 pyramidal neurons (Uemura et al., 1991). The postischemic protein synthesis deficit in these cells is well known (Dienel et al., 1980, Thilmann et al., 1986), and appears to be sufficiently prolonged even after 2 min ischemia (Araki et al., 1990) that it precludes translation of the very transiently expressed c-*fos* mRNA (Onodera et al., 1989; Nowak et al., 1990). The differences in time course and frequency of Fos protein expression in other hippocampal neuron populations (Fig. 2) can be similarly interpreted, with a shift toward later expression in CA3 and dentate hilus in comparison with the less vulnerable dentate granule cells. More rapid recovery of protein synthesis after the 2 min insult would also result in a higher proportion of positively stained neurons. In dentate granule cells it would appear that protein synthesis recovery does not limit Fos protein accumulation, and a tendency toward more prolonged Fos expression after the 5 min insult is suggested.

The importance of protein synthesis recovery in determining the functional consequences of induced gene expression is of general relevance in pathological conditions such as ischemia, and may thereby directly influence the evolution of neuronal damage. Using experimental paradigms comparable to those employed above it was demonstrated that 2 min ischemia allowed the accumulation of immunoreactive hsp70 in CA1 neurons (Kirino et al., 1991) that after more severe insults express the mRNA but not the protein (Vass et al., 1988; Nowak, 1991). Analogous results have been reported for hsp70 immunoreactivity following ischemia of varied duration in the rat (Simon et al., 1991), with neurons of dentate hilus and CA1 showing positive staining after short insults but not after prolonged occlusions that result in severe neuronal injury.

Hsp70 induction following short priming insults has been suggested to contribute to functional alterations in CA1 neurons that can result in an induced tolerance to subsequent more severe ischemia (Kitagawa et al., 1990; Kirino et al., 1991). There are methodological complications that limit interpretation of these results, since there can be a prolonged delay in detection of immunoreactive hsp70 even in cell populations that show relatively rapid protein synthesis recovery (Vass et al., 1988; Nowak 1991). Hyperthermia has also been suggested to induce ischemic tolerance (Chopp et al., 1989; Kitagawa et al., 1991), but generally results in a distribution of induced hsp70 that does not include the protected neurons (Sprang and Brown, 1987; Blake et al., 1990; Marini et al., 1990). In our own preliminary studies we have failed to detect significant hsp70 staining in CA1 after 2 min insults that produce tolerance, while reproducible expression of the protein is found after 2.5 min ischemia (S. Suga and T. S. Nowak, Jr., unpublished observations).

With regard to Jun staining it is notable that immunoreactive protein was detected in CA1 neurons of animals subjected to 2 min but not to 5 min ischemia (Fig. 3). This provides clear evidence of functional protein synthesis recovery in CA1 within this recirculation interval after the short ischemic insult, and implies that the encoding mRNA is expressed with a longer time course than that of c-*fos*. The major significance of this finding is the demonstration that conditions of moderate ischemia associated with induction of ischemic tolerance also increase the expression of a known transcription factor. It remains to be determined which member(s) of the Jun family are detected in these studies. Nevertheless, these experiments have identified a mechanism that could account for widespread activation or repression of specific genes after brief ischemic insults. The mRNA for another class of transcription factor also has been shown to be induced after ischemia (Abe et al., 1991), although functional expression and cellular localization remain to be established. Future studies may be expected to further expand the range of known transcriptional changes that comprise the neuronal response to injury, to establish the signal transduction mechanisms that account for these various changes in gene expression, and to clarify the possible roles of specific induced proteins in determining cell death, survival or acquisition of tolerance.

REFERENCES

Abe, K., Kawagoe, J., Sato, S., Sahara, M. and Kogure, K., 1991, Induction of the 'zinc finger' gene after transient focal ischemia in rat cerebral cortex, Neurosci. Lett., 123:248.

Araki, T., Kato, H., Inoue, T., and Kogure, K., 1990, Regional impairment of protein synthesis following brief cerebral ischemia in the gerbil, Acta Neuropathol., 79:501.

Blake, M. J., Nowak, T. S., Jr. and Holbrook, N. J., 1990, In vivo hyperthermia induces expression of HSP70 mRNA in brain regions controlling the neuroendocrine response to stress, Mol. Brain Res., 8:89.

Choi, H.-K., Li, B., Lin, Z., Huang, L. E. and Liu, A. Y.-C., 1991, cAMP and cAMP-dependent protein kinase regulate the human heat shock protein 70 gene promoter activity, J. Biol. Chem., 266:11858.

Chopp, M., Chen, H., Ho, K.-L., Dereski, M. O., Brown, E., Hetzel, F. W. and Welch, K. M. A., 1989, Transient hyperthermia protects against subsequent forebrain ischemic cell damage in the rat, Neurology, 39:1396.

Dienel, G. A., Pulsinelli, W. A. and Duffy, T. E., 1980, Regional protein synthesis in rat brain following acute hemispheric ischemia, J. Neurochem., 35:1216.

Ferriero, D. M., Soberano, H. Q., Simon, R. P. and Sharp, F. R., 1990, Hypoxia-ischemia induces heat shock protein-like (hsp72) immunoreactivity in neonatal rat brain, Dev. Brain Res., 53:145.

Gonzalez, M. F., Shiraishi, K., Hisanaga, K., Sagar, S. M., Mandabach, M. and Sharp, F. R., 1989, Heat shock proteins as markers of neural injury, Mol. Brain Res., 6:93.

Herrera, D. G. and Robertson, H. A., 1990, Application of potassium chloride to the brain surface induces the c-fos proto-oncogene: reversal by MK-801, Brain Res., 510:166.

Kirino, T., Tsujita, Y. and Tamura, A., 1991, Induced tolerance to ischemia in gerbil hippocampal neurons, J. Cereb Blood Flow Metab, 11:299.

Kitagawa, K., Matsumoto, M., Tagaya, M., Hata, R., Ueda, H., Niinobe, M., Handa, N., Fukunaga, R., Kimura, K., Mikoshiba, K. and Kamada, T., 1990, "Ischemic tolerance" phenomenon found in brain, Brain Res., 528:21.

Kitagawa, K., Matsumoto, M., Tagaya, M., Kuwabara, K., Hata, R., Handa, N., Fukunaga, R., Kimura, K. and Kamada, T., 1991, Hyperthermia-induced neuronal protection against ischemic injury in gerbils, J. Cereb. Blood Flow Metab., 11:449.

Marini, A. M., Kozuka, M., Lipsky, R. L. and Nowak, T. S., Jr., 1990, 70-Kilodalton heat shock protein induction in cerebellar astrocytes and cerebellar granule cells in vitro: comparison with immunocytochemical localization after hyperthermia in vivo, J. Neurochem., 54:1509.

Morgan, J. I. and Curran, T., 1991, Stimulus-transcription coupling in the nervous system: Involvement of the inducible proto-oncogenes fos and jun, Annu. Rev. Neurosci., 14:421.

Nowak, T. S., Jr., 1990, Protein synthesis and the heat shock/stress response after ischemia, Cerebrovasc. Brain Metab. Rev., 2:345.

Nowak, T. S., Jr, 1991, Localization of 70 kDa stress protein mRNA induction in gerbil brain after ischemia, J. Cereb. Blood Flow Metab., 11:432.

Nowak, T. S., Jr., Ikeda, J. and Nakajima, T., 1990, 70 Kilodalton heat shock protein and c-fos gene expression following transient ischemia, in: "Proceedings, 17th Princeton Conference on Cerebrovascular Diseases," J. N. Davis, M. J. Alberts, L. B. Goldstein, eds., Stroke, 21(Suppl. III):107.

Nowak, T. S., Jr. and Osborne, O. C., 1991, Threshold ischemic duration for stress protein induction in gerbil brain (Abstract), Stroke, 22:131.

Nowak, T. S., Jr., Osborne, O. C. and Ikeda, J., 1991, Role of altered gene expression in development of neuronal changes after ischemia, in: "Maturation Phenomena in Cerebral Ischemia," U. Ito and I. Klatzo, eds., Springer-Verlag, Berlin, (in press).

Onodera, H., Kogure, K., Ono, Y., Igarashi, K., Kiyota, Y. and Nagaoka, A., 1989, Proto-oncogene c-fos is transiently induced in the rat cerebral cortex after forebrain ischemia, Neurosci. Lett., 98:101.

Sharp, F. R., Lowenstein, D., Simon, R. and Hisanaga, K., 1991, Heat shock protein hsp72 induction in cortical and striatal astrocytes and neurons following infarction, J. Cereb. Blood Flow Metab., 11:621.

Sheng, M. and Greenberg, M. E., 1990, The regulation and function of c-fos and other immediate early genes in the nervous system, Neuron, 4:477.

Simon, R. P., Cho, H., Gwinn, R. and Lowenstein, D. H., 1991, The temporal profile of 72-kDa heat-shock protein expression following global ischemia, J. Neurosci., 11:881.

Sonnenberg, J. L., Mitchelmore, C., Macgregor-Leon, P. F., Hempstead, J., Morgan, J. I. and Curran, T., 1989, Glutamate receptor agonists increase the expression of Fos, Fra, and AP-1 DNA binding activity in the mammalian brain, J. Neurosci. Res., 24:72.

Sprang, G. K. and Brown, I. R., 1987, Selective induction of a heat shock gene in fiber tracts and cerebellar neurons of the rabbit brain detected by in situ hybridization, Mol. Brain. Res., 3:89.

Szekely, A. M., Barbaccia, M. L., Alho, H. and Costa, E., 1989, In primary cultures of cerebellar granule cells the activation of N-methyl-D-aspartate-sensitive glutamate receptors induces c-fos mRNA expression, Mol. Pharmacol., 35:401.

Thilmann, R., Xie, Y., Kleihues, P. and Kiessling, M., 1986, Persistent inhibition of protein synthesis precedes delayed neuronal death in postischemic gerbil hippocampus, Acta Neuropathol., 71:88.

Uemura, Y., Kowall, N. W. and Beal, M. F., 1991, Global ischemia induces NMDA receptor-mediated c-fos expression in neurons resistant to injury in gerbil hippocampus, Brain Res., 542:343.

Uney, J. B., Leigh, P. N., Marsden, C. D., Lees, A. and Anderton, B. H., 1988, Stereotaxic injection of kainic acid into the striatum of rats induces synthesis of mRNA for heat shock protein 70, FEBS Lett., 235:215.

Vass, K., Berger, M. L., Nowak, T. S., Jr., Welch, W. J. and Lassmann, H., 1989, Induction of stress protein HSP70 in nerve cells after status epilepticus in the rat, Neurosci. Lett., 100:259.

Vass, K., Welch, W. J. and Nowak, T. S., Jr., 1988, Localization of 70 kDa stress protein induction in gerbil brain after ischemia, Acta Neuropathol., 77:128.

Welsh, F. A., Moyer, D. J. and Harris, V. A., 1991, Regional expression of mRNAs for heat shock protein-70 and c-fos following focal ischemia in rat brain (Abstract), J. Cereb. Blood Flow Metab., 11:S213.

CORRELATION BETWEEN SECOND-MESSENGER ACTIVITIES AND CEREBRAL BLOOD FLOW IN THE BASAL GANGLIA AND HIPPOCAMPUS AFTER UNILATERAL CAROTID ARTERY OCCLUSION IN GERBILS

K. Tanaka, F. Gotoh, Y. Fukuuchi, S. Gomi, S. Takashima, B. Mihara, T. Shirai, S. Nogawa and E. Nagata

Department of Neurology, School of Medicine, Keio University, Tokyo 160, Japan

INTRODUCTION

We reported previously that cerebral ischemia for 6 h induced a significant reduction of forskolin (FK) binding over extensive brain regions, whereas the phorbol 12,13-dibutyrate.(PDBu) binding was relatively well preserved in various cerebral structures[1]. These findings suggest that the adenylate cyclase (AC) system may be vulnerable to ischemic insult, while the protein kinase C (PKC) system may be relatively resistant to ischemia. The quantitative autoradiographic method employed previously[1] permitted us to measure the regional FK and PDBu bindings and cerebral blood flow (CBF) in the same brain, so that the interrelationship between these three parameters could be precisely analyzed in each cerebral region. The basal ganglia including the caudate-putamen (striatum) and hippocampus are known to be highly vulnerable to ischemia[2], and these structures formed the focus of the present study on cerebral ischemia. The aim of this communication is to report the correlation between the second-messenger activities and CBF in the basal ganglia and hippocampus in 6-hour ischemia of the gerbil brain.

MATERIALS AND METHODS

Animal Preparation

Twenty-seven Mongolian gerbils of either sex weighing 70-90 g were used. The details of the animal preparation were as described previously[1]. In brief, the right common carotid artery was doubly ligated with a silk thread under light anesthesia with 1.5% halothane in order to induce 6-hour cerebral ischemia. At 60 min after the ligation, the behavior of the animals was evaluated in order to calculate the stroke index[3]. Eight animals which scored more than 5 on the stroke index (SI) were subjected to further investigation as the Ischemia Group (SI=15.8+5.1). In another seven animals which formed the Sham Group, the right common carotid artery was only exposed. In both groups, at 70 min after the ligation or sham operation, polyethylene catheters were inserted into the femoral artery and vein. Following such surgical procedures, the gerbils were immobilized in a plaster cast on a board.

Measurement of Local Cerebral Blood Flow

At 6 h after the ligation or sham surgery, CBF was determined by the iodoantipyrine method[4] as described elsewhere[1]. Triple sets of 6 brain sections, each at predetermined levels (the caudate-putamen, globus-pallidus, and hippocampus), from each animal were utilized for autoradiographic measurement of the CBF and for labelling procedures with [3H]forskolin (FK) and [3H]phorbol 12,13-dibutyrate (PDBu), respectively. In addition, 2 sections each at the same levels were stained with thionine for anatomical identification and alignment of the autoradiograms as described below.

Measurement of [3H]FK and [3H]PDBu Binding

The cut brain sections assigned for measurement of the FK and PDBu binding were labeled with 10 nM [3H]FK and 2.5 nM [3H]PDBu, respectively, as described previously[1]. The nonspecific binding was simultaneously determined in the presence of an unlabeled ligand. All data presented in this paper are for the specific binding.

Data Analysis

Quantitative densitometric analysis of the autoradiograms was performed with the computerized digital image processing system developed in our laboratory[5]. The geometrical information concerning the region of interest (ROI) for each anatomical structure was registered in the image processor. According to these procedures, ROI was identical for measurement of the three parameters (CBF, FK and PDBu binding) in each cerebral structure of each animal.

The data for the above parameters in both hemispheres obtained from the animals of both groups were pooled, and the interrelationship between these parameters was analyzed by the linear or non-linear least-squares method.

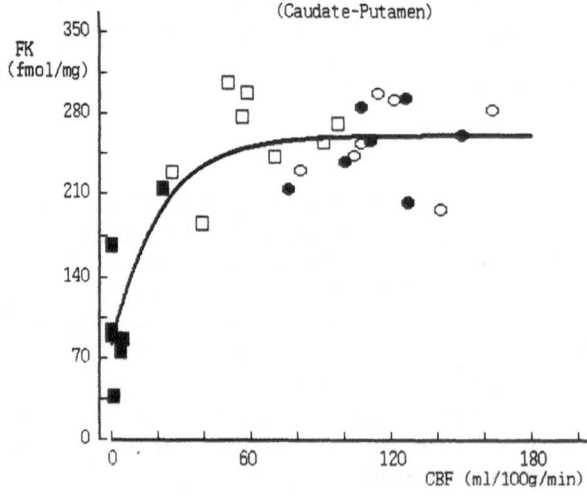

Figure 1. Relationship between cerebral blood flow (CBF) and forskolin (FK) binding in the caudate-putamen. Filled and empty squares represent data obtained from the ligated and non-ligated sides of the Ischemia Group, respectively. Filled and empty circles represent the data obtained from the operated and non-operated sides of the Sham Group, respectively.

RESULTS

1) CBF vs. FK Binding

An exponential relationship was observed between CBF and the FK binding in the basal ganglia including the caudate-putamen (Fig. 1), globus-pallidus and accumbens nucleus. In these regions, the FK binding decreased precipitously as CBF fell below 20 ml/100g/min. In contrast, the FK binding remained fairly constant at CBF above 40-60 ml/100g/min. On the other hand, a linear correlation was noted in various structures of the hippocampus (CA_1 and CA_3 regions (Fig. 2) and dentate gyrus). Accordingly, there appears to be no apparent CBF threshold for reduction of the FK binding in the hippocampus.

2) CBF vs. PDBu Binding

A significant correlation between CBF and the PDBu binding was observed only in the caudate-putamen (Fig. 3) and dentate gyrus (Fig. 4) of the hippocampus. In both regions, the correlation was linear with relatively low values for the correlation coefficients, and no apparent CBF threshold for reduction of the PDBu binding was noted.

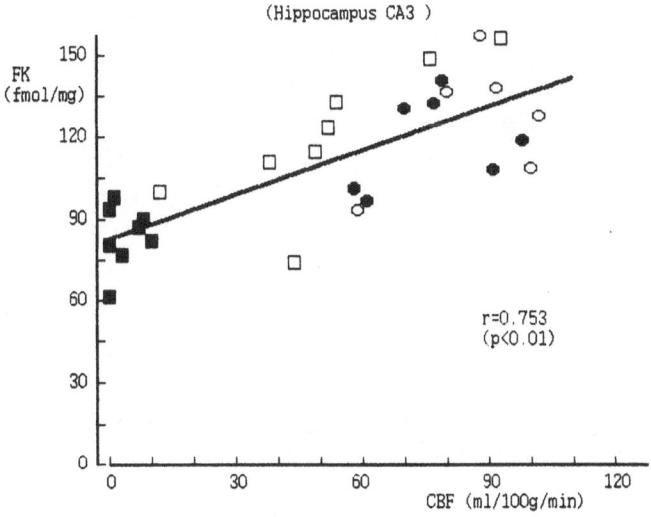

Figure 2. Relationship between cerebral blood flow (CBF) and forskolin (FK) binding in the CA_3 region of the hippocampus. Each data symbol is as in Fig. 1.

DISCUSSION

The present study clearly demonstrated regional specificities in the alteration of the FK binding during cerebral ischemia. The basal ganglia including the caudate-putamen, globus-pallidus and accumbens nucleus showed an ischemic threshold of 20 ml/100g/min, below which the FK binding began to decrease precipitously. A similar relationship between CBF and the FK binding has also been noted in the cerebral cortices as reported previously[1]. On the other hand, in the hippocampus (CA_1 and CA_3 regions and dentate gyrus), the FK binding appeared to be flow-dependent when CBF ranged between 0 and 110 ml/100g/min, and a positive linear relationship was observed between CBF and the FK binding.

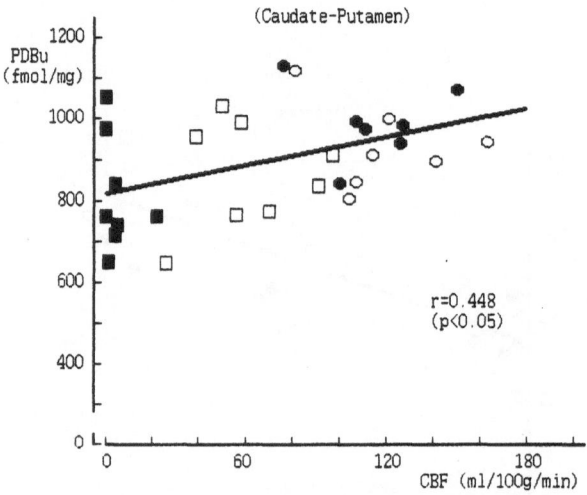

Figure 3. Relationship between cerebral blood flow (CBF) and phorbol 12,13-dibutyrate (PDBu) binding in the caudate-putamen. Each data symbol is as in Fig. 1.

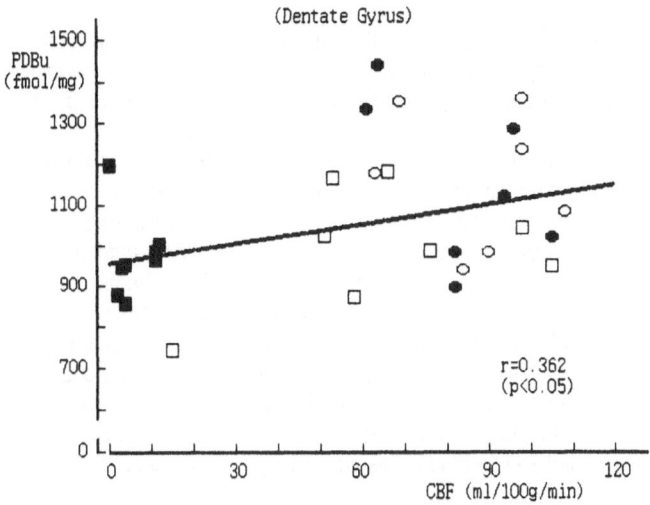

Figure 4. Relationship between cerebral blood flow (CBF) and phorbol 12,13-dibutyrate (PDBu) binding in the dentate gyrus of the hippocampus. Each data symbol is as in Fig. 1.

The FK binding basically correlates with the activated functional state of the adenylate cyclase system provoked by certain neurotransmitters or modulators[1]. Our results therefore suggest that the signal transmission via the adenylate cyclase system may begin to deteriorate significantly in the basal ganglia and cerebral cortices at a flow level of 20 ml/100g/min. Indeed, electrical failure of the cerebral cortex generally begins to manifest itself when CBF decreases below 15-20 ml/100g/min[6]. Similarly, impairment of the brain energy metabolism, such as reduction of the high-energy phosphate content and accumulation of lactate, has been reported to be initiated when CBF fell below approximately 20 ml/100g/min[7,8]. The existence of CBF thresholds for these phenomena appears to be compatible with our observations concerning the FK binding in the above regions.

The caudate-putamen and globus-pallidus receive glutamatergic afferent fibers from the cerebral cortex and subthalamic nucleus, respectively[9]. The microdialysis technique has revealed that the extracellular glutamate concentration increased substantially at a CBF level below 20 ml/100g/min[10], and a massive increase in extracellular glutamate level has been demonstrated in the caudate-putamen during severe ischemia[11]. These findings suggest that an ischemic flow threshold may exist for functional derangement of the glutamatergic neurons with an excessive release and impaired cellular uptake of glutamate, and this flow threshold is apparently consistent with that for the reduction of the FK binding observed in the present study. Glutamate is widely known to be neurotoxic at high concentrations[12], and intrastriatal injection of kainic acid, a glutamate analog, has been found to cause a substantial loss of striatal dopamine-sensitive adenylate cyclase activity[13]. Based on these findings, glutamate may, at least, be one of the factors responsible for causing the threshold-type relationship between CBF and the FK binding observed in the present study.

The basal ganglia have other input and output systems including the dopaminergic nigrostriatal pathway. A profound increase in extracellular concentration of striatal dopamine has been noted during severe ischemia[11], and excessive release of dopamine is suggested to be important for the development of ischemic cell damage. The accumbens nucleus is also known to be innervated by the mesolimbic dopaminergic system[14]. In addition, global forebrain ischemia has been reported to increase the extracellular serotonin level significantly in the striatum[15]. However, it is currently unknown whether or not ischemic flow thresholds similar to that for glutamate also exist for the increase in extracellular dopamine and serotonin levels.

As in the caudate-putamen, a massive release of excitatory amino acids[16], noradrenaline[17], dopamine[18] and serotonin[15] was noted in the hippocampus at the early stage of ischemia. The relatively low basal activity of the AC system in the hippocampus as compared to that in the caudate-putamen might obscure the CBF threshold for the reduction of the FK binding.

In contrast to the FK binding, the PDBu binding exhibited a statistically significant correlation with CBF only in the caudate-putamen and dentate gyrus, and the correlation was relatively loose and linear in each region. The PDBu binding obtained in the present study can be considered to represent a selective measure of the membrane-associated PKC, namely the activated form of PKC[19]. The data indicate that the PKC system appears to be ischemia-resistant as compared to the AC system. Severe brain ischemia is known to induce a marked increase in the cerebral contents of diacylglycerol and arachidonic acid[20], both of which are potent activators of PKC[21]. Such production of PKC activators as well as a massive release of neuro-transmitters (excitatory amino acids, noradrenaline, etc.) which ordinarily stimulate the phosphoinositide system, may cancel out the effect of the Ca^{2+}-activated proteolysis during cerebral ischemia, resulting in a relatively well-maintained PDBu binding as observed in the present study.

The different patterns of changes in second-messenger activities in response to reduction of the CBF in each cerebral region might closely reflect the regionally specific pathophysiology of the ischemic tissue injury. A molecular analysis of the components of the second-messenger systems such as each key enzyme and guanine nucleotide-binding protein during ischemia will be required in order to gain a better understanding of the data obtained in the present study.

REFERENCES

1. Tanaka K, Gotoh F, Gomi S, et al. Autoradiographic analysis on second-messenger systems and local cerebral blood flow in ischemic gerbil brain. J Cereb Blood Flow Metab 1991;11:283-292
2. Pulsinelli WA. Selective neuronal vulnerability:Morphological and molecular characteristics. In: Kogure K, Hossmann K-A, Siesjo BK, et al. eds. Molecular Mechanisms of Ischemic Brain Damage, Progress in Brain Research, Vol 63, Amsterdam, Elsevier Science Publishers, 1985:29-37
3. McGraw CP. Experimental cerebral infarction. Effects of pentobarbital in mongolian gerbils. Arch Neurol 1977;34:334-336
4. Sakurada O, Kennedy C, Jehle J, et al. Measurement of local cerebral blood flow with iodo[^{14}C]antipyrine. Am J Physiol 1978;234:H59-H66
5. Tanaka K, Gotoh F, Ishihara N, et al. Autoradiographic analysis of second-messenger systems in the gerbil brain. Brain Res Bull 1988;21:693-700
6. Astrup J. Thresholds in cerebral ischemia-The ischemic penumbra. Stroke 1981;12:723-725
7. Crockard HA, Gadian DG, Frackowiak RS, et al. Acute cerebral ischemia: Concurrent changes in cerebral blood flow, energy metabolites, pH, and lactate measured with hydrogen clearance and ^{31}P and ^{1}H nuclear magnetic resonance spectroscopy. II. Changes during ischaemia. J Cereb Blood Flow Metab 1987;7:394-402
8. Naritomi H, Sasaki M, Kanashiro M, et al. Flow thresholds for cerebral energy disturbance and Na^+ pump failure as studied by in vivo ^{31}P and ^{23}Na nuclear magnetic resonance spectroscopy. J Cereb Blood Flow Metab 1988;8:16-23
9. Graybiel AM. Neurotransmitters and neuromodulators in the basal ganglia. Trends Neurosci 1990;13:244-254
10. Shimada N, Graf R, Rosner G, et al. Ischemic flow threshold for extracellular glutamate increase in cat cortex. J Cereb Blood Flow Metab 1989;9:603-606
11. Globus MY-T, Busto R, Dietrich WD, et al. Effect of ischemia on the in vivo release of striatal dopamine, glutamate, and γ-aminobutyric acid studied by intracerebral microdialysis. J Neurochem 1988;51:1455-1464
12. Choi DW. Methods for antagonizing glutamate neurotoxicity. Cerebrovasc Brain Metab Rev 1990;2:105-147
13. Kebabian JW, Calne DB. Multiple receptors for dopamine. Nature 1979;277:93-96
14. Dreher JK, Jackson DM. Role of D_1 and D_2 dopamine receptors in mediating locomotor

activity elicited from the nucleus accumbens of rats. Brain Res 1989;487:267-277

15.   Sarna GS, Obrenovitch TP, Matsumoto T, et al. Effect of transient cerebral ischaemia and cardiac arrest on brain extracellular dopamine and serotonin as determined by in vivo dialysis in the rat. J Neurochem 1990;55:937-940

16.   Hagberg H, Lehmann A, Sandberg M, et al. Ischemia-induced shift of inhibitory and excitatory amino acids from intra- to extracellular compartments. J Cereb Blood Flow Metab 1985;5:413-419

17.   Globus MY-T, Busto R, Dietrich WD, et al. Direct evidence for acute and massive norepinephrine release in the hippocampus during transient ischemia. J Cereb Blood Flow Metab 1989;9:892-896

18.   Bhardwaj A, Brannan T, Martinez-Tica J, Weinberger J. Ischemia in the dorsal hippocampus is associated with acute extracellular release of dopamine and norepinephrine. J Neural Transm [GenSect] 1990;80:195-201

19.   Olds JL, Anderson ML, McPhie DL, et al. Imaging of memory-specific changes in the distribution of protein kinase C in the hippocampus. Science 1989;245:866-869

20.   Ikeda M, Yoshida S, Busto R, et al. Polyphosphoinositides as a probable source of brain free fatty acids accumulated at the onset of ischemia. J Neurochem 1986;47:123-132

21.   Huang K-P. The mechanism of protein kinase C activation. Trends Neurosci 1989;12:425-432

INCREASED PHORBOL ESTER BINDING AND DECREASED FORSKOLIN BINDING IN ISCHAEMIC CORTEX AFTER SUBDURAL HAEMATOMA IN THE RAT: AN IN VITRO AUTORADIOGRAPHIC STUDY

Yasuhiro Kuroda, Deborah Dewar, Ross Bullock

University Department of Neurosurgery, and Wellcome
Surgical Institute, Garscube Estate, Bearsden Road
Glasgow, Scotland

INTRODUCTION

An acute intracranial haematoma develops in about 45% of severely head injured patients, and dramatically worsens the prognosis, because secondary ischaemic brain damage develops.[1] Although the frequency of ischaemic damage is much more difficult to assess in patients who survive an intracranial haematoma, only about one third of patients will recover fully even though surgery is nowadays performed within a few hours of injury, in most cases.

We have recently devised a rat model of acute subdural haematoma, producing a zone of focal ischaemic damage, affecting 14% of the hemisphere under the haematoma, which closely replicates the damage seen after a subdural haematoma in humans.[2] We have shown that this ischaemic damage is reduced by 54% with pretreatment using an NMDA antagonist.[3]

In the study reported here, we have employed this model to measure [3]H Forskolin, ^{3}H phorbol ester and beta adrenergic and muscarinic cholinergic receptor binding, to evaluate acute changes in receptor - coupled second messenger systems. Changes in intracellular second messenger systems may be crucial in determining the effect of early drug therapy after acute ischaemia.

MATERIALS AND METHODS

Ten adult male Sprague-Dawley rats weighing 425-570 g. were studied. The rats were divided into two groups; subdural haematoma (n=6) and sham operation (n=4). Anaesthesia was induced, (nitrous oxide - 70%, oxygen - 30% halothane - 2%) and tracheostomy and controlled ventilation were initiated; animals were fully monitored, and normothermia maintained.

The Role of Neurotransmitters in Brain Injury, Edited by
M. Globus and W.D. Dietrich, Plenum Press, New York, 1992

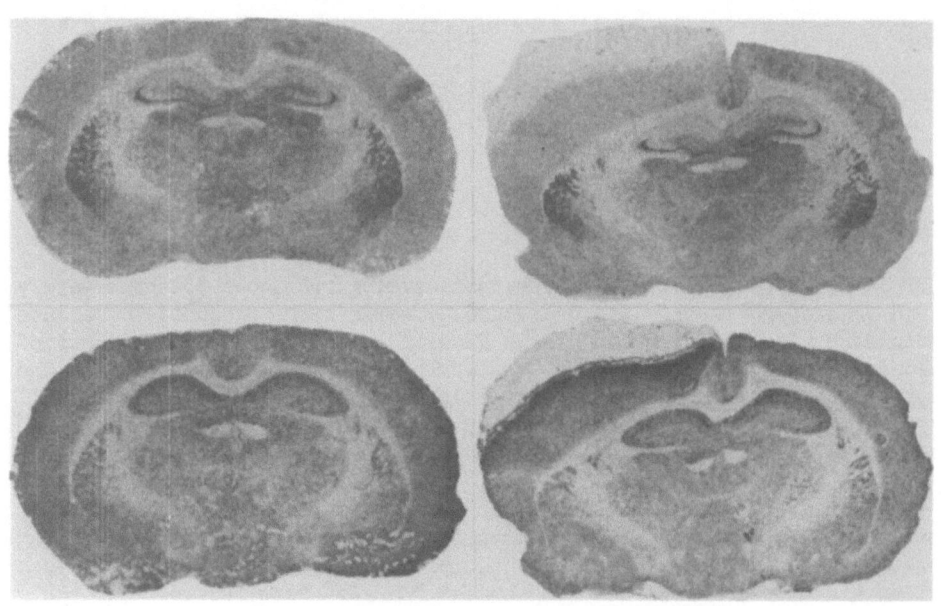

Figure 1.   Top row.   ³H Forskolin binding in a sham-operated control
(left) and two hours after subdural haematoma (right). Note
the decrease in binding in cortex under the haematoma.
(arrow)
Bottom row.   ³H Phorbol ester binding in a sham-operated
control (left) and two hours after subdural haematoma
(right). Note increased binding under the haematoma. (arrow)

An acute subdural haematoma was made in six animals, using the method we have reported previously.[2] Autologous venous blood was injected into the subdural space via a parietal burr hole, and after injection, the induction needle was cut off and sealed and the scalp was sutured. In the sham group, the same procedure was performed without injection of blood.

After two hours the anaesthetised animals were decapitated and the brain was quickly removed and frozen at -42°C. Coronal forebrain sections 20 um thick were cut on a cryostat and mounted onto subbed slides. Five sections selected from each anatomical structure were processed; four sections for binding studies; one for histology (stained with hematoxylin and eosin).

Second messenger binding

Adenylate cyclase (AC) and stimulatory G proteins (Gs) were studied using [^{3}H] forskolin. Sections were incubated in 50 mM Tris-HCl (pH 7.5) containing 10 mM $MgCl_2$ and 20 nM [^{3}H] forskolin (38 Ci/mmol), for 20 minutes. Adjacent sections were incubated in 20 uM forskolin to define non-specific binding.

Protein kinase C (PKC) was quantified using [^{3}H]phorbol 12,13 dibutyrate (PDBu). Sections were incubated in 50 mM Tris-HCl (pH 7.7) containing 1 mM $CaCl_2$ and 2.5 nM [^{3}H]PDBu (19.1 Ci/mmol) for 90 minutes. Adjacent sections were incubated in the presence of 2 uM PDBu to define non-specific binding.

Muscarinic cholinergic and Beta Adrenergic receptor binding

To establish whether receptors which operate through protein kinase C and adenylate cyclase were structurally intact two hours after an acute subdural haematoma, binding studies were performed using ^{3}H quinuclidyl benzylate (^{3}H QNB) and ^{3}H dihydroalprenolol (^{3}H DHA). For ([^{3}H]QNB) binding, sections were incubated in 50 mM Phosphate buffer (pH 7.4) and 2 nM [^{3}H]QNB (41.5 Ci/mmol) for 120 minutes. Atropine 2 um was used to define non-specific binding. [^{3}H] DHA binding was studied by incubating sections in 170 mM Tris-HCl (pH 7.7) containing 10 mM $MgCl_2$ and 2 nM [^{3}H] DHA (83 Ci/mmol) for 40 minutes. Two uM propranolol was used to define non-specific binding. The slides were apposed to Hyperfilm-[^{3}H] (Amersham) along with a set of [^{3}H]-microscales (Amersham) for five days (QNB), seven days (PDBu), 14 days (DHA) and 25 days (Forskolin), respectively. Autoradiograms were analysed by computer-assisted densitometry (Quantimet 970, Cambridge Instruments) comparing with the histological section.

RESULTS

There were no significant differences in mean physiological variables between groups.

The patterns of ^{3}H forskolin and [^{3}H]PDBu binding seen at two hours in control and subdural haematoma animals, are shown in Figure 1. for controls, the binding density for ^{3}H forskolin was highest in the caudate nucleus, and for [^{3}H]PDBu it was highest in the superficial cortex. A significant reduction (72%) in ^{3}H forskolin binding was seen in parietal cortex, (most marked in layers IV-V) when comparison is

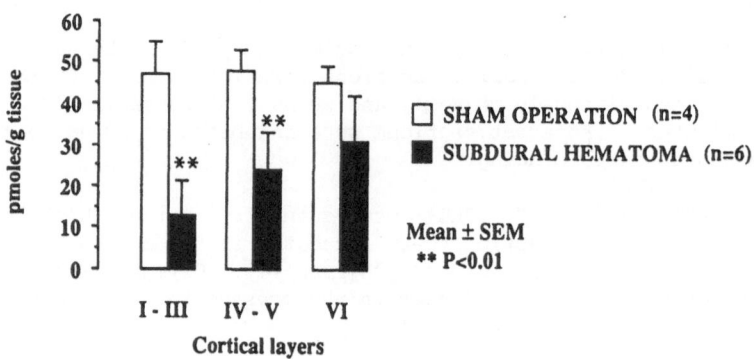

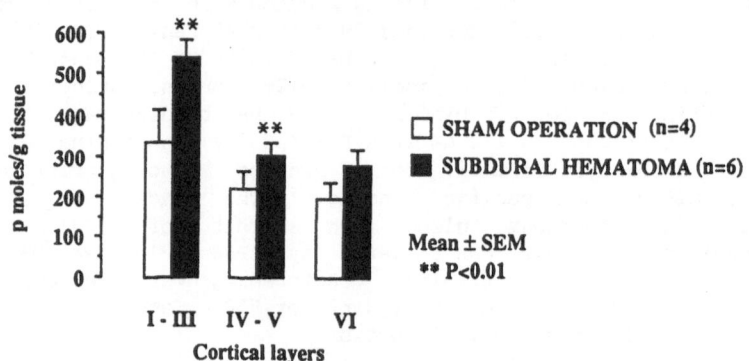

Figure 2.   Top.   ³H Forskolin binding in cortex after subdural haematoma.
Bottom.   ³H Phorbol ester binding in cortex after subdural haematoma.   (unpaired t-test)

made with controls.(Figure 2)   In all other grey matter structures studied, ^{3}H forskolin binding was increased. Increases were most marked for limbic structures.

[3]H PDBu binding was significantly increased (62%) in all layers of the ischaemic parietal cortex under the subdural haematoma and also in posterior cingulate and contralateral parietal cortex, layer VI.(Fig.2)

The highest density of [^{3}H]QNB binding was observed in cerebral cortex and hippocampus (CA1 and molecular layer of dentate gyrus). In the sham group there were no significant differences between sham operated and subdural haematoma groups. The highest density of [^{3}H]DHA binding in controls was observed in the cortex and caudate nucleus. There were no significant differences between sham and subdural animals.

DISCUSSION

Severe head injury currently kills and disables more young people than any other disease category in the USA. Recent animal studies show that outcome after both fluid percussion trauma to the brain, and subdural haematoma can be profoundly influenced by receptor mediated processes.[3,6,7]

Although certain receptors directly regulate ion channels (eg. the NMDA receptor) others, such as muscarinic, B_1 adrenergic and metabotropic glutamate receptors, act via intracellular second messenger systems, such as the adenylate cyclase/G protein system, and protein kinase C.[5,8,9] The functional integrity of these second messenger systems may thus crucially determine outcome and may be a major limiting factor determining the efficacy of drug therapy.

The relationship of both [3]H Forskolin and [3]H Phorbol ester binding to enzymic activity of the adenylate cyclase complex and protein kinase C is not yet fully resolved.[8,9,10] However, several authors have shown that [3]H Forskolin binding represents the activated adenylate cyclase/quanine nucliotide binding protein subunit.[8,10] In this study, [3]H Forskolin binding was significantly reduced in the densely ischaemic cortex under the haematoma, at two hours. This accords with the findings after both focal and global ischaemia, in other models.[8,10]

During activation of protein kinase C, translocation of the active site from the cytosol to cell membrane takes place, and this highly calcium dependent process occurs early after an ischaemic event.[9,10] This is associated with increased [3]H Phorbol ester binding.[11] In this study, a significant increase of 62% in Phorbol binding in the ischaemic cortex occurred.(Figures 1 and 2) In the hippocampus, Phorbol binding increased by 46% (molecular layer). This accords with the findings of Onodera et al) who demonstrated increased Phorbol binding early after ischaemia in tissues which were destined to undergo delayed ischaemic necrosis.[11] This supports the hypothesis that a glutamate mediated process may induce calcium flux within neurons and glia beneath the subdural haematoma, and in the hippocampus, in this model.[7]

REFERENCES

1.   R Bullock, GM Teasdale.   Head Injuries - Surgical Management:
          Traumatic intracranial haematomas.   In: Vinken and Bruyn's
          Handbook of Clinical Neurology: Vol. 24 - Head Injury.

Ed. Braakman R. Elsevier Science Publishers, Amsterdam. Chapter 10. pp.249-298, (1991)

2. JD Miller, R Bullock, DI Graham, M-H Chen, GM Teasdale: Ischaemic brain damage in a model of acute subdural haematoma. Neurosurgery. 27: 433-439, (1990)

3. M-H Chen, R Bullock, DI Graham, JD Miller, J McCulloch. Ischaemic neuronal damage after acute subdural haematoma in the rat: Effects of pretreatment with a glutamate antagonist. J Neurosurg. 74: 944-950, (1991)

4. DR Gehlert, TM Dawson, HI Yamamura, JK Wamsley: Quantitative autoradiography of [^{3}H]forskolin binding sites in the rat brain. Brain Res. 361: 351-360, (1985)

5. PF Worley, JM Baraban, EB De Souza, SH Snyder: Mapping second messenger systems in the brain: Differential localisations of adenylate cyclase and protein kinase C. Proc. Natl. Acad. Sci. USA 83: 4053-4057, (1986)

6. Y Katayama, DP Becker, T Tamura, D Hovda: Massive increases in extracellular potassium and the indiscriminate release of glutamate following concussive brain injury. J. Neurosurg. 73: 889-900, (1990)

7. R Bullock, SP Butcher, M-H Chen, L Kendall, J McCulloch: Correlation of the extracellular glutamate concentration with the extent of blood flow reduction after subdural haematoma in the rat. J. Neurosurg. 74: 794-802, (1991)

8. H Onodera, K Kogure: Mapping second messenger systems in the rat hippocampus after transient forebrain ischaemia: in vitro [^{3}H]forskolin and [^{3}H]inositol 1,4,5-trisphosphate binding. Brain Res. 487: 343-349, (1989)

9. M Cardell, H Bingren, T Wieloch, J Zivin, T Saitoh. Protein kinase C is translocated to cell membranes during cerebral ischaemia. Neurosci. Lett. 119L 228-232, (1990)

10. K Tanaka, F Gotoh, S Gomi, S Takeshima, B Mihara: Autoradiographic analysis on second messenger systems and local cerebral blood flow in the ischaemic gerbil brain. J. Cereb Blood Flow Metab. 11: 283-291, (1991)

11. H Onodera, T Araki, K Kogure. Protein kinase C activity in the rat hippocampus after forebrain ischaemia: autoradiographic analysis by [^{3}H]phorbol 12,13-dibutyrate. Brain Res. 481: 1-7, (1989)

FREE FATTY ACID AND DIACYLGLYCEROL LEVELS ARE RELATED TO

CEREBRAL O_2 DURING SEIZURES

L.L. Rihn, F. Visioli, E.B. Rodriguez de Turco, N.R. Kreisman, and
N.G. Bazan

Department of Physiology, Tulane University School of Medicine, and
LSU Eye Center and Neuroscience Center, Louisiana State University
School of Medicine, New Orleans, LA 70112, USA

INTRODUCTION

Seizure-induced Changes in Free Fatty Acid and Diacylglycerol Levels

It is well established that levels of free fatty acids (FFA) and diacylglycerols
(DAG) are elevated during seizures and that these are derived primarily from
membrane phospholipids that are degraded via activation of phospholipases A_1, A_2,
and C (Bazan, 1970; Siesjö et al, 1982; Bazan et al., 1983; 1986; Rodriguez de Turco,
1986). Arachidonic acid, which is the primary fatty acid released during seizures, can
be metabolized further by cyclooxygenase and lipoxygenase to generate prostaglandins,
leukotrienes, and thromboxanes. These factors can alter intracellular signalling,
generate free radicals and hydroperoxides, and damage cell membranes. Reacylation
of FFA and DAG to phospholipids requires energy in the form of ATP from oxidative
metabolism. Therefore, it would be expected that a deficiency of oxygen delivery
during seizures should enhance levels of FFA and DAG. The objective of the present
investigation was to determine if levels of FFA and DAG are related to changes in
cerebral oxygenation during the course of recurrent seizures. Furthermore, an attempt
was made to determine the relative contribution of reacylation of FFA and reutilization
of DAG to the levels observed during and between seizures (i.e., during ictal and
interictal periods, respectively).

Changes in Cerebral Blood Flow and Oxygen Delivery During Recurrent Seizures

Seizures are accompanied by profound increases in cerebral metabolic rate
(Meldrum and Nilsson, 1976). If ischemia is to be prevented, cerebral blood flow
(CBF) and oxygen delivery must increase to meet the enhanced metabolic requirements of brain tissue. It has been confirmed repeatedly that CBF and oxygen delivery
increase more than does metabolic rate during individual seizures, as indicated by a
rise in cerebral tissue PO_2 and an increase in the ratios of NAD^+ to NADH and
oxidized to reduced cytochrome aa_3 (Plum et al., 1968; Jöbsis et al., 1971; Caspers and
Speckmann, 1972; Mayevsky and Chance, 1975; Vern et al., 1976; Hempel et al., 1980;

Kreisman et al., 1981b; Pinard et al., 1984; Kreisman et al., 1991). During serial seizures, however, cerebral blood flow and oxygen delivery do not always meet metabolic demand, as indicated by a seizure-associated fall in brain tissue PO_2 and a decrease in the ratio of oxidized to reduced cytochrome aa_3 (Caspers and Speckmann, 1972; Kreisman et al, 1981b; 1991). Kreisman et al. (1991) showed that a critical blood flow increase of 175-200% of control levels was required during seizures to prevent a relative hypoperfusion of the cerebral cortex (Fig. 1). In the present experiments, changes in levels of free fatty acids and diacylglycerols were measured during ictal and interictal episodes which were associated with changes in cerebral oxygenation as described above.

METHODS

Male Wistar rats (250-350 g) were anesthetized with sodium pentobarbital (50 mg/kg i.p.), intubated, and paralyzed with d-tubocurarine (10 mg/kg i.p.). The animals were respired with a mixture of 30% O_2 and 70% N_2 using a positive pressure ventilator, which was adjusted in stroke volume and rate to yield blood gas values for PaO_2 between 100 and 140 mm Hg, $PaCO_2$ between 35 and 40 mm Hg, and pH between 7.35 and 7.45. Body temperature was maintained at 36-37°C with a heating pad. Heparinized femoral arterial and venous cannulae were inserted for measurement of blood pressure and injection of drugs or fluids, respectively. Supplemental doses of pentobarbital were administered periodically to maintain a surgical level of anesthesia. The EEG was monitored from the dura with two small pin jacks, which were fitted into small holes drilled into the skull.

Optical measurements of relative changes in the oxidation/reduction state of cytochrome aa_3 were made in vivo at 605-590 nm using a dual-wavelength reflectance spectrophotometer as described previously (Kreisman et al., 1981a). Optical measurements were made through the intact skull from a field 3.2 mm across and approximately 1-2 mm deep.

Pentylenetetrazol (PTZ; in a 10% solution with pH adjusted to 7.40) was injected intravenously as a bolus of 0.02 ml every 20 s, until an individual seizure was evoked. Auditory stimulation was used in conjunction with PTZ to help trigger generalized seizures. This procedure was repeated at intervals of 5-7 min to evoke up to 15 seizures. Withdrawal of 1-3 ml of arterial blood during seizures and addition of 3-5 cm of positive-end expiratory pressure helped to avoid secondary complications, such as prolonged increases in blood pressure and pulmonary edema.

Brain enzymes were deactivated rapidly by head-focussed microwave irradiation in situ. The brain was rapidly removed and cortex was dissected and homogenized in chloroform:methanol (2:1, by vol) with an Ultraturrax Polytron. Lipids were extracted according to Folch (Folch et al., 1957) and free fatty acids and diacylglycerols were separated by thin-layer chromatography using a system of hexane:ether:acetic acid (45:55:1.3, by vol) as eluent. Spots were visualized by spraying the plates with 2'7'-dichlorofluorescein (0.005% in methanol), then the spots were scraped, and the fatty acids were derivatized to methylesters (FAMEs) using BF_3 in methanol (Aveldaño and Bazan, 1973). FAMEs were quantified by gas-liquid chromatography.

Rats were divided into three experimental groups. One group (n=6) received surgery but was not given PTZ. The second group sustained seizures as described above but was sacrificed in the interictal period preceding the third seizure (n=6) or during the third seizure (n=6). The third group was sacrificed after 2 h of serial seizures during the interictal period (n=9) or during the final seizure (n=15).

RESULTS

Seizure-Associated Increases in Cerebral Cortical Blood Flow and Oxygenation Attenuate as Seizures Recur

Early seizures were accompanied by large increases in CBF and cortical oxygen levels (Fig. 1). The seizure-associated rises in CBF were the result of both an increase in mean arterial blood pressure (MABP) and a decrease in cerebrovascular resistance (not shown). As seizures recurred, the associated increases in MABP diminished in magnitude, resulting in attenuated increases in both CBF and cortical oxygenation. In 65% of the animals, cortical oxygenation continued to increase with subsequent seizures. In the remaining 35% of the animals, cortical oxygenation decreased during later seizures because the associated rise in CBF failed to exceed 200% of control values (e.g. Fig. 1). The major factor contributing to the attenuation of seizure-associated rises of CBF was attenuation, and often reversal, of the MABP responses.

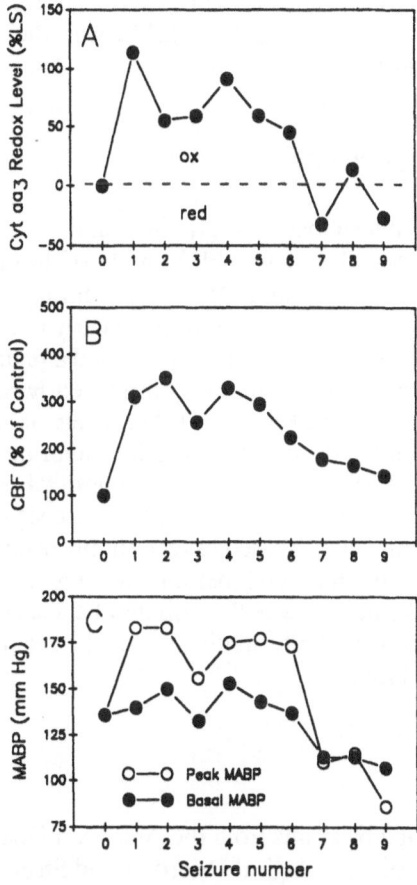

Fig. 1. Example of changes in the oxidation/reduction level of cytochrome aa₃ (A), CBF (B), and MABP (C) during serial seizures induced with pentylenetetrazol (PTZ) in a rat anesthetized with sodium pentobarbital (see Kreisman et al., 1991). Abbreviations: CBF, cerebral blood flow; MABP, mean arterial blood pressure, LS, labile signal; ox, oxidation; red, reduction. Seizure number zero represents the pre-seizure control period.

Cortical FFA and DAG Levels Change in Relation to Both Time and Degree of Oxygenation

Early seizures were accompanied by a large increase in both FFA and DAG levels. FFA levels were elevated to a lesser degree during interictal periods, suggesting that ictally generated FFAs were partially removed or reacylated following each seizure. DAG levels remained relatively stable both ictally and interictally during the early seizure period.

During late seizures, the rise in FFA levels was diminished, compared to levels observed during early seizures. Also, FFA levels returned to control levels during the late interictal periods. FFA levels rose more during late seizures accompanied by decreased cerebral oxygenation than during late seizures accompanied by increased cerebral oxygenation.

Cortical DAG levels were also sensitive to levels of oxygenation. During later seizures, DAGs increased during both ictal and interictal periods only if cortical oxygenation fell. In contrast, DAG levels remained stable during later seizures if cortical oxygenation increased.

The greatest changes in both FFAs and DAGs were in arachidonic and stearic acids. This suggests that breakdown of membrane inositol phospholipids, which are known to be enriched with these two fatty acids, was the primary source of the increases in total FFA and DAG.

SUMMARY AND CONCLUSIONS

Seizures activate phospholipases A_1, A_2, and C, causing breakdown of membrane phospholipids and elevation of FFA and DAG levels in rat cerebral cortex. Differences were observed in FFA and DAG levels during recurrent seizures. FFA levels increased during the ictal period but returned nearly to control values during the interictal period. In contrast, DAG levels rapidly reached a plateau and failed to return to basal levels during the interictal period. During early seizures, phospholipase activation and the resulting rise in FFA and DAG levels occurred despite adequate cerebral oxygenation. Higher acyltransferase activity and/or an increased CBF may be responsible for the lowered FFA levels during the interictal period.

However, during late seizures, these processes were significantly affected by the oxygenation state of the brain: a decreased oxygen supply resulted in elevation of FFA and DAG levels during the interictal period, possibly as a result of a slowed reutilization or removal. We hypothesize that this may be a consequence of a mismatch between the need for energy, in the form of ATP for the reacylation process, and the delivery of oxygen to the brain.

ACKNOWLEDGMENT

This research was supported by grants from the American Heart Association (89-902) and the National Institute of Neurological Disorders and Stroke, National Institutes of Health (NS23002).

REFERENCES

Auer, R.N., and Siesjö, B.K., 1988, Biological differences between ischemia, hypoglycemia, and epilepsy, Ann. Neurol. 24: 600-704.

Aveldaño, M.I., and Bazan, N.G., 1973, Fatty acids composition and level of diacylglycerol and phosphoglycerides in brain and retina, Biochim. Biophys. Acta 296: 1-9.

Bazan, N.G., 1970, Effects of ischemia and electroconvulsive shock on free fatty acid pool in the brain, Biochim. Biophys. Acta 218: 1-10.

Bazan, N.G., Birkle, D.L., Tang, W., and Reddy, T.S., 1986, The accumulation of free arachidonic acid, diacylglycerols, prostaglandins, and lipoxygenase reaction products in the brain during experimental epilepsy, Adv. Neurol. 44:879-902.

Bazan, N.G., Rodriguez de Turco, E.B., and Morelli de Liberti, S.A., 1983, Free arachidonic acid and membrane lipids in the central nervous system during bicuculline-induced status epilepticus, Adv. Neurol. 34: 305-310.

Caspers, H., and Speckmann E.J., 1972, Cerebral pO_2, pCO_2, and pH changes during convulsive activity and their significance for spontaneous arrest of seizures, Epilepsia 13: 699-725.

Folch, J., Lees, M., and Sloane Stanley, G.H., 1957, A simple method for the isolation and purification of total lipids from animal tissues, J. Biol. Chem. 226: 497-509.

Hempel, F.G., Kariman K., and Saltzman H.A., 1980, Redox transitions in mitochondria of cat cerebral cortex with seizures and hemorrhagic hypotension, Am. J. Physiol. 238: H249-H256.

Jöbsis, F.F., O'Connor M., Vitale A., and Vreman H., 1971, Intracellular redox changes in functioning cerebral cortex. I. Metabolic effects of epileptiform activity, J. Neurophysiol. 34: 735-749.

Kreisman, N.R., Sick, T.J., LaManna, J.C., and Rosenthal, M., 1981a, Local tissue oxygen tension-cytochrome aa_3 redox relationship in rat cerebral cortex in vivo, Brain Res. 218: 161-174.

Kreisman, N.R., LaManna, J.C., Rosenthal, M., and Sick, T.J., 1981b, Oxidative metabolic responses with recurrent seizures in rat cerebral cortex: Role of systemic factors, Brain Res. 218:175-188.

Kreisman, N.R., Rosenthal, M., Sick T.J., and LaManna, J.C., 1983, Oxidative metabolic responses during recurrent seizures are independent of convulsant, anesthetic, or species, Neurology 33: 861-867.

Kreisman, N.R., Magee, J.C., and Brizzee, B.L., 1991, Relative hypoperfusion in rat cerebral cortex during recurrent seizures, J. Cereb. Blood Flow Metab. 11:77-87.

Mayevsky, A., and Chance B., 1975, Metabolic responses of the awake cerebral cortex to anoxia, hypoxia, spreading depression and epileptiform activity, Brain Res 98: 149- 165.

Meldrum, B.S., and Nilsson, B., 1976, Cerebral blood flow and metabolic rate early and late in prolonged epileptic seizures induced in rats by bicuculline, Brain 99: 523-542.

Pinard, E., Tremblay E., Ben-ari Y., and Seylaz J., 1984, Blood flow compensates oxygen demand in the vulnerable CA3 region of the hippocampus during kainate-induced seizures, Neuroscience 13: 1039-1049.

Plum, F., Posner J.B., and Troy B., 1968, Cerebral metabolic and circulatory responses to induced convulsions in animals, Arch. Neurol. 18: 1-13.

Rodriguez de Turco, E.B., 1986, Drugs affecting membrane lipid catabolism: the brain free fatty acids effect, in: Phospholipids Research and the Nervous System, Biochemical and Molecular Pharmacology, L.A. Horrocks et al., eds., Liviana Press, Padova, pp.57-66.

Siesjö, B.K., Ingvar, M., and Westerberg, E., 1982, The influence of bicuculline-induced seizures on free fatty acid concentrations in cerebral cortex, hippocampus, and cerebellum, J. Neurochem., 39: 796-802.

Vern, B., Schuette W.H., Whitehouse W.C., and Matsuga N., 1976, Cortical oxygen consumption and NADH fluorescence during metrazol seizures in normotensive and hypotensive cats, Exp. Neurol. 52: 83-98.

TRAUMATIC BRAIN INJURY ALTERS CYCLIC AMP

SECOND MESSENGER SYSTEMS IN RAT BRAIN

Bruce G. Lyeth and Ji Y. Jiang

Division of Neurosurgery
Medical College of Virginia
Richmond, Virginia 23298

INTRODUCTION

Neurochemical alterations related to abnormal agonist-receptor interactions mediate important components of brain pathophysiology associated with TBI (Hayes et al., 1986, in press) The mechanical forces applied to the brain by a traumatic insult may initiate a widespread neuronal depolarization which produces a large non-specific release of neurotransmitters. Membrane depolarization may be a phenomenon common to all forms of mechanical trauma whether focal or generalized and may contribute to long term changes in brain information flow (Hayes et al., 1989). The resulting abnormal activation of receptors could produce changes in intracellular signal transduction pathways resulting in transient, long lasting, or irreversible alterations in cell function. Functional deficits arising from TBI could be similar regardless of whether information flow pathways were disrupted by sublethal disruptions in cell signaling or interrupted by cell death or axotomy (Lyeth at al., 1990).

A number of receptor systems are involved in TBI mechanisms including: glutamate, acetylcholine, opioids, and catecholamines. While some neurotransmitters directly modulate ion channels, others act through second messenger systems such as cyclic AMP, cyclic GMP, calcium, and diacylglycerol. The magnitude to which a neurotransmitter influences cellular function depends to a large extent upon the degree of receptor-linked signal amplification produced by second messenger systems (Worley et al., 1986). Thus, determining the localization and magnitude of TBI-induced alterations in second messenger systems associated with pertinent neurotransmitters may reveal important information about the roles of these systems in TBI.

Opioids and catecholamines are known to either inhibit or activate the enzyme adenylate cyclase which is responsible for the synthesis of the intracellular messenger, cyclic AMP. We used autoradiographic techniques with [^{3}H]-forskolin to examine adenylate cyclase activation in brain regions in the mid-dorsal coronal plane of the rat following moderate fluid percussion TBI.

The Role of Neurotransmitters in Brain Injury, Edited by
M. Globus and W.D. Dietrich, Plenum Press, New York, 1992

MATERIALS AND METHODS

Male Sprague-Dawley rats, weighing 300 to 325 gm, were surgically prepared for fluid-percussion brain injury under sodium pentobarbital anesthesia (54 mg/kg, i.p.) by attaching a hollow injury tube (a modified Leur-loc syringe hub, 2.6 mm inside diameter) epidurally over a 4.8 mm craniectomy on the sagittal suture midway between the lambda and bregma.

Twenty-four hours after surgery rats were delivered a moderate fluid percussion brain injury (2.2 atmospheres) or sham injury under methoxyflurane anesthesia.

At 3 hours after injury (n=7) or sham injury (n=7) rats were intubated and ventilated with 2% isoflurane in 70% N_2O / 30% O_2 and the brain frozen <u>in situ</u> with liquid nitrogen. The brains were removed in a cold glove box (-20°C), blocked, and mounted on cryostat chucks.

Autoradiographic localization of forskolin binding was performed according to the method of Worley et al.(1986). Ten micron coronal brain sections from the mid-dorsal hippocampus (Bregma -3.8) were cut on a cryostat and thaw-mounted onto gelatin coated slides. Sections were incubated in 50 mM tris HCl (pH 7.7), 100 mM NaCl, 5 mM $MgCl_2$, 10 nM [^3H]-forskolin (40.0 ci/mmol) for 10 minutes at room temperature. Sections were then washed 2 min twice in buffer and dipped in distilled H_2O, dried under a stream of room temperature dry air, and apposed to LKB ultrafilm for 4 weeks at room temperature. Non specific binding was calculated in the presence of unlabeled 10uM forskolin. Optical density of silver grains on ultrafilm were quantified by computer assisted microdensitometry.

RESULTS

[^3H]-forskolin binding in hippocampal formation and cortex was significantly lower in TBI rats compared to sham-injured rats at 3 hours after injury.

Table 1

Brain Region	Percent Change from Control
Hippocampus	
CA1 stratum oriens	-53% **
CA1 stratum pyramidale, stratum radiatum, & stratum lacunosm moleculare	-51% **
CA3 stratum pyramidale	-23% *
CA4 stratum pyramidale	-39% **
Dentate Gyrus	-44% **
Cortex	-25% **

Changes in [^3H]-forskolin binding in rat brain sections (Bregma -3.8) compared to sham-injured controls. * p< 0.03, ** p< 0.01 Students t-test

DISCUSSION

Moderate fluid percussion TBI produced large significant decreases in [³H]-forskolin binding in the mid-dorsal hippocampal formation and in the overlying cortex at 3 hours after injury. These data suggest that TBI produces marked alterations in intracellular signal transduction systems utilizing cyclic AMP in these regions.

Cyclic AMP is a second messenger coupled with several neurotransmitter systems including: alpha$_2$ adrenergic, beta adrenergic, dopamine, serotonin, and adenosine. Catecholaminergic systems have recently been show to play important roles in experimental TBI pathophysiology. For example, d-amphetamine administered 3 hours after fluid percussion TBI enhanced behavioral recovery (Romhanyi et al., 1990) and metabolic recovery (Tandian et al., 1990). In contrast to catecholamines, opioids are negatively coupled to cyclic AMP. Endogenous opioids have also been implicated in mechanisms of TBI injury. Activation of kappa opioid receptors has been shown to exacerbate TBI pathophysiology (McIntosh et al., 1987) while activation of mu opioid receptor subtypes have protective effects (Hayes et al., 1990).

In this study we focussed our analysis on cAMP changes in the hippocampal formation. A number of studies implicate the hippocampus (and especially the CA1 sector) as one of the most vulnerable brain region to experimental TBI. For example, TBI increases dorsal hippocampi CA1 pyramidal neuron vulnerability to a controlled post-traumatic ischemic insult even when delayed up to 24 hours after TBI (Jenkins et al., 1988, 1989). Moreover, long-term potentiation of the hippocampal Schaeffer collateral/CA1 system is also altered following TBI (Miazaki et al., 1989) suggesting altered synaptic function in the CA3-CA1 system. Lastly, changes in NMDA (Miller et al., 1990) and opioid (Perry et al., in press) receptor binding are significantly altered at 3 hours after moderate TBI in the CA1. Thus, pathophysiological changes within the CA1 sector of the hippocampus appear to be a prominent feature of TBI.

Forskolin binds to adenylate cyclase at a low and high affinity binding site. It is believed that neurotransmitter receptor activation stimulates a G$_s$-protein which in turn stimulates adenylate cyclase and unmasks the high affinity binding site for forskolin on the catalytic subunit (Nelson and Seamon, 1988). Thus presumably, an increase in forskolin binding would suggest a recent activation of adenylate cyclase coupled neurotransmitter receptors. However, one must be cautious in ascribing forskolin binding as solely representing adenylate cyclase activity. One must also consider that forskolin also labels glucose transporter protein (Sergate and Kim, 1985) and that forskolin binding sites in the CA1 may not be regulated by G proteins (Poat et al., 1988).

The decreased forskolin binding in the hippocampal formation, an area that appears to be particularly vulnerable to TBI insults, and the cortex suggests that normal cell signaling processes are perturbed for at least 3 hours after injury. The decreased binding may be related to any of several mechanisms: 1. diminished release of neurotransmitters that normally activate adenylate cyclase, 2. altered coupling efficiency of neurotransmitter-cyclic AMP 2nd messenger systems, 3. down regulation of receptors positively coupled to cyclic AMP, 4. increased release of neurotransmitters that normally inhibit adenylate cyclase (e.g. opioids), 5. diminished synthesis or enhanced breakdown of constituents of adenylate cyclase. In any case, the decreased binding observed in the hippocampus following TBI may reflect cellular dysfunction involving the cAMP second messenger system which may contribute to pathophysiological processes associated with injury.

ACKNOWLEDGEMENTS

This research was supported by Grants NS 21587 from The National Institutes of Health and Grant H133B80029 from the National Institute on Disability and Rehabilitation Research, the U.S. Department of Education.

REFERENCES

Hayes R.L., Stonnington. H.H., Lyeth, B.G., Dixon, C.E., and Yamamoto, T., 1986, Metabolic and neurophysiological sequelae of brain injury: A cholinergic hypothesis, J. CNS Trauma 3:163-173.

Hayes, R.L., Lyeth, B.G., and Jenkins, L.W., 1989, Neurochemical mechanisms of mild and moderate head injury: Implications for treatment, in: "Mild Head Injury," H.S. Levin, H.M. Eisenberg, and A.L. Benton, eds., Oxford University Press, New York.

Hayes, R.L., Lyeth, B.G., Jenkins, L.W., Zimmerman, R.S., McIntosh, T.K., Clifton, G.L., and Young, H.F., 1990, Possible protective effect of endogenous opioids in traumatic brain injury, J. Neurosurgery 72:252-261.

Hayes, R.L., Jenkins, L.W., Lyeth, B.G., (in press), Neuropharmacological mechanisms of traumatic brain injury: Acetylcholine and excitatory amino acids, J. Neurotrauma.

Jenkins, L.W., Lyeth, B.G., LeWelt, W., Moszynski, K., DeWitt, D.S., Balster, R.L., Miller, L.P., Clifton, G.L., Young, H.F. and Hayes R.L. Combined pre-trauma scopolamine and phencyclidine attenuates post-traumatic increased sensitivity to delayed secondary ischemia, J. Neurotrauma 5(4):303-315, 1988.

Jenkins, L.W., Moszynski, K., Lyeth, B.G., Lewelt, W., DeWitt, D.S., Allen, A., Dixon, C.E., Povlishock, J.T., Majewski, T.J., Clifton, G.L., Young, H.F., Becker, D.P., and Hayes, R.L., 1989, Increased vulnerability of the mildly traumatized brain to cerebral ischemia: The use of controlled secondary ischemia as a research tool to identify common or different mechanisms contributing to mechanical and ischemic brain injury, Brain Res. 477:211-224.

Lyeth, B.G., Jenkins, L.W., Hamm, R.J., Dixon, C.E., Phillips, L.L., Clifton, G.L., Young, H.F., Hayes, R.L., 1990, Prolonged memory impairment in the absence of hippocampal cell death following traumatic brain injury in the rat, Brain Res. 526:249-258.

McIntosh, T.K., Hayes, R.L., DeWitt, D.S., Agure, V., and Faden, A.I., 1987, Endogenous opioids may mediate secondary damage after experimental brain injury, Am. J. Physiol. 253:E565-E574.

Miazaki, S., Newlon, P., Goldberg, S., Jenkins, L., Lyeth, B., Katayama, Y., and Hayes, R.L., 1989, Cerebral concussion suppresses hippocampal long-term potentiation (LTP) on rats, in: "Intracranial Pressure VII," A. Marmarou and J. Hoff, eds, Springer-Verlag, New York.

Miller, L.P., Lyeth, B.G., Jenkins, L.W., Oleniak, L., Hamm, R.J., Phillips, L.L., Clifton, G.L., and Hayes, R.L., 1990, Excitatory amino acid receptor subtype binding following traumatic brain injury, Brain Res. 526:103-107.

Nelson, C.A. and Seamon, K.B., 1988, Binding of 3[H]forskolin to solubilized preparations of adenylate cyclase, Life Sci 42:1375-1383.

Poat, J.A., Cripps, H.E., and Iversen, L.L., 1988, Differences between high-affinity forskolin binding sites in dopamine-rich and other regions of rat brain, Proc. Natl. Acad. Sci. USA 85:3216-3220.

Perry, D.C., Lyeth, B.G., Miller, L.P., Getz, R.L., Jenkins, L.W., and Hayes, R.L., (in press), Effects of traumatic brain injury in rats on bonding to forebrain opiate receptor subtypes, Mol. Chem. Neuropath..

Sergate, S. and Kim, H.D., 1985, Inhibition of 3-O-methyl-glucose transport in human erythrocytes by forskolin, J. Biol. Chem. 260:14677-14682.

Romhanyi, R.S., Tandian, D., Hovda, D.A., Kawamata, T., Yoshino, A., Cristescu, S.V., and Becker, D.P., 1990, Catecholaminergic stimulation enhances recovery of function following concussive brain injury, Proc. 8th Annual Neurotrau. Soc..

Tandian, D., Romhanyi, R.S., Hovda, D.A., Yoshino, A., Kawamata, T., Balady, N.F., and Becker, D.P., 1990, Amphetamine enhances both behavioral and metabolic recovery following fluid percussion brain injury, Proc. 8th Annual Neurotrau. Soc..

Worley, P.F., Baraban, J.M., Souza, E.B., and Snyder, S.H., 1986, Mapping second messenger systems in the brain: differential localization of adenylate cyclase and protein kinase C, Proc. Natl. Acad. Sci. USA 83:4053-4057.

EXPRESSION OF BASIC FGF IN PERIINFARCTED BRAIN TISSUE AND MODIFICATION OF POSTISCHEMIC THALAMIC DEGENERATION BY EXOGENOUS BASIC FGF

Kazuo Yamada, Eiji Kohmura, Junji Taguchi, Akira Kinoshita, Tateo Sakaguchi, and Toru Hayakawa

Department of Neurosurgery, Osaka University Medical School
1-1-50 Fukushima, Fukushima-ku, Osaka 553, Japan

INTRODUCTION

Thalamus of the infarction side becomes atrophic after middle cerebral artery (MCA) occlusion of rats. Because the atrophy is caused mainly by retrograde degeneration, lack of growth factors for neurons might be related to this event. We recently found that basic fibroblast growth factor (bFGF) supports survival of cultured thalamic, cortical and hippocampal neurons in vitro (Yamada et al., in press). We therefore analyzed expression of bFGF in the periinfarcted brain tissue by immunostaining and Western blotting. With this result, we then examined whether intracisternal injection of bFGF modified the thalamic degeneration.

EXPRESSION OF BASIC FGF

Immunostaining for bFGF. Wistar rats were subjected to MCA occlusion and sacrificed 1, 4, 7, 14 and 21 days later by perfusion and fixation. Brains were embedded in paraffin, thin-sliced and immunostained by anti-bFGF antibody (x500),which recognizes the epitope located within the first 9 amino acid residues at the amino terminal of the bFGF (Seno et al, 1989). The ABC-*Elite* system was used for visualization.

The bFGF immunoreactivity was not detectable in one day after occlusion. It appeared at 4 days after occlusion and lasted till 21 days. Reactive astrocytes were strongly positive for bFGF throughout the period, though density of bFGF-positive astrocyte became maximal at 7 days after occlusion. Some neurons of the periinfarcted cortex became positive for bFGF at 4 to 21 days after occlusion. The highest density of bFGF positive neurons were attained at 4 days after occlusion (Figure 1). In the thalamus however, no neurons nor astrocytes became positive for bFGF immunoreactivity (Figure 1).

Abbreviations used: bFGF, basic fibroblast growth factor; MCA, middle cerebral artery; GFAP, glial fibrillary acidic protein.

The Role of Neurotransmitters in Brain Injury, Edited by
M. Globus and W.D. Dietrich, Plenum Press, New York, 1992

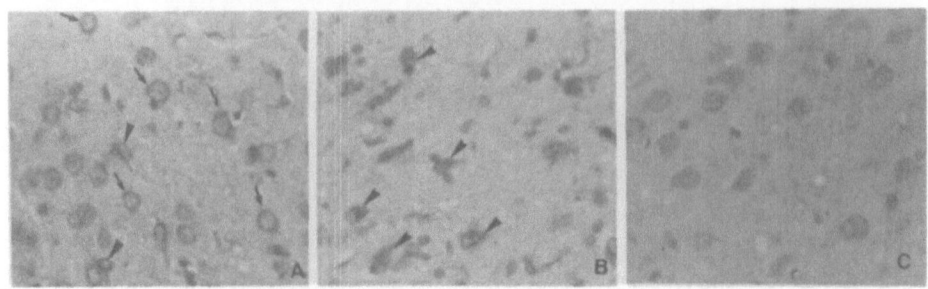

Figure 1. Immunoreactivity of bFGF in the MCA occlusion model.
(A) Cortex (4 days after ischemia); (B) white matter (4 days);
(C) thalamus (7days); arrows, bFGF-positive neurons; arrowheads,
bFGF-positive astrocytes.

Western blotting for bFGF. In 7 and 14 days after occlusion,
tissues were sampled from periinfarcted cortex and thalamus. The
contralateral corresponding site was also sampled for comparison.
They were homogenated with sample buffer and ultracentrifuged at
$100,000g$ for 60 minutes. The supernatant was applied to Molcut
system, which deleted large particles with molecular weight of more
than 100,000. The extract was applied to the SDS-PAGE gel with
concentration of 12.5%. They were electrophorased for 60 minutes
and transferred to the nitrocellulose membrane for 2 hours. The
membrane was stained with anti-bFGF antibody (x1000) and ABC-*Elite*
system.

In 7 days after occlusion, bFGF was located in high concentration in the lesion-side and contralateral cortex. The bFGF was
however, not expressed in the both side of the thalamus. The authentic form of bFGF was precipitated at 16kD, whereas bFGF from
cortical samples has precipitation around 31 and 45kD. Therefore,
bFGF might present in the rat brain as diploid or triploid form. I
14 days after occlusion, bFGF was expressed only in the cortex but
not in the thalamus. The amount of bFGF was higher in the periinfarcted cortex than the contralateral side (Figure 2).

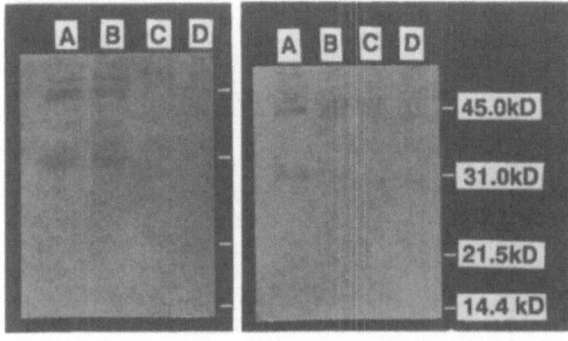

Figure 2. Western blotting
of periinfarcted brain for
bFGF.(*left*) 7 days after
occlusion; (*right*) 14 days
(lane A) lesion-side cor-
tex; (B) contralateral
cortex; (C) lesion-side
thalamus; (D) contralatera
thalamus.

PREVENTION OF THALAMIC DEGENERATION BY BASIC FGF

Method of bFGF administration. The bFGF used in this experiment was human recombinant bFGF mutein, CS23, which was provided k
Takeda Chemical Industries Ltd., Osaka, Japan (Kurokawa et al.,
1987). Two cysteine residues (Cys[69] and Cys[87]) of the native bFGF
were replaced by serine residues to stabilize the molecules (Seno
et al., 1988). Purity of the product was assured by polyacrylamide

gel electrophoresis, and the biological activity was evaluated by mouse 3T3 cells *in vitro* and by angiogenesis on the chick embryo chorioallantoic membrane (Seno et al., 1988).

The MCA of the Wistar rats was occluded, and animals were divided in two groups. One group received intracisternal injection of bFGF according to the protocol (Figure 3), and the other group received vehicle solution.

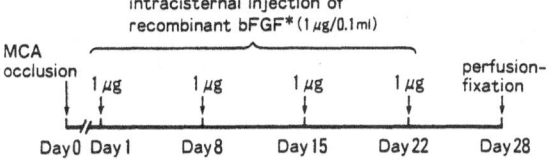

Figure 3. Protocol for intracisternal injection.

Macroscopical and microscopical observation. The animals were perfused and fixed in 4 weeks after occlusion and histological evaluation was made by tissue sections at the levels of caudoputamen and thalamus. Planimetric measurement was employed for measuring the size of each hemisphere at caudate level or size of the posterior ventral thalamus.

The size of infarction was evaluated by atrophy rate at caudate level, and it was not statistically different in both groups (Figure 4 *left*). The atrophy of the thalamus was evaluated by comparing cross sectional area of posterior ventral thalamus and data were expressed as percent of the contralateral side. The saline-treated group showed 25% reduction of the posterior ventral thalamus whereas bFGF group showed only 7% reduction (Figure 4 *right*).

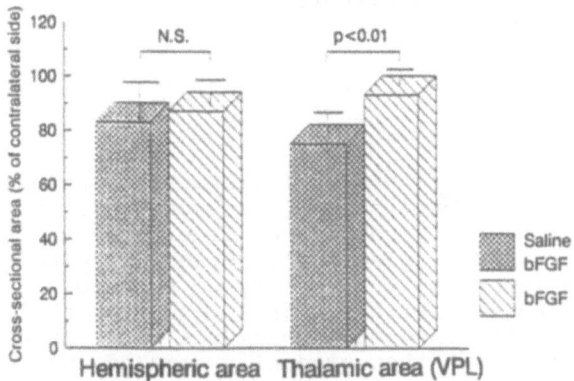

Figure 4. Macroscopical measurement of hemispheric size at caudate (*left* and VPL nucleus of the thalamus (*right*)

Microscopical observation showed shrinkage and disappearance of thalamic neurons in the vehicle treated animals, whereas bFGF-treated group showed preservation of the thalamic neurons (Figure 5). The computerized analysis of the cell size substantiated the observation (Yamada et al., 1991).

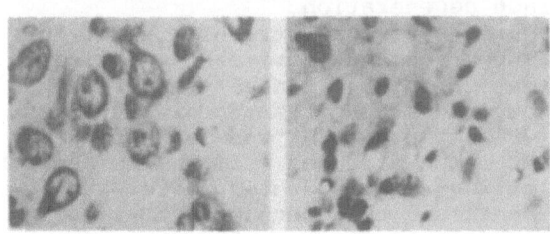

Figure 5. Microscopical appearance of thalamic neurons treated with bFGF (*left*) or vehicle saline (*right*).

The mechanism why bFGF is effective for prevention of thalamic
degeneration remains unknown. Therefore, we injected bFGF or vehicle saline to the normal Wistar rats by the same schedule as treatment protocol shown in Figure 3.

In 3 days after injection of bFGF, neurons of the cingulate
cortex became positive for c-Fos at 3 days after injection (Figure
6). Therefore bFGF may have direct effect to the neurons. Moreover
bFGF has effect to the astrocytes. In 4 weeks after injection of
bFGF, density of GFAP positive astrocyte increased significantly
(Yamada et al., 1991). The astrocytes of the bFGF injected animals
is more fibrillary-shaped with having longer process than those of
normal-saline treated group.

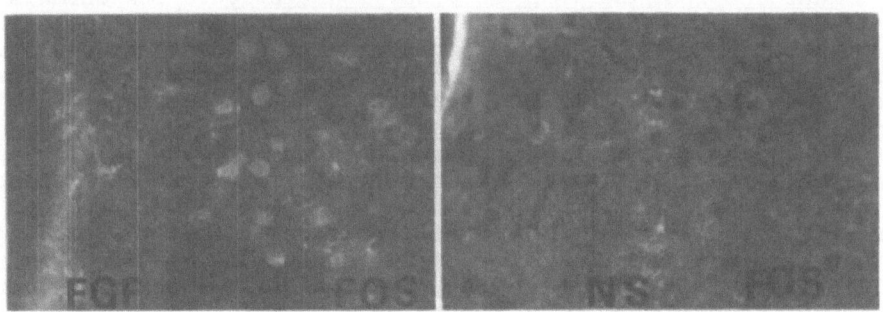

Figure 6. Fos immunoreactivity 3 days after intracisternal injection of bFGF (*left*) or vehicle saline (*right*)

DISCUSSION

Neuroanatomical study indicates that thalamic neurons have a
dense projection to the somatosensory cortex (Jacobson and Trojanowski, 1975). When the somatosensory cortex was widely removed by
surgical ablation, thalamic neurons selectively shrank and died of
retrograde degeneration (Barron et al., 1973). As we pointed out i
the previous communication (Kataoka et al., 1989), axons of the
thalamic neurons were damaged by MCA occlusion of the rat. The cel
bodies of the thalamic neurons were then gradually damaged and the
died of retrograde degeneration.

Retrograde degeneration of the same kind was observed in the
septo-hippocampal cholinergic neurons, in which transection of the
fimbria-fornix caused death of the septal cholinergic neurons.
Intraventricular administration of NGF prevented the degeneration
(Hefti, 1986). It is therefore believed that with proper administration of a neurotrophic factor, neurons might be rescued from
axonal injury without retrograde degeneration.

We selected bFGF as a candidate for neurotrophic factor for
thalamic neurons, because it supported the survival of cortical,
hippocampal and thalamic neurons *in vitro*. It also prevented the
death of these neurons after hypoxic stress (Yamada et al., in
press). The second reason why we selected bFGF was lack of bFGF
expression in the thalamus of the MCA-occluded animals. As shown t
immunostaining and Western blotting, periinfarcted cortex has
neurons and astrocytes which contained bFGF immunoreactivity. In
the thalamus however, bFGF immunoreactivity was missing in any

periods after cortical infarction. The third reason why we selected bFGF is the generous supply of the stabilized mutein of recombinant human bFGF. This recombinant form is stable and keeps equivalent biological activity to native bFGF. This mutein partially reversed retrograde death of septal neurons after fimbrial transection, when it was soaked in Gelfoam and applied to the transected area (Suno et al., personal communication). We selected percutaneous intracisternal administration as the route of drug delivery, because this technique has been maintained in our laboratory and wide distribution of drugs to the ventriculo-cisternal system is well documented (Arita et al., 1988).

As we expected, intracisternal administration of bFGF prevented retrograde degeneration of the thalamic neurons after MCA occlusion (Yamada et al., 1991). This is clearly demonstrated by macroscopic measurement of thalamic atrophy and by microscopical evaluation and computerized measurement of the cell size of thalamic neurons. The mechanism by which bFGF prevented retrograde degeneration of the thalamic neurons must be discussed. The first possibility is the effect of bFGF as a neurotrophic factor. The bFGF supported the survival of cultured thalamic neurons and reversed hypoxic neuronal injury *in vitro*. Therefore, bFGF can be a neurotrophic factor for thalamic neurons when their axons are injured by ischemia. The normal Wistar rats which received an intracisternal injection of bFGF showed a wide range of GFAP-positive astrocytes in the tissue facing the ventriculo-cisternal system. Therefore, bFGF might diffuse into the brain facing the ventriculo-cisternal system and it activated astrocytes. The area of astrocytic reaction correlated with the thalamocortical pathways. The data suggest that intracisternal bFGF has a direct supportive effect on the injured thalamic neurons and prevents retrograde degeneration.

The second possibility is an indirect effect of bFGF on neurons through reactive astrocytes. The bFGF has a rather wide range of target cells. In the cortical ablation model, bFGF-immunoreactivity was detected in the glial cells at the ablation edge, and such an increase in local bFGF concentration might be beneficial for glial and capillary proliferation, and for neuronal sprouting (Finklestein et al., 1988). In the present infarction model, bFGF-activated astrocytes may have a supportive effect on the neurons suffering from ischemic axonal injury. It is well known that astrocytes secrete NGF (Furukawa et al., 1987), and reactive gliosis might produce NGF for collateral sprouting of cholinergic neurons (Gage et al., 1988). The supportive effect of astrocytes on the injured thalamic neurons may therefore be due both to secretion of some factor and to mechanical contact with the neurons. Although further study is needed for analysis of the mechanism, the results obtained in this study may have important implications for the use of bFGF in the treatment of slowly-progressing ischemic injury.

ACKNOWLEDGEMENT

This work is supported by a Grant-in-Aid for Scientific Research (63570679, 02670628, 02807132) from the Ministry of Education and a Research Grant for Cardiovascular Disease (2C-2) from the Ministry of Health and Welfare. We thank Miss Reiko Fujita for secretarial support and Miss Naoko Kuratani, Chiaki Imai and Ikuyo Tani for technical assistance.

REFERENCES

Arita N, Ushio Y, Hayakawa T, Nagatani M, Huang T-Y, Izumoto S, Mogami H (1988) Intrathecal ACNU: a new therapeutic approach against malignant leptomeningeal tumors. *J Neuro-Oncol* 6:221-226

Barron KD, Means ED, Larsen E (1973) Ultrastructure of retrograde degeneration in thalamus of rat. 1. Neuronal somata and dendrites. *J Neuropath Exp Neurol* 32:218-244

Finklestein SP, Apostolides PJ, Caday CG, Prosser J, Philips MF, Klagsbrun M (1988) Increased basic fibroblast growth factor (bFGF) immunoreactivity at the site of focal brain wounds. *Brain Res* 460:253-259

Furukawa S, Furukawa Y, Satoyoshi E, Hayashi K (1987) Regulation oi nerve growth factor synthesis/secretion by catecholamine in cultured mouse astroglial cells. *Biochem Biophys Res Commun* 147:1048-1054

Gage FH, Olejniczak P, Armstrong DM (1988) Astrocytes are importani for sprouting in the septohippocampal circuit. *Exp Neurol* 102:2-13

Hefti F (1986) Nerve growth factor promotes survival of septal cholinergic neurons after fimbrial transection. *J Neurosci* 6:2155-2162, 1986

Jacobson S, Trojanowski JQ (1975) Corticothalamic neurons and thalamocortical terminal fields: An investigation in rat using horseradish peroxidase and autoradiography. *Brain Res* 85:385-401

Kataoka K, Hayakawa T, Yamada K, Mushiroi T, Kuroda R, Mogami H (1989) Neuronal network disturbance after focal ischemia in rats. *Stroke* 20:1226-1235

Kurokawa T, Sasada R, Iwane M, Igarashi K (1987) Cloning and expression of cDNA encoding human basic fibroblast growth factor. *FEBS Lett* 213:189-194

Seno M, Sasada R, Iwane M, Sudo K, Kurokawa T, Ito K, Igarashi K (1988) Stabilizing basic fibroblast growth factor using protein engineering. *Biochem Biophys Res Commun* 151:701-708

Seno M, Iwane M, Sasada R, Moriya N, Kurokawa T, Igarashi K (1989) Monoclonal antibodies against human basic fibroblast growth factor *Hybridoma* 8:209-221

Yamada K, Kinoshita A, Kohmura E, Sakaguchi T, Taguchi J, Kataoka K, Hayakawa T (1991) Basic fibroblast growth factor prevents thalamic degeneration after cortical infarction. *J Cereb Blood Flow Metabol* 11:472-478

Yamada K, Kohmura E, Kinoshita A, Taguchi J, Sakaguchi T, Tsuruzon K, Hayakawa T (in press) Modification of ischemic neuronal injury by basic fibroblast growth factor. Ito U and Klatzo I (eds). *Maturation Phenomenon*. Springer, Tokyo

Chapter 6

Blood Brain Barrier in Brain Injury

THE BLOOD-BRAIN BARRIER IN BRAIN INJURY:

AN OVERVIEW

John T. Povlishock[1] and W. Dalton Dietrich[2]

[1]Department of Anatomy, Medical College of Virginia/Virginia Commonwealth
University and
[2]Departments of Neurology and Anatomy, University of Miami, School of Medicine

INTRODUCTION

The blood-brain barrier functionally resides in the cerebral vascular endothelium which creates a dynamic regulatory interface between the blood and brain tissue front. As has been long recognized, the cerebral vascular endothelium, unlike endothelial cells found in other vascular beds, lacks numerous fenestrae and vesicles and further adjacent endothelial cells are joined by tight junctions creating a continuous endothelial membranous interface between the blood and brain tissue. As would be anticipated, this continuous endothelial sheet permits the passage of only those solutes that are lipid soluble, of low molecular weight and not highly ionized. Nutrients, on the other hand, which are polar compounds cannot easily cross through such a continuous membrane. This potential problem, however, has been obviated by the existence of multiple carrier systems that allow for the facilitated diffusion of the various nutrients from blood to brain front. To date, seven carrier-mediated systems have been identified allowing for the passage of glucose, monocarboxylic acids, neutral amino acids, basic amino acids, purine bases, amines and nucleosides (Pardridge, 1985). These carriers are believed to exist on both the luminal and abluminal endothelial membranes, allowing for effective flux of nutrients from blood to brain and brain to blood. In addition to these facilitated transport mechanisms, active transport has also been demonstrated in relation to the blood-brain barrier. However, in this case, this energy requiring process is believed confined to the abluminal endothelial surface. Here it is envisioned that active transport functions to remove potassium against a concentration gradient from the brain microenvironment to the blood front. Similarly, it has also been speculated that damaging excitatory amino acids such as glutamate are selectively sequestered and moved from the brain to blood front via such active transport mechanisms (Pardridge, 1979).

To date, dysfunction of the blood-brain barrier has been assessed through the analyses of altered nutrient transport in the belief that barrier perturbation with attendant altered nutrient transport could adversely impact upon brain metabolism. Because of both technical and conceptual shortcomings, however, studies of this nature have been limited and sometimes controversial (Betz et al, 1973, 1974; Hertz et al. 1981; Daniel et al. 1981; Gjedde and Crone, 1975). More commonly, contemporary pathological assessment of blood-brain barrier dysfunction has focused on more dramatic barrier changes and, as such, has examined altered cerebral vascular permeability to various endogenous serum proteins or exogenous tracers via light and electron microscopy or various forms of quantitative autoradiography (Balin et al, 1986, Broadwell et al, 1985; Dietrich et al, 1987, 1988; Ellison et al, 1986; Noble and Wrathall, 1988; Povlishock et al, 1978; Wei et al., 1986). Altered cerebral vascular permeability has been described in a host of pathological states and through the use of various endogenous and exogenous tracers, provocative information has evolved on not only the precise sites of altered barrier permeability, but also its temporal duration, and its subcellular correlates (Beggs and Waggener, 1976; Dietrich et al. 1987, 1988a, 1988b; Kuroiwa et al, 1985; Lossinsky et al 1987; Nag and Harik 1987; Nag et al 1977; Noble and Wrathall 1988; Noble and Ellison, 1989; Petito 1980; Petito et al, 1982; Povlishock et al 1978, 1980; Wei et al., 1986; Westergard 1977, 1980).

In the following paragraphs, we will attempt to review this information and provide new insight into the morphological substrates of altered barrier status while fully explicating the significance of such barrier change for brain function.

Substrates of Blood-Brain Barrier Dysfunction

As noted above the use of various endogenous and exogenous tracers has allowed many investigators to appreciate that a myriad of insults can disrupt the blood-brain barrier resulting in an extravasation of the tracer employed. Primarily, through the use of horseradish peroxidase and serum proteins ultrastructural studies have attempted to detect the precise subcellular routes of such macromolecular tracer passage. Although, in some of the most destructive processes, direct mechanical damage of the endothelial membranes and/or endothelial necrotic changes correlate with the passage of these macromolecules from the blood to brain front (Dietrich et al 1987; Petito et al 1982), this represents the most extreme condition, and clearly, these events are not operant with more subtle perturbations of the barrier. With the more subtle perturbations of blood-brain barrier status that occur with various experimental and neuropathological insults, most advocate a transcytosis of macromolecules through the blood-brain barrier involving one of three potential routes: 1.) Dramatically increased transendothelial vesicular transport; 2.) Creation of transendothelial tubular-vesicular channels, and 3.) Cleaving of interendothelial tight junctions. Although the literature exploring each of these potential routes of transendothelial or paraendothelial passage is extensive, no clear consensus has evolved that any of these routes are the actual and/or exclusive biological conduits of altered cerebral vascular permeability. In the case of interendothelial tight junctional cleaving as advocated by some investigators (Brightman et al, 1973, Nagy et al., 1979a, b, c), it has been suggested that inattention to perfusion pressure as well as the type and quality of fixation can contribute to artifactual changes in tight junctional integrity. In this regard, however, cryofixation, has recently demonstrated that cleaving of the interendothelial channels is a real event that is most likely involved in altered permeability. At present, this issue is being further critically evaluated in order to appreciate its full biological significance (Nagy et al, 1988). In the case of transendothelial vesicular transport, other investigators have advocated that various fluid-phase tracers (such as horseradish peroxidase) can, under certain pathological and experimental conditions, be sequestered by an increased number of luminal pits. These are assumed to internalize the tracer as vesicles and move it to the abluminal endothelial front where the tracer is dispersed via abluminal pits to the subendothelial basal lamina (Beggs and Waggener 1976, Dietrich et al., 1987; Povlishock et al 1978 and Westergard 1977, 1980). Despite the fact that most concur that many neuropathological and experimental states are associated with increased pit/vesicle number within the cerebral vascular endothelium, some have begun to question the overall role of transendothelial vesicular transport in the transcytosis of macromolecules from the blood to brain front (Broadwell 1989). In the case of fluid phase tracers such as horseradish peroxidase, it has been recently noted that they can, in a nonspecific fashion, be incorporated into the endothelial cell where they do not undergo further transport but rather undergo degradation in the lysosomal compartment (Broadwell, 1989). Thus, based upon these suggestions, many now question whether macromolecules can actively move through the endothelium as previously purported. Further complicating this issue is the recent observation by several investigators that the of primary fixative influences the total number of vesicles visualized (Nagy et al 1988). With increasing concentrations of glutaraldehyde, increased numbers of vesicles and pits have been recognized, whereas with decreasing concentrations, proportionally few have been seen (Nagy et al 1988). These issues cloud the previous biological interpretations of pit/vesicular activity and also impact upon the previous interpretations attached to the identification of fused pits and vesicles forming continuous transendothelial channels. Primarily relying on high voltage electron microscopy and thick sections, a case has been advanced to support the concept the such transendothelial channels provide conduits for tracer passage; yet, here again, the evidence in support of this is both limited and not highly compelling (Lossinsky et al, 1989).

What is disquieting in all of the above discussion is that, to date, despite exhaustive efforts, our understanding of the subcellular mechanisms of macromolecular transport associated with altered barrier status has not significantly improved. Admittedly, the passage of tracers (macromolecules) via cleaved interendothelial tight junctions still constitutes a potentially viable route; yet, the actual biological significance of consistantly increased pit/vesicular number remains to be ascertained. Although pit and vesicle number may be increased in various pathological states, they may merely reflect a general form of endothelial perturbation and not directly constitute conduits for increased vascular permeability (Povlishock et al 1980). In appreciating the subcellular correlates of altered blood-brain barrier status, it is important to recall some of the limitations associated with routine ultrastructural research. Simply stated, the images obtained with fixation do not faithfully replicate *in vivo* forms and more importantly, the two dimensional images generated also contribute to a lack of appreciation of those changes ongoing in a three-dimensional plane. In regards to the issue of altered macromolecular tracer passage, it is instructive to note that even in a case of local neuronal flooding with tracers such as horseradish peroxidase, the precise transmembrane route has yet to be ascertained (Povlishock et al 1979). In fact, when such neuronal flooding occurs, ultrastructural analysis has failed to demonstrate any evidence of cell membrane perturbation or disruption. Rather, the suggestion has been advanced that perhaps current ultrastructural approaches are incapable of revealing the subtle membrane change that correlates with the transmembrane diffusion of the tracer (Povlishock et al 1979). This also may the case in the passage of macromolecules occurring with alterations in the blood-brain barrier. Perhaps, direct transendothelial diffusion of the tracer is a major route of passage; yet this route of passage may not be easily detected by the ultrastructural approaches currently in use. Obviously, this issue requires further evaluation.

Significance of Altered Blood-Brain Barrier Status

Dogma has long dictated that alteration of the blood-brain barrier impacts upon brain function. Alterations in barrier status, particularly involving the passage of serum proteins, have long been linked to the subsequent generation of vasogenic edema (Klatzo et al 1980; Klatzo 1987). Additionally, alterations in blood-brain barrier status have been assumed to involve perturbations of nutrient transport which, in turn, also adversely impacts upon brain function. Although in various neuropathologic and experimental conditions, examples exist to support the development of vasogenic edema and altered nutrient transport, in most cases, a more direct relation between barrier change and subsequent brain dysfunction has yet to be clearly elucidated. Admittedly, within foci of altered barrier status, damaged brain parenchyma has been recognized; yet, in most cases, it has been unclear whether these events are causally related or rather the one and the same tissue response to the primary insult. Recently, however, there has been mounting experimental evidence that alteration of the barrier may provide for a direct delivery of damaging substances to the already perturbed brain. Some have recently alluded to this possibility (Sokral et al.), and indeed, have suggested that the altered blood-brain barrier may allow for the influx of normally excluded substances such as neurotransmitters or vasoactive compounds. That altered barrier permeability may directly impact upon brain parenchymal function is now supported both directly and indirectly by several experimental investigations. Specifically, in the case of ischemic brain injury, it is well-recognized that a direct correlation exists between the sites of permeability change and neuronal damage in addition to local neuronal inundation with intravascular tracers (Dietrich et al 1991). Importantly, the number and distribution of inundated and damaged neurons appear to bear a relation to the overall degree of barrier perturbation. Further, with the exacerbate of barrier change via hyperthermia, there appears a concomitant increase in the extent and number of damaged neurons. In the case of traumatic brain injury, there also appears to be a direct correlation between barrier change and brain parenchymal dysfunction. Mild traumatic brain injury has long been recognized to result in transient opening of the blood-brain barrier which occurs in foci linked to ensuring morbidity. Interestingly, when this traumatically induced permeability change was attenuated by hypothermia (Jiang et al 1992), not only did these brain regions show significant reductions in local tracer concentration, but more importantly, the morbidity typically associated with dysfunction of these regions was significantly reduced (unpublished findings). Collectively, then, these studies provide compelling evidence that a causal relationship exists between blood-brain barrier change and ensuing neuronal structural and functional perturbation. In relation to the theme of this conference, it is instructive to recall that perhaps blood-brain barrier perturbation may contribute to the delivery of various transmitters including the excitatory neurotransmitters to already injured brain parenchyma. Perhaps, the perturbed barrier will allow for the influx of glutamate from the blood to brain front. Also, as the blood-brain barrier under normal circumstances, works to remove glutamate from the brain to blood front (Pardridge 1979), it is possible that this simultaneously increased permeability to glutamate, coupled with an inability to clear glutamate, may further contribute to local glutamate elevations with their damaging consequences. Clearly, these issues require continued investigation. Admittedly, in terms of neuroexcitation, primary agonist receptor interaction is most likely the predominant theme, but it is not implausible to assume that transmitters derived from the blood may be interacting in continued and agonist-receptor interactions which, in turn, translate into enduring abnormal secondary signal transduction, leading to continued morbidity.

REFERENCES

Balin B, Broadwell R, Saleman M (1987) Evidence against tubular profiles contributing to the formation of transendothelial channels through the blood-brain barrier. *J Neurocytol* 16:721-728

Beggs J, and Waggener J (1976) Transendothelial vesicular transport of protein following compression injury to the spinal cord. *Lab Invest.* 34:428-439.

Betz A, Gilboe D, Yudilevich D and Drewes L (1973) Kinetics of unidirectional glucose transport into the isolated dog brain. *Am J Physiol* 225:586-592

Brightman M, Hori M, Rapoport S, Reese T, Westergard E (1973) Osmotic opening of tight junctions in cerebral endothelium. *J Comp Neurol* 152:317-326

Broadwell R (1989) Transcytosis of macromolecules through the blood-brain barrier: a cell biological perspective and critical appraisal. *Acta Neuropathol (Berl)* 79:117-128

Cremer J and Cunningham V (1979) Effects of some chlorinated sugar derivatives on the hexose transport system of the blood-brain barrier. *Biochem J* 180:677-679

Daniel P, Love E, Moorhouse S and Pratt O (1981) The effect of insulin upon the influx of tryptophan into the brain of the rabbit. *J Physiol* 312:551-556

Dietrich D, Busto R, Watson B, Scheinberg P and Ginsberg M (1987) Photochemically induced cerebral infarction II. Edema and blood-brain barrier disruption. *Acta Neuropathol (Berl)* 72:326-334

Dietrich D, Prado M and Watson B (1988a) Photochemically stimulated blood-brain factors induced blood-brain barrier alterations in rats. *Stroke* 19:857-862

Dietrich D, Prado R, Watson B and Nakayama H (1988b) Middle cerebral artery thrombosis: Acute Blood Brain Barrier Consequences. *J Neuropath Exp Neurol* 47:443-451

Dietrich D, Busto R, Halley M and Valdes I (1990) The importance of brain temperature in alterations of blood-brain barrier following cerebral ischemia. *J Neuropath and Exp Neurol* 49:486-497

Dietrich D, Halley M, Valdes I and Busto R (1991)Interrelationships between increased vascular permeability and acute neuronal damage following temperature controlled brain ischemia in rats. *Acta Neuropathol (Berl)* 81:152-163

Ellison M, Povlishock JT, and Hayes RL (1986) Examination of the blood-to-brain transfer of α-Aminoisobutyric acid and horseradish peroxidase: Regional alteration in blood-brain barrier function following acute hypertension. *J Cerebral Blood Flow and Metab* 6:471-480.

Gjedde A, and Crone C (1975) Induction processes in blood-brain transfer of ketone bodies during starvation. *Am J Physiol* 229:1165-1169

Hertz M, Paulson O, Barry D, Christiansen J and Svendsen P (1981) Insulin increases glucose transfer across the blood-brain barrier in man. *J. Clin Invest* 67:597-604

Jiang J, Lyeth B, Kapasi M and Povlishock JT (1991) Moderate hypothermia reduces blood-brain barrier disruption following traumatic brain injury In: Proceedings of the Neurotransmitter Satellite Symposium of Brain 1991. (In press)

Klatzo I (1987) Pathophysiological aspects of brain edema. *Acta Neuropathol (Berl)* 72:236-239

Klatzo I, Chui E, Fujiwara K, Spatz M (1980) Resolution of vasogenic edema. *Adv Neuro* 28:359-379

Lossinky A, Moretz R, Carp R, Wisniewski H (1987) Ultrastructural observations of spinal cord lesions and blood-brain barrier changes in scrapie-infected mice. *Acta Neuropathol (Berl)* 3:43-52

Lossinsky A, Song M, Wisniewski H (1989) High-voltage electron microscopic studies of endothelial cell tubular structures in the mouse blood-brain barrier following brain trauma. *Acta Neuropathal* 77:489-493

Nag S, Harik S (1987) Cerbrovascular permeability to horseradish peroxidase in hypertensive rat: effects of unilateral locus ceruleus lesion. *Acta Neuropathol (Berl)* 73:247-353

Nag S, Robertson D, Dinsdale G (1977) Cerebral cortical changes in acute experimental hypertension. An ultrastructure study. *Lab Invest* 36:150-160

Nagy Z, Pappius H, Mathieson G, Hüttner I (1979a) Opening of tight junctions in cerebral endothelium. I. Effect of hyperosmolar mannitol infused through the internal carotid artery. *J Comp Neurol* 185:569-578.

Nagy Z, Mathieson G, Hüttner I (1979b) Opening of tight junctions in cerebral endothelium. II. Effect of pressure-pulse induced acute arterial hypertension. *J Comp Neurol* 185:579-586

Nagy Z, Mathieson G, Hüttner I (1979c) Blood-brain barrier opening to horseradish peroxidase in acute arterial hypertension. *Acta Neuropathol* 48:45-53

Noble L, Wrathall J (1988) Blood-spinal cord barrier disruption proximal to a spinal cord transection in the rat: time course and pathways associated with protein leakage. *Exp Neurol* 99:567-578

Noble L, Ellison J (1989) Effect of transection on the blood-spinal cord barrier of the rat after isolation from the descending sources. *Brain Res* 487:299-310

Pardridge W (1985) Cerebral vascular permeability status in brain injury In: Central Nervous System Trauma Status Report Becker D and Povlishock J (eds) William Byrd Press, pp 503-512

Pardridge W (1979) Regulation of amino acid availability to brain: Elective control mechanisms for glutamate. In: Glutamic Acid: Advances in Biochemistry and Physiology. Filer L et al (eds), Raven Press

Petito C, Levy D (1980) The importance of cerebral arterioles in alterations of the blood-brain barrier. *Lab Invest* 43:262-268

Petito C, Pulsinelli W, Jacobson G, Plum F (1982) Edema and vascular permeability in cerebral ischemia: Comparison between ischemic neuronal damage and infraction. *J Neuropathol Exp Neurol* 41:423-436

Povlishock JT, Becker D, Sullivan H, and Miller, J (1978) Vascular permeability alterations to horseradish peroxidase in experimental brain injury. *Brain Res* 153:223-239

Povlishock JT, Becker D, Miller J, and Dietrich W (1979) The morphopathologic substrates of concussion? *Acta Neuropathol (Berl)* 47:1-12

Povlishock JT, Kontos H, Rosenblum W, Becker D and Jenkins L (1980) A scanning electron microscopic analysis of the intraparenchymal brain vasculature subsequent to systemic hypertension. *Acta Neuropathol (Berl)* 51:203-213

Sokrab T, Johansson BB, Halimo H, and Olsson Y (1988) A transient hypertensive opening of the blood-brain barrier can lead to brain damage. *Acta Neuropathol (Berl)* 75:557-565

Wei E, Kontos H, Ellison M and Povlishock J (1986) Oxygen radicals in arachidonate-induced increased blood-brain barrier permeability to proteins. *Am J Physiol* 251:H693-H699

Westergard E (1977) The blood-brain barrier to horseradish peroxidase under normal and experimental conditions. *Acta Neuropathol (Berl)* 39:181-187

Westergard E (1980) Ultrastructural permeability properties of cerebral microvasculature under normal and experimental conditions after application of tracers. *Adv Neurol* 28:55-74

ULTRASTRUCTURAL STUDIES OF ENDOTHELIUM IN

NMDA-INDUCED EXCITOTOXICITY

Sukriti Nag

Department of Pathology (Neuropathology)
Queen's University, Kingston, Ontario. Canada K7L 3N6

INTRODUCTION

Glutamate and aspartate serve as excitatory neurotransmitters in the central nervous system and have been implicated in the pathophysiology of brain damage in a variety of pathological states such as anoxia-ischemia, hypoglycemia and seizures[1,2]. In addition, excitotoxic mechanisms have been proposed to play a role in the pathogenesis of chronic degenerative diseases including Huntington's disease, senile dementia of the Alzheimer type and amyotrophic lateral sclerosis[3,4].

Our previous study[5] demonstrated that continuous intrathecal infusion of N-methyl-D-aspartate (NMDA) for 2-4 weeks produced toxicity in spinal neuronal systems. Of interest was the nature of the vascular changes occurring in this model. The present study was undertaken to determine whether breakdown of the blood-brain barrier occurs in this model and when, during the course of neurotoxicity, breakdown of the blood-brain barrier occurs. This was studied in rats following a short-term intrathecal infusion of NMDA in the region of the lumbar enlargement of the cord. Horseradish peroxidase (HRP) was used as a tracer to detect changes in vascular permeability to protein.

MK-801, (+)-5-methyl-10, 11-dihydro-5-dibenzo[a, d] cyclohepten-5, 10-imine, a water-soluble, selective, noncompetitive, open channel blocker at the NMDA receptor site crosses the blood-brain barrier and has a half-life of ~ 1 h[6]. One group of rats were pretreated with MK-801 to determine its effect on the observed histological changes produced in the spinal cord by the NMDA infusion.

MATERIALS AND METHODS

Fifteen Wistar-Furth rats weighing 200-250 g were anesthetised by an i.p. injection of sodium amytal (80 mg/kg). Polyethylene cannulas (PE 50) were inserted into the femoral artery for measurement of the blood pressure, and the femoral vein for the adminstration of test substances. Following a skin incision over the lower back, silastic tubing was inserted into the lumbar subarachnoid space through a slit in the dura between lumbar 5 and 6 vertebrae. The tube was passed rostrally for a distance of about 3 cm to allow test solutions to infuse the lumbar enlargement of

The Role of Neurotransmitters in Brain Injury, Edited by
M. Globus and W.D. Dietrich, Plenum Press, New York, 1992

the cord. This tubing was connected to an infusion pump. Body temperatures were maintained at 37^0C by means of a rectal probe connected to a proportional feedback temperature controller connected to an infrared lamp.

Experimental Groups

Group I: Three rats were infused for one hour with NMDA (Sigma Chemical Co., St. Louis, MO) 30 ug/min.
Group II: Three rats were infused for one hour with NMDA 60 ug/min.
Group III: Three rats were pretreated with MK-801 (Research Biochemicals Incorp., Natick, MA) 1 mg/Kg i.v., 5 min prior to the infusion of 60 ug/min NMDA.
Group IV : Three control rats were infused with phosphate -buffered saline (PBS) for 1 hour instead of NMDA.
Group V: Three additional control rats were pretreated with MK-801 1 mg/Kg i.v. 5 min. prior to the infusion of saline.

At 50 min all rats were injected intravenously with 250 mg/Kg HRP, Type II (Sigma Chemical Co., St. Louis, MO) dissolved in physiological saline. Ten minutes later a thoracotomy was performed and rats were perfused with fixative solutions via a cannula inserted in the ascending aorta. Rats were perfused for 3 minutes with a solution containing 1.25% paraformaldehyde in 0.1 M cacodylate buffer (pH 7.4) containing 4.35% sucrose. This was followed by perfusion of a solution which contained 1.25% paraformaldehyde and 1.25% glutaraldehyde in the same buffer. The lumbar enlargement of the cord was cut into two halves. Half the cord was sectioned at 50 micron intervals and the other half at 200 micron intervals using a tissue chopper.

All spinal cord slices were processed for the demonstration of HRP reaction product as described previously[7]. Tissue blocks containing either the anterior or posterior horn were processed for electron microscopy. Blocks were post-fixed in Dalton's fixative for 30 min. This fixative contained 1 part of 4% potassium dichromate, 1 part of 3.4% NaCl and 2 parts of 4% osmium tetroxide in 0.1 M cacodylate buffer. Blocks were then stained en bloc with uranyl acetate in 0.05 M sodium hydrogen maleate-NaOH buffer (pH 7.0) for 1 hour. This was followed by dehydration and embedding in JEMBED 812 resin. Semithin sections were stained with 1 % toluidine blue. Ultrathin sections were stained with lead citrate and examined with a Hitachi H500 electron microscope at 75 kV.

RESULTS

The resting mean maximum systolic blood pressure of test and control animals was in the normal range. The blood pressure of rats was unaffected by the NMDA infusion or the injection of MK-801.

Behavioral Changes

About 30 min after the onset of the NMDA infusion rats displayed spontaneous movements of their tails. This was more marked in rats receiving the higher dose of NMDA. In addition some animals displayed a ripple-like movement of the skin of the lower back.

Nervous Tissue Changes

Control rats infused with PBS and those pretreated with MK-801 prior to the PBS infusion did not show any morphological changes in the anterior or posterior horns by light (Fig. 1a) or electron microscopy (Table 1).

TABLE 1

Number of rats in the different experimental
groups showing tissue changes.

		NERVOUS TISSUE	VESSELS
NMDA			
30 ug/min	(3)	3	1
60 ug/min	(3)	3	3
MK801 + NMDA	(3)	0	0
Controls			
saline	(3)	0	0
MK801 + saline	(3)	0	0

The total number of rats in each group is shown
in parentheses.

All rats infused with 30 ug/min NMDA showed mild vacuolation of the
neuropil in the region of the substantia gelatinosa of the posterior horn
of the cord (Table 1). The degree of changes were more marked in rats
infused with 60 ug/min NMDA. In these rats vacuolation of the neuropil
extended beyond the substantia gelatinosa into the adjacent laminae of the
posterior horn (Fig. 1b). In addition these changes extended in continuity

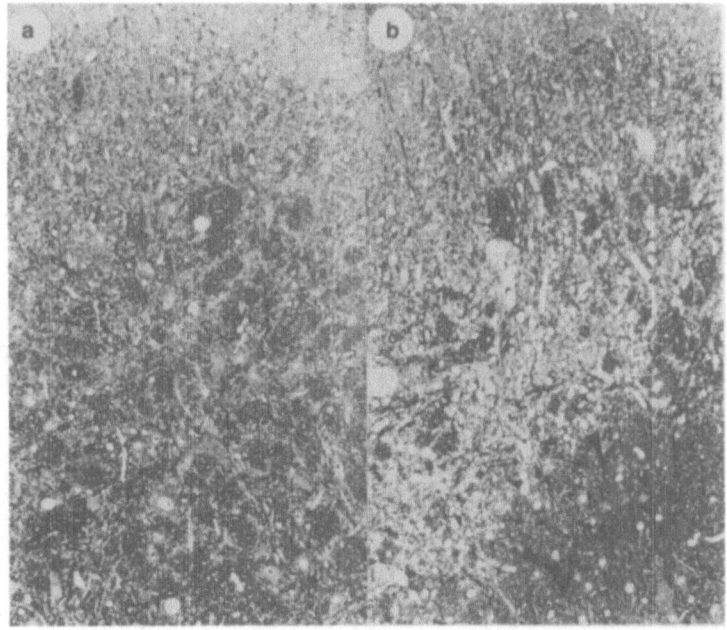

Fig. 1a. Posterior horn of a control rat showing no pathological changes.
b.   Posterior horn of a rat infused with NMDA showing vacuolation of the
neuropil in the substantia gelatinosa and the adjacent lamina. Neurons in
the region are necrotic (arrowheads). Toluidine blue stain.

to the junction of the grey and white matter at the lateral and anterior aspects of the cord. Ultrastructural examination showed that the vacuolation was due to swelling of dendrites (Fig. 2). Some dendrites showed absence of organelles due to a severe degree of swelling, while other dendrites showed a few residual mitochondria and neurofilaments.

The neurons in these regions were necrotic showing cytoplasmic shrinkage and nuclear pyknosis (Fig. 1b), while other neurons displayed intracytoplasmic vacuoles (Fig. 2).

The frequency of nervous tissue changes was less in the anterior than the posterior horns.

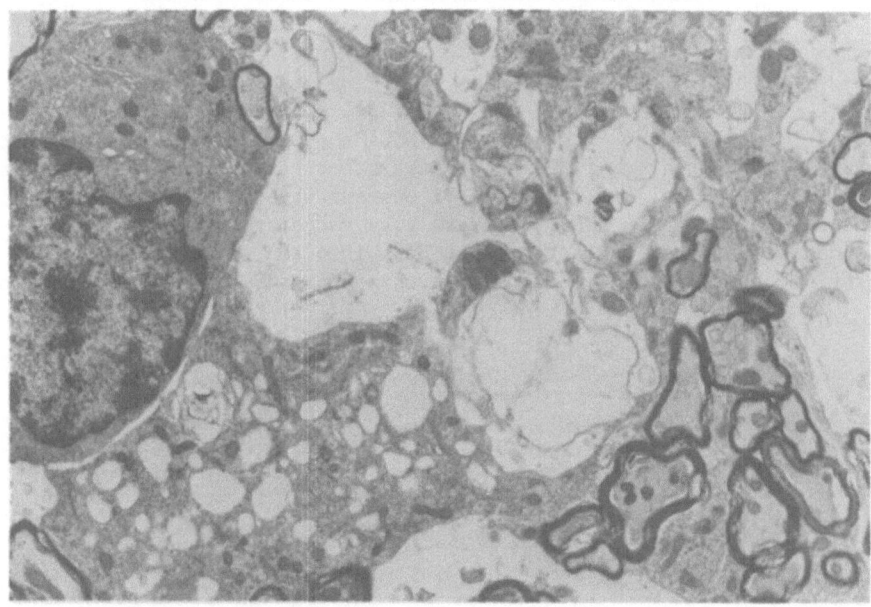

Fig 2. Electron micrograph showing the posterior horn of a rat infused with 30 ul NMDA/min. The dendrites in the area are swollen with electron-lucent cytoplasm. The oligodendrocyte in the upper left hand corner of the micrograph appears to be normal. The lower part of the micrograph shows a necrotic neuron with intracytoplasmic vacuoles of varying size.

Vascular Changes

Vessels within the substance of the spinal cord of control rats did not show increased permeability to HRP. Occasional pial arterioles were permeable to HRP and showed tracer in their walls.

One of the 3 rats infused with 30 ug/min NMDA showed mild blood-brain barrier breakdown with presence of HRP in 12 vessels in the 30 spinal cord slices examined. Although increased permeability affected all types of vessels, it was principally capillaries and venules that showed increased permeability. There was extravasation of HRP from many vessel walls into the surrounding neuropil forming focal collections of HRP in the grey matter. Permeable vessels were present in the areas with vacuolation of the neuropil. However, they appeared to be more common in the adjacent viable tissue.

Extensive breakdown of the blood-brain barrier was observed in all rats infused with 60 ug/ml NMDA. Tracer was observed in vessel walls and in continuity in the extracellular spaces of the surrounding neuropil (Fig. 3a). Endothelium of permeable vessels showed increased numbers of pinocytotic vesicles many of which contained HRP (Fig. 3b). In addition tracer was observed in abluminal pinocytotic vesicles and in continuity in the adjacent basement membrane. Tracer was observed focally in intercellular junctions, but was not observed occupying the entire length of junctions from the luminal to the abluminal end.

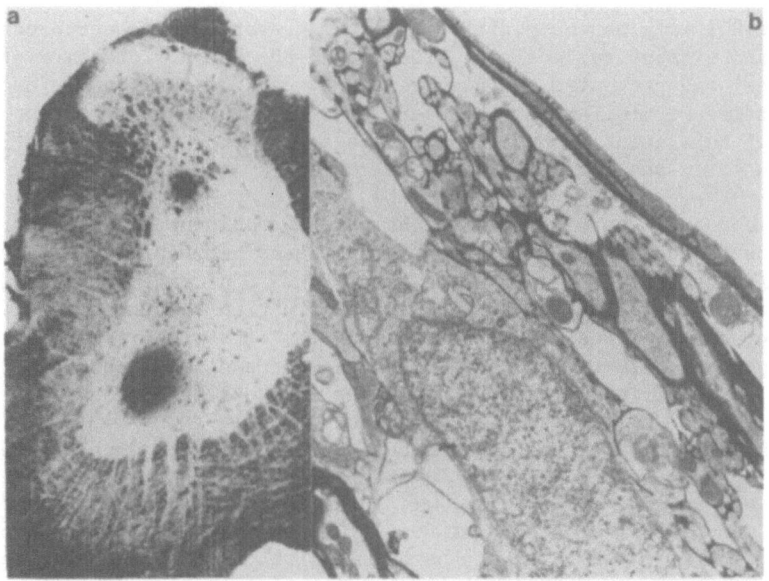

Fig. 3a. Unstained slice of spinal cord from a rat infused with NMDA showing two focal areas of HRP extravasation in the grey matter adjacent to the areas showing NMDA-induced tissue changes. b. Electron micrograph showing a segment of venular wall with HRP in the wall and in continuity in the extracellular spaces of the surrounding neuropil. Note the increased numbers of pinocytotic vesicles in the endothelium.

DISCUSSION

This study demonstrates toxicity in spinal neuronal systems in response to a short intrathecal infusion of NMDA. Breakdown of the blood-brain barrier in the grey matter of spinal cord was a consistent finding in rats infused with the higher dose of NMDA. These changes were specific for NMDA since they did not occur in rats pretreated with MK-801, a non-competitive antagonist of NMDA.

Nervous tissue changes were present in both the anterior and posterior horns. However, they were more frequent in the posterior horn. The latter may be related to the fact that a higher concentration of NMDA was present in the region of the posterior horn which was in the direct vicinity of the tip of the infusion tube. Although, NMDA receptors have been mapped in the region of the substantia gelatinosa and the anterior horn[8], there may be a differential density of NMDA receptors in the gray matter of the adult rat spinal cord which may account for the greater toxicity observed in the posterior horn.

This study demonstrates the typical histological findings associated with excitotoxin-induced brain injury, namely, swelling and destruction of dendrites and neurons but survival of presynaptic terminals and nonneuronal cells. Olney[9] reported that glutamate-induced lesions whether in the retina or brain and regardless of species, involved rapid swelling of neuronal dendrites and cell bodies followed by acute degenerative changes in intracellular organelles and coarse clumping of nuclear chromatin. The reaction was very rapid with onset of dendritic swelling being detectable in 15-30 min and phagocytosis of the necrotic neuronal cell bodies beginning as early as 3 hr after s.c. adminstration of glutamate[10].

MK-801 is a non-competitive NMDA antagonist with a neuroprotective effect and without serious side effects. MK-801 had a protective effect in this model and prevented the occurrence of the NMDA-induced neurotoxicity. This finding is similar to the observations of others of a neuroprotective effect of this agent in excitotoxic damage occurring in ischemia[11], hypoglycemia[12,13], and in response to NMDA exposure[14,15].

Blood brain-barrier alterations were a prominent feature in rats infused with the higher dose of NMDA. Increased permeability affected all types of vessels. However, capillary and venular permeability was more prominent. This is in keeping with the observation that the reactivity of different parts of the vascular system varies with arterioles playing a more important role in conditions such as hypertension[16] and capillary and venular permeability being more prominent in other conditions.

Experimental studies have demonstrated that passage of protein and protein tracers across cerebral endothelium could occur by transcytotic passage of pinocytotic vesicles, via interendothelial junctions and via damaged endothelial plasma membranes[17]. One or more of these mechanisms play a major role depending on the pathological state being studied. In the present study the principal finding was increased numbers of pinocytotic vesicles in endothelium suggesting that transcytotic passage of vesicles resulted in protein extravasation. Similar findings have been observed in the early stages of blood-brain barrier breakdown in other models[16,18,19].

In the present study vascular alterations occurred following the development of nervous tissue changes suggesting that increased blood-brain barrier alterations develop late in excitotoxicity and lead to edema in these lesions.

ACKNOWLEDGEMENTS

Thanks are expressed to Mrs. Verna Norkum and Mr. Blake Gubbins for skilled technical assistance.

This work was supported by the Heart and Stroke Foundation of Ontario.

REFERENCES

1. C. W. Cotnam, R. J. Bridges, J. S. Taube, A. S. Clark, J. W. Geddes, and D. T. Monaghan: The role of the NMDA receptor in central nervous system plasticity and pathology, J of NIH Research 1:65-74 (1989).

2. S. M. Rothman, and J. W. Olney: Glutamate and the pathophysiology of hypoxic-ischemic brain damage, Ann Neurol 19:105-111 (1986).

3. B. Meldrum, and J. Garthwaite: Excitatory amino acid neurotoxicity and neurodegenerative disease, Trends in Pharmacol Sciences 11:379-387 (1990).

4. S. Sahai: Glutamate in the Mammalian CNS. Eur Arch Psychiatry Clin Neurosci 240:121-133 (1990).

5. S. Nag, and R. J. Riopelle: Spinal neuronal pathology associated with continuous intrathecal infusion of N-methyl-D-aspartate in the rat, Acta Neuropathol (Berl) 81:7-13 (1990).

6. H. B. Hucker, J. E. Hutt, S. D. White, B. H. Arison, and A. G. Zacchei: Disposition and metabolism of (+)-5-methyl-10,11-dihydro-5H-dibenzo[a,d]cyclohepten-5,10-imine in rats, dogs and monkeys, Drug Metab Dispos 11:54-58 (1983).

7. S. Nag: Localisation of calcium-activated adenosine-triphosphatase (Ca^{2+}-ATPase) in intracerebral arterioles in acute hypertension, Acta Neuropathol (Berl) 75:547-553 (1988).

8. D. T. Monaghan, and C. W. Cotman: Distribution of N-methyl-aspartate-sensitive L-[^{3}H] glutamate-binding sites in rat brain, J Neurosci 5:2909-2919 (1986).

9. J. W. Olney: Excitotoxic amino acids: Research applications and safety implications. In "Glutamic acid:Advances in Biochemistry and Physiology", L. J. Filer (Ed) Raven Press, New York, pp 287-319 (1979).

10. J. W. Olney, L. G. Sharpe, and R. D. Feigin: Glutamate-induced brain damage in infant primates, J Neuropath Exp Neurol 31:464-488 (1972).

11. R. P. Simon, J. H. Swan, T. Griffiths, and B. S. Meldrum: Blockade of N-methyl-D-aspartate receptors may protect against ischemic damage in the brain, Science 226:851-852 (1984).

12. M. P. Papagapiou, and R. N. Auer: Regional neuroprotective effects of the NMDA receptor antagonist MK-801 (Dizocilpine) in hypoglycemic brain damage, J Cereb Blood Flow Metab 10:270-276 (1990).

13. T. Weiloch: Hypoglycemia-induced neuronal damage prevented by an N-methyl-D-aspartate antagonist, Science 230:681-683 (1985).

14. J. W. Olney, M. Price, K. S. Salles, J. Labruyere, and G. Friedrich: MK-801 powerfully protects against N-methyl aspartate neurotoxicity, Eur J Pharmacol 141:357-361 (1987).

15. A. C. Foster, R. Gill, J. A. Kemp, and G. N. Woodruff: Systemic adminstration of MK-801 prevents N-methyl-D-aspartate-induced neuronal degeneration in rat brain, Neurosci Lett 76:307-311 (1987).

16. S. Nag, D. M. Robertson, and H. B. Dinsdale: Quantitative estimate of pinocytosis in experimental acute hypertension, Acta Neuropathol (Berl) 46:107-116 (1979).

17. M. W. Brightman, K. Zis, and J. Anders: Morphology of cerebral endothelium and astrocytes as determinants of the neuronal microenvironment, Acta Neuropathol (Berl) Suppl VIII, 21-33 (1983).

18. S. Nag, and S. I. Harik: Cerebrovascular permeability to horseradish peroxidase in hypertensive rats: effects of unilateral locus ceruleus lesion, Acta Neuropathol (Berl) 73:247-253 (1987).

19. E. Westergaard: The blood-brain barrier to horseradish peroxidase under normal and experimental conditions, Acta Neuropathol (Berl) 39:181-187 (1977).

NMDA RECEPTORS MEDIATE ACTIVATION OF POLYAMINE SYNTHESIS

AND BLOOD-BRAIN BARRIER BREAKDOWN AFTER COLD INJURY

Harold Koenig[*], Jerome J. Trout, Alfred D. Goldstone, and Chung Y. Lu

Neurology Service, VA Lakeside Medical Center; and Departments of
Neurology and Cellular, Molecular and Structural Biology, and the Feinberg
Cardiovascular Research Institute, Northwestern University Medical School
Chicago, IL 60611 USA

Cold injury (CI) of cerebral cortex is an extensively used model for focal blood-brain barrier (BBB) breakdown and spreading vasogenic brain edema (Klatzo et al., 1981). CI also induces changes in neuronal function and metabolism, including epileptic discharges and slow (delta) wave activity (Pappius & McCann, 1969) and a depression of local cerebral glucose metabolism (Pappius, 1981). We found that focal CI of rat cerebral cortex induces a biphasic increase in polyamine (PA) levels and the activity of their regulatory synthetic enzyme ornithine decarboxylase (ODC), which converts ornithine to putrescine, in perilesional cortex involving both capillaries and brain cells (Koenig et al., 1989a). In the first phase, ODC activity and PA levels increase transiently between 1-5 min after CI. A secondary rise in ODC activity and PA levels occurs after a lag period of 4 h which lasts for more than 72 h. Agents which directly or indirectly inhibit CI-induced stimulation of ODC activity and PA accumulation also prevent BBB breakdown, monitored by fluorescein and horseradish peroxidase (HRP) leakage, while exogenous putrescine reverses the effects of ODC inhibition and restores BBB breakdown. These agents include the specific suicide ODC inhibitor α-difluoromethylornithine (DFMO), (Metcalf et al., 1978; Koenig et al., 1983, 1989a; Trout et al., 1986), verapamil, dexamethasone, and aspirin (Koenig et al., 1989a).

We recently found that gerbil brain ODC activity and PA levels are regulated by NMDA receptors during global forebrain ischemia and reperfusion (Koenig et al., 1990), and that the ODC/PA signaling cascade is involved in NMDA receptor transduction (Siddiqui et al., 1988; Koenig et al., 1989b, 1991a). Since the extracellular concentration of glutamate, the major excitatory neurotransmitter and NMDA receptor agonist in mammalian brain(Monaghan et al., 1989), is frequently

---

[*]Harold Koenig passed away June 26, 1992

The Role of Neurotransmitters in Brain Injury, Edited by
M. Globus and W.D. Dietrich, Plenum Press, New York, 1992

elevated after cerebral injury, including focal CI (Baethmann et al., 1989), we tested the hypothesis that NMDA receptors may mediate some effects of CI through the use of the noncompetitive NMDA receptor antagonist (+)5-methyl-10-11-dihydro-5H-dibenzo(a,d)cyclohepten-5,10-imine (MK-801) (Wong et al., 1986). We also conducted in vitro studies in isolated rat cerebral capillaries.

METHODS

Young adult female Sprague-Dawley rats were anesthetized with pentobarbital (50 mg/kg). Two 3-mm openings were made in the right parietal bone with a dental drill, the exposed dura was overlaid with plastic wrap, and metal probes 2 mm in diameter precooled in liquid nitrogen were applied at both sites for two 30-sec periods separated by a 30 sec interval (Koenig et al., 1983, 1989a, Trout et al., 1986). Unoperated and warm probe-treated (sham operated) animals served as controls. For measurement of ODC activity, brains were rapidly chilled by brief immersion of the decapitated heads in liquid nitrogen 2 min after initiating CI. For assessing BBB integrity, the micromolecular tracers sodium fluorescein (1 ml/kg of a 10% solution) or α-[1-^{14}C]aminoisobutyrate ([^{14}C]AIB) (50 μCi/kg), and the macromolecular tracer HRP (100 mg/kg) were administered into the caudal vein 5 min before CI. Animals were killed 3.5 min after initiating CI. A cardiac blood sample was removed at sacrifice to normalize tracer uptake in brain to blood tracer concentration. Fluorescein, [^{14}C]AIB, HRP, and ODC activity were measured as previously described (Koenig et al., 1989a). For semiquantitative analysis of HRP transcytosis across cerebral capillary endothelium, HRP was injected intravenously immediately before CI, and brains were perfuse-fixed transcardially with 3% glutaraldehyde in 0.1 M sod. cacodylate buffer, pH 7.3. Frontal sections through lesions were cut at 50-100 μM with a Vibratome, incubated for peroxidase activity, and processed for light and electron microscopy as previously described (Trout et al., 1986). The first hundred capillaries in four brains per treatment were scored for the presence of luminal pits and HRP-labeled vesicles. The following treatment protocols were generally used: (1) unoperated or sham operated (warm probe-treated) controls; (2) CI controls; (3) MK-801, 1,2 and 10 mg/kg injected i.p. 30 min before CI.

Isolated rat cerebral capillaries were used as an in vitro model of the BBB, as described earlier (Koenig et al., 1989c). Capillaries were isolated according to Kobayashi et al. (1981). Electronmicroscopy revealed that capillaries and small numbers of arterioles/venules were the major constituents. Contamination by neurons, neuronal processes, and glia was not detected, although astrocytic membrane fragments still attached to capillaries were occasionally observed. Rates of endocytosis, hexose, and ^{45}Ca transport were studied by monitoring the temperature-sensitive uptake of HRP, 2-[^{3}H]deoxyglucose ([^{3}H]DG), and ^{45}Ca^{2+} in isolated vessels incubated in an artificial cerebrospinal fluid at 37° C under 95% O_2-5% CO_2 with shaking for 2 min as previously described (Koenig et al., 1989c). Data were expressed as mean $\pm$ SEM values. Results were analyzed statistically by a one-way analysis of variance. Fischer's least significant difference test was used to assess the statistical significance of differences between individual group means. Statistical significance was set at p<.05.

RESULTS

It can be seen in Table 1 that CI produced a prompt (< 2 min) increase in ODC activity and an early leakage of [^{14}C]AIB, fluorescein (not shown) and HRP into perilesional cortex, as described earlier (Trout et al., 1986; Koenig et al., 1989a). MK-801, in all doses tested (1-10 mg/kg), prevented the increase in ODC activity and the leakage of tracers in the perilesional tissue.

We previously showed that BBB breakdown and leakage of HRP following CI is mediated by increased endocytosis and vesicular transport (transcytosis) across the capillary endothelium (Trout et al. 1986). We therefore examined the effect of MK-801 on the enhanced transcytosis of HRP produced by CI. It can be seen in Table 1 that CI markedly enhanced the percentage of capillaries displaying increased endocytosis aand vesicular transport, identified by the presence of endocytic pits and HRP-labeled vesicles. Pretreatment with MK-801 abolished this response, indicating involvement of NMDA receptors in BBB breakdown.

To assess the possibility that NMDA receptors could be present on brain capillaries proper, we examined the efffects of NMDA on ODC activity and on select transport processes in acutely isolated rat cerebral capillaries. The basal ODC activity in brain capillaries was enriched 7- to 10-fold over the cortical homogenate, suggesting that polyamines play an important role in the BBB (Koenig et al., 1989c). NMDA (50 μM) evoked a 2- to 3-fold increase in capillary ODC activity within 2 min (Table 2). This effect was blocked by the competitive NMDA receptor antagonist 2-amino-5-phosphonovalerate (AP5, 200 μM) (Watkins & Olverman, 1987), and the ODC inhibitor DFMO (5-10 mM). In addition, ODC activation was blocked by removing extracellular Ca^{2+} and by the Ca^{2+} channel antagonist nisoldipine (0.1 μM). Thus, the NMDA-induced stimulation of ODC activity in brain capillaries is NMDA receptor-dependent and requires Ca^{2+} entry via dihydropyridine-sensitive Ca^{2+} channels.

DISCUSSION

A major finding of this investigation is that the CI-induced activation of polyamine synthesis and BBB breakdown in rat cerebral cortex involves NMDA receptors, as these responses are coordinately blocked by the specific NMDA receptor antagonist MK-801. It is known that CI markedly increases the brain extracelllular glutamate level in edema fluid (Baethmann et al., 1989), which doubtless activates NMDA receptors. The rapidity of these MK-801-sensitive capillary responses to CI raises the possibility that glutamate could be acting directly on NMDA receptors associated with brain capillaries. This inference is supported by the finding that NMDA rapidly stimulates ODC activity in isolated rat cerebral capillaries by activating NMDA receptors. Moreover, NMDA coincidentally stimulates the uptake of ^{45}Ca, DG and HRP in a NMDA receptor- and temperature-dependent manner, and this uptake is inhibited by the ODC inhibitor, DFMO, suggesting that it requires polyamine synthesis. The NMDA receptor-mediated stimulation of ODC activity and capillary transport processes is dependent on Ca^{2+} entry via dihydropyridine-sensitive, voltage-gated Ca^{2+} channels. Such channels have been previously detected in rat brain capillaries by Dooley et al. (1987).

Electron cytochemical study has demonstrated that NMDA stimulates endocytic uptake and vesicular transport of HRP in isolated cerebral capillaries (unpublished data). This confirms the biochemical data suggesting that NMDA receptors at the abluminal plasma membrane of brain capillary endothelium regulate the transcapillary transport of proteins, and inferentially of Ca^{2+} and hexose, by their respective transport systems. We have recently identified NMDA receptors in rat cerebral capillary membranes by radioligand binding assays (Koenig et al., 1991). Thus capillary membranes display NMDA-displaceable L-[^{3}H]glutamate binding activity (K_d, 50-150 nM; B_{max}, 175-300 fmol/mg protein). L-Glutamate plus glycine-stimulatable, AP5-sensitive binding of MK-801 to capillary membranes has also been detected. These data suggest that brain capillaries possess NMDA receptors similar to those present on neuronal membranes (Monaghan et al., 1989).

These findings support the view that extracellular glutamate is an important modulator of capillary transport, and, under pathological conditions, can breach the BBB barrier by excessive activation of capillary NMDA receptors. Astrocytes in synaptic zones play an important role in removing extracellular glutamate via the Na^{2+}-dependent glutamate transport system (Nicholls & Atwell, 1990). We suggest that perivascular astrocytes, whose vascular processes constitute a cytoplasmic investment of brain capillaries, play a similar role in rapidly removing glutamate from the pericapillary spaces.

ACKNOWLEDGEMENTS

Supported by Veterans Administration Research Service, National Institutes of Health grants NS 18047 and HL-26835 and grants from the Searle Family Center for Neurological Diseases and Feinberg Cardiovascular Research Institute, Northwestern University Medical School.

REFERENCES

Baethmann, A., Maier-Hauff, K., Schurer, L., et al., 1989, Release of glutamate and of free fatty acids in vaogenic brain edema. J. Neurosurg. 70:578-591.

Dooley, D.J., Mahlmann, H., Brenner, O., and Osswald, H., 1987, Characterization of the dihydropyridine binding sites of rat neocortical synaptosomes and microvessels. J. Neurochem. 49:900-904.

Klatzo, I., Chui, E., and Fujiwara, K., 1981, Aspects of the blood-brain barrier in brain edema, in: "Brain Edema," de Vlieger, M., de Lange, J.A., and Beks, J.W.F., eds., John Wiley, New York, pp. 11-18.

Kobayashi, H., Memo, M., Spano, P.F., and Trabucchi, M., 1981, Identification of β-adrenergic receptor binding sites in rat brain microvessels, using [^{125}I]-iodohydroxybenzypindol, J. Neurochem. 36:1383-1388.

Koenig, H., Goldstone, A.D., and Lu, C.Y., 1983, Blood-brain barrier breakdown in brain edema following cold injury is mediated by microvascular polyamines. Biochem. Biophys. Res. Comm. 116:1039-1048.

Koenig, H., Goldstone,, A.D., and Lu, C.Y., 1989a, Blood-brain barrier in cold-injured brain is linked to a biphasic stimulation of decarboxylase activity and polyamine synthesis:both are coordinately inhibited by verapamil, dexamethasone and aspirin, J. Neurochem. 52:101-109.

Koenig, H., Iqbal, Z., Goldstone, A.D., Siddiqui, F., and Lu, C.Y., 1989b, NMDA receptors are coupled to ornithine decarboxylase and use polyamines to modulate Ca^{2+} fluxes and transmitter release in cortical synaptosomes, FASEB 3:A588.

Koenig, H., Goldstone, A.D., Lu, C.Y., and Trout, J.J., 1989c, Polyamines and Ca^{2+} mediate hyperosmolal opening of the blood-brain barrier:in vitro studies in isolated rat cerebral capillaries, J. Neurochem. 52:1135-1142.

Koenig, H., Goldstone, A.D., Lu, C.Y., and Trout, J.J., 1990, Brain polyamines are controlled by N-methyl-D-aspartate receptors during ischemia and recirculation, Stroke 21(Suppl. III):III-98-III-102.

Koenig, H., Goldstone, A.D., Trout, J.J., Lu, C.Y., and Iqbal, Z.,, 1991, NMDA receptors on brain capillaries mediate cold injury-induced activation of polyamine synthesis and blood-brain barrier breakdown, Abstr., Soc. Neurosci. 18, in press.

Metcalf, B..W., Bey, P., Danzin, C., Jung, M.J., Casara, P., and Vevert, J.P., 1978, Catalytic irreversible inhibition of mammalian ornithine decarboxylase (EC 4.1.1.17) by substrate and product analogues, J. Amer. Chem. Soc. 100:2551-2553.

Monaghan, D.T., Bridges., R.J., and Cotman, C.W., 1989, The excitatory amino acid receptors:their classes, pharmacology, and distinct properties in the function of the central nervous system, Ann. Rev. Pharmacol. Toxicol. 29:365-402.

Nicholls, D. and Attwell, D., 1990, The release and uptake of excitatory amino acids, Trends Pharmacol. Sci. 11, 462-468.

Pappius, H.M. and McCann, W.P., 1969, Effects of steroids on cerebral edema in cats, Arch. Neurol. 20:207-216.

Pappius, H.M., 1981, Local cerebral glucose utilization in thermally traumatized brain, Ann. Neurol. 9:484-491.

Siddiqui, F., Iqbal, Z., Koenig, H., Goldstone, A.D., and Lu, C.Y., 1988, Polyamine dependence of NMDA receptor-mediated Ca^{2+} fluxes and transmitter release from rat hippocampus, Abstr., Soc. Neurosci. 14:1048..

Trout, J.J., Koenig, H., Goldstone, A.D., and Lu, C.Y., 1986, Blood-brain barrier breakdown by cold injury. Polyamine signals mediate acute stimulation of endocytosis, vesicular transport, and microvillus formation in rat cerebral capillaries, Lab. Invest. 55:622-631..

Watkins, J.C. and Olverman, H.J., 1987, Agonists and antagonists for excitatory amino acid receptors, Trends Neurosci. 10:265-272.

Wong, E.F., Kemp, J.A., Priestley, T., Knight, A.R., Woodruff, G.N., and Iverson, L.L., 1986, The anticonvulsant MK-801 is a potent N-methyl-D-aspartate antagonist, Proc. Natl. Acad. Sci. USA 83:7104-7108.

MODERATE HYPOTHERMIA REDUCES BLOOD-BRAIN

BARRIER DISRUPTION FOLLOWING TRAUMATIC BRAIN INJURY

Ji Y. Jiang, Bruce G. Lyeth, Manisha Kapasi[1], and John Povlishock[1]

Division of Neurosurgery
[1]Department of Anatomy
Medical College of Virginia
Richmond, Virginia 23298

INTRODUCTION

Several studies indicate that hypothermia may modulate pathological responses to experimental traumatic brain injury (TBI). For example, induction of hypothermia (30°C) prior to fluid percussion injury in rats reduces mortality (Clifton et al., 1991) and induction of hypothermia (30° or 33°C) 5 minutes after injury reduces motor deficits (Clifton et al., 1991). The mechanisms of hypothermic protection in TBI are presently not known. Recent data, however, now indicate that hypothermia (30°C) significantly reduces the increase in acetylcholine (ACh) concentration in cerebrospinal fluid (CSF) of TBI rats (Lyeth et al., 1991).

In cerebral ischemia, mild-to-moderate hypothermia also improves neurologic outcome (Hoffman et al., 1991) as well as reduces histologic and biochemical sequelae (Busto et al., 1987; 1989a,b,; Chopp et al., 1989; Clifton et al., 1989; Dempsey et al., 1987; Gingsberg et al., 1989). Additionally, hypothermia (30-33°C) reduces and hyperthermia (39°C) exacerbates the BBB disruption associated with cerebral ischemia in rats (Dietrich et al., 1990).

It has long been recognized that TBI results in alteration of BBB (for reviews see: Pardridge, 1985; Povlishock, 1985). Mild injury produces subtle endothelial perturbations while injuries of increasing intensity also produce hemorrhage and frank vascular disruption which obviously contribute to such barrier change. The factors involved in such BBB perturbation include not only direct mechanical stimulation of the microvessels involved, but perhaps also neurogenic and hydrostatic influences.

Alterations in BBB function associated with TBI (Povlishock and Lyeth 1989; McIntosh 1989a) could allow the abnormal passage of blood-borne neurotransmitters (e.g. ACh, glutamate) into the brain which could summate with vesicular release of brain neurotransmitters (Robinson et al., 1990).

The Role of Neurotransmitters in Brain Injury, Edited by
M. Globus and W.D. Dietrich, Plenum Press, New York, 1992

Excitatory agonist-receptor interactions mediate at least same of the pathophysiological consequences of traumatic brain injury (TBI) (Hayes et al., 1988; Lyeth et al., 1988; Faden et al 1989; McIntosh et al., 1989b; Robinson et al., 1990). Thus, if hypothermia attenuates BBB disruption following TBI, then one possible mechanism of hypothermia protection may be the blunting of blood-borne toxic substances from entering the brain parenchyma.

In the present study, we examined the effects of moderate hypothermia upon TBI-induced alterations in cerebral vasculature permeability to circulating serum proteins.

MATERIALS AND METHODS

All animals were anesthetized, intubated, ventilated with 2% isoflurane in 70%N_2O/30%O_2, and surgically prepared 24 hours prior to injury. A hole of 4.8 mm diameter was trephined into the skull over the right parietal cortex midway between the sagittal suture and lateral ridge, and midway between lambda and bregma. Two steel screws were placed 1 mm rostral to bregma and 1 mm caudal to lambda. A rigid plastic injury tube (modified Leur-loc syringe hub, 2.6 mm inner diameter) was placed over the exposed dura and bonded with cyanoacrylate adhesive and dental acrylic. After the acrylic had hardened, the injury tube was plugged with Gelfoam sponge, the scalp was sutured closed, and the animal returned to its home cage.

Brain temperature was monitored by a probe in the temporalis muscle with a thermocouple thermometer (Jiang et al., 1991). Brain temperatures of 30°C was achieved by ice pack application to the body. The ice pack was removed, when brain temperature had fallen to 32°C. Brain temperatures continued to decrease to 30°C. Normothermic brain temperature of 37°C was maintained with use of a heating lamp under general anesthesia after injury. The target brain temperatures (±0.2°C) were maintained for one hour in both groups by continuing general anesthesia at room temperature and applying ice packs or heating lamps as needed.

A hypothermia group (n=4) was cooled to 30°C prior to injury and maintained at 30°C for 60 minutes after injury. A normothermia group (n=4) was maintained at 37°C during this same period. Animals were injured at 2.10-2.25 atmospheres by a fluid percussion model as soon as the appropriate target temperature was achieved. Seventy minutes after injury rats were perfused transcardially under deep sodium pentobarbital anesthesia (50 mg/kg) and immunocytochemically prepared for light microscopic visualization of endogenous serum albumin which is normally excluded from entering the brain by the BBB.

Rats were perfused with 0.9% saline followed by 4% paraformaldehyde and 0.2% glutarylaldehyde in Millonig's phosphate buffer. The brains were sectioned (50 μm), incubated in 3% Triton X, and then incubated in primary antibody, Goat anti-rat albumin IgG (1: 15,000) for 16-18 hours. The sections were next incubated in the biotinylated secondary antibody, rabbit anti-goat IgG (1:200) for one hour, followed by incubation in an avidin-biotin (Vectastain) peroxidase complex for 45 minutes. Finally, the sections were immersed with an equal mixture of 0.02% hydrogen peroxide and 0.1% DAB (Diaminobenzidene) for the visualization of the reaction product. The sections were picked up on glass

TABLE 1. MK-801 blocks the rapid activation of ODC, leakage of fluorescein and HRP, and transcapillary transcytosis of HRP induced by CI in perilesional cortex.

TREATMENT	ODC activity (pmol/h/mg prot.)	Fluorescein uptake (µg/mg prot.)	HRP uptake (ng/mg prot.)	% of capillaries w/luminal pits	% of capillaries w/HRP-labeled vesicles
Sham control	10.2 ± 1.2	.226 ± .04	1.00 ± .30	15.4 ± 1.5	13.3 ± 2.7
CI control	20.3 ± 2.8*	.552 ± .03**	3.14 ± .60*	76.3 ± 1.3***	68.7 ± 1.5***
MK-801 (1 mg/kg)	9.0 ± 1.9†	.234 ± .04††	1.18 ± .16†	16.7 ± 6.8†††	10.7 ± 2.7†††

Data are means ± SEM (n=3-5). *,**,***: p<.05, .01, .001 (vs sham control). †,††,†††: p<.05, .01, .001 (vs CI control).

TABLE 2. NMDA rapidly stimulates ODC activity and uptake of ^{45}Ca, $[^{14}C]DG$, and HRP in rat cerebral capillaries in vitro. Effect of inhibitors.

TREATMENT	ODC activity (pmol/h/mg prot.)	^{45}Ca uptake (nmol/mg prot.)	$[^{14}C]DG$ uptake (nmol/mg prot.)	HRP uptake (ng/mg prot.)
Control	177 ± 12.2	4.52 ± 1.86	9.53 ± 1.93	185 ± 10.2
NMDA (50 µM)	329 ± 25.0**	25.10 ± 4.37*	22.90 ± 2.47*	603 ± 43.0***
AP5 (200 µM)	192 ± 23.0†	8.70 ± 3.84†	14.60 ± 1.32†	228 ± 88.0†
DFMO (10 mM)	117 ± 22.0††	8.70 ± 2.62†	8.63 ± 1.34††	247 ± 47.0††
Nisoldipine (0.1 µM)	138 ± 9.2††	12.70 ± 2.17†	8.30 ± 2.63†	165 ± 49.0††
CNQX (10 µM)	294 ± 44	n.d.	n.d.	n.d.
Ca^{2+}-free	173 ± 43†	n.d.	n.d.	n.d.

Acutely isolated capillaries were preincubated without or with the various inhibitors for 3 min. At zero time 50 µM NMDA and the substrates ^{45}Ca (1 µCi/ml), $[^{14}C]DG$ (0.1 mM, 4 µCi/ml), and HRP (1 mg/ml) were added and incubations were terminated by chilling in ice at 2 min. Preincubation and incubation were at 37° C under 95% O_2-5% CO_2 with shaking. 0° C uptake values were subtracted from 37° C values to correct for nonspecific uptake. For additional details, see Koenig et al. (1989c). Data are means ± SEM (n=4). *,**,***: p<.05, .01, .001 (vs control). †,††,†††: p<.05, .01, .001 (vs NMDA).

slides, dehydrated, and cover slipped for subsequent light
microscopic analysis.

RESULTS

In normothermia brain injured rats (Figure 1A), increased
vascular permeability to serum albumin was observed throughout
the cortical gray and white matter as well as in the
underlying hippocampi at the site of injury and both rostral
and caudal to the site of injury in both hemispheres. BBB
permeability was greater in the hemisphere under the injury
site.

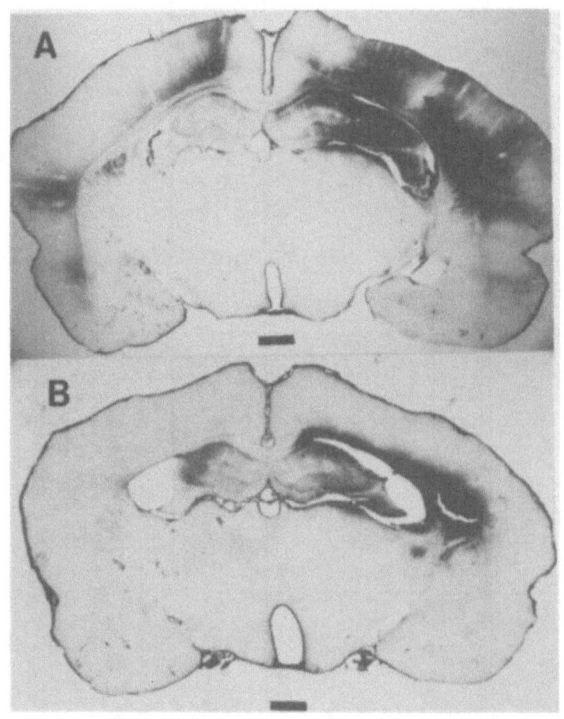

Figure 1.   (A) Normothermia (37°C) TBI rat. Permeability
changes were found throughout the cortical grey and white
matter and in the underlying hippocampi; (B) Hypothermia
(30°C) TBI rat. Permeability changes were confined to the
grey-white interface with no involvement in the overlying
cortices and reduced involvement in the hippocampi. Dimension
bar = 1mm

In contrast to normothermic TBI rats, the permeability to
serum albumin in hypothermic rats was greatly reduced (Figure
1B). Visualization of the reaction product was confined to
the grey-white interface between cortex and hippocampi.
Additionally, serum albumin was not observed in the overlying
cortices and was greatly reduced in the hippocampi.

DISCUSSION

Our findings of increased BBB permeability following moderate fluid percussion TBI are consistent with previous findings (McIntosh et al., 1989a; Povlishock and Lyeth 1989) demonstrating, after TBI, the passage from blood to brain of various substances that are normally confined to the vasculature. We observed intervascular permeability of endogenous rat albumin visualized by the use of antibodies targeted against the rat albumin in both hemispheres at the plane of injury as well as at sites both rostral and caudal to the injury site. The source of this exudation typically involves the passage through intact microvessels, however, sometimes we observed isolated foci of petechial hemorrhage which would also clearly contribute to permeability change. Moderate hypothermia significantly reduced the BBB permeability to endogenous rat albumin following TBI.

Alterations in BBB permeability may contribute to TBI excitotoxic processes. Recent studies indicate that excitatory agonist-receptor interactions may mediate at least some of the pathophysiological consequences of TBI. Excessive activation of muscarinic cholinergic and/or NMDA glutamate receptors significantly contribute to this process (Hayes et al., 1988; Lyeth et al., 1988; Faden et al 1989; McIntosh et al., 1989b). Perturbations in BBB integrity could also contribute to the activation of these receptors by allowing the abnormal passage of blood borne exogenous neurotransmitters into the brain summating with brain neurotransmitters released from vesicular pools. For example, Robinson et al. found that selective depletion of peripheral ACh prior to TBI in the rat with A-5, a bis quaternatery amine derivative of hemicholinium-3, reduced injury-induced elevations of ACh in the CSF by 39% and also significantly reduced functional deficits associated with injury (Robinson et al., 1990). Their results strongly suggest that BBB perturbations resulting from TBI allow blood plasma constituents such as ACh to gain access to the brain and influence injury processes. Furthermore, we have recently found that hypothermia (30°C) significantly reduces the increase in ACh concentration in CSF of TBI rats (Lyeth et al., 1991). Thus, the reduction of BBB permeability by hypothermia may curtail one source of damaging neurotransmitters following injury.   Our observations are also consistent with previous reports of hypothermic protection in ischemic brain injury (Dietrich et al., 1990). Dietich and his associates have found that mild-to moderate hypothermia (30-33°C) significantly reduced BBB permeability associated with forebrain ischemia as compared to the normothermic (36°C) ischemic brain injured rats. Additionally, BBB permeability was more widespread in the hyperthermic (39°C) ischemic brain injured rats (Dietrich et al., 1990).

The mechanisms of by which hypothermia provides protection of BBB function observed following TBI are not known. Multiple factors may be involved in the observed hypothermic protection including: direct effects upon transendothelial passage of albumin; reduced rate of diffusion of albumin once it has crossed the BBB; decreased hydrostatic pressure via reduced vascular hypertension. Regardless of the mechanism involved, the hypothermic reduction of albumin tracer into and through the brain may exert a significant protective effect via the exclusion or reduction of passage of various harmful substances normally confined to the vasculature.

ACKNOWLEDGEMENTS

This research was supported by Grant NS 21587 from The National Institutes of Health and Grant H133B80029 from the National Institute on Disability and Rehabilitation Research, the U.S. Department of Education.

REFERENCES

Busto, R., Dietrich, W.D., Globus, M.Y.T., Valdes, I., Scheinber, P., and Ginsberg, M.D., 1987, Small differences in intraischemic brain temperature critically determine the extent of ischemic neuronal injury. J Cereb Blood Flow Metab 7:729-738.

Busto, R., Dietrich, W.D., Globus, M.Y.T., Castella, Y., and Ginsberg, M.D. 1989a, Postischemic moderate hypothermia inhabits CA1 hippocampal ischemic neuronal injury. J Cereb Blood Flow Metab 9(suppl 1):266.

Busto, R., Globus, M.Y.T., Dietrich, D., Martinez, E., Valdes, I., and Ginsberg, M.D., 1989b, Effect of mild hypothermia on ischemia-induced release of neurotransmitters and free fatty acids in rat brain. Stroke 20:904-910.

Chopp, M., Knight, R., Tidwell, C.D., Helpern, J.A., Brown, E., and Welch, K.M.A., 1989, The metabolic effects of mild hypothermia on global cerebral ischemia and recirculation in the cat: comparison to normothermia and hyperthermia. J Cereb Blood Flow Metab 9:141-148.

Clifton, G.L., Taft, W.C., Blair, R.E., Choi, S.F., and DeLorenzo, R.J., 1989, Conditions for pharmacologic evaluation in the gerbil model of forebrain ischemia. Stroke 20:1545-1552.

Clifton, G.L., Jiang, J.Y., Lyeth, B.G., Jenkins, L.W., Hamm, R.J., and Hayes, R.L., 1991, Marked protection by moderate hypothermia after experimental traumatic brain injury. J Cereb Blood Flow Metab 11:114-121.

Dempsey, R., Combs, D.J., Maley, M.E., Cowen, D.E., Roy, M.W., and Donaldson, D.L., 1987, Moderate hypothermia reduces postischemic edema development and leukotriene production. Neurosurgery 21:177-181.

Dietrich, W.D., Busto, R., Halley, M., and Valdes, I., 1990, The importance of brain temperature in alterations of the blood-brain barrier following cerebral ischemia. J Neuropathol Exp Neurol 49:486-497.

Faden, A.I., Demediuk, P., Panter, S.S., and Vink, R., 1989, The role of excitatory amino acid and NMDA receptors in traumatic brain injury. Science 244:798-800.

Gingsberg, M.D., Busto, R., Castella, Y., Valdes, I., and Loor, J. 1989, The protective effect of moderate intra-ischemic brain hypothermia is associated with improved postischemic glucose utilization and blood flow. J Cereb Blood Flow Metab 9(suppl. 1):380.

Hayes, R.L., Jenkins, L.W., Lyeth, B.G., Balster, R.L., Robinson, S.E., Miller, L.P., Clifton, G.L., and Young, H.F., 1988, Pretreatment with phencyclidine, an N-methyl-D aspartate receptor antagonist, attenuates long-term behavioral deficits in the rat produced by traumatic brain injury. J Neurotrauma 5:287-302.

Hoffman, W.E., Werner, C., Baughman, V.L., Thomas, C., Miletich, D.J., and Albrecht, R.F., 1991, Postischemic treatment with hypothermia improves outcome from incomplete cerebral ischemia in rats. J Neurosurg Anesthesiol 3:34-38.

Jiang, J.Y., Lyeth, B.G., Clifton, G.L., Jenkins, L.W., Hamm, R.J., and Hayes, R.L., 1991, Relationship between body and brain temperature in traumatically brain-injured rodents. J Neurosurg 74:492-496.

Lyeth, B.G., Dixon, C.E., Jenkins, L.W., Hamm, R.J., Alberico, A., Young. H.F., Stonnington, H.H., and Hayes, R.L., 1988, Effects of scopolamine treatment on long-term behavioral deficits following concussive brain injury to the rat. Brain Res 452:39-48.

Lyeth, B.G., Jiang, J.Y., Robinson, S.E., Guo, H., Jenkins, L.W., and Hayes, R.L., 1991, Hypothermia blunts acetylcholine increase in CSF in traumatically brain injured rats. Soc Neurosci Abstr (in press).

McIntosh, T.K., Vink, R., Noble, L., Yamakami, I., Soares, H., and Faden, A.L., 1989a, Traumatic brain injury in the rat: characterization of a lateral fluid-percussion model. Neuroscience 28:233-244.

McIntosh, T.K., Vink, R., Soares, H., Hayes, R.L., and Simmon, R., 1989b, Effects of N-Methyl-D-Aspartate receptor blocker MK-801 on neurologic function after experimental brain injury. J Neurotrauma 6:247-259.

Pardridge, W.M., 1985, Cerebral vascular permeability status in brain injury, in: Central Nervous System Trauma Status Report, D.P. Becker, and J.T. Povlishock, eds., NINCDS & NIH, Washington, D.C..

Povlishock, J.T., 1985, The morphopathologic responses to experimental head injuries of varying severities, in: Central Nervous System Trauma Status Report, D.P. Becker, and J.T. Povlishock, eds., NINCDS & NIH, Washington, D.C..

Povlishock, J.T., and Lyeth, B.G., 1989, Traumatically induced blood-brain barrier disruption: A conduit for the passage of circulating excitatory neurotransmitters. Soc Neurosci Abstr 15:1113

Robinson, S.E., Martin, R.M., Davis, T.R., Gyenes, C.A., Ryland, J.E., and Enters, E.K., 1990, The effect of acetylcholine depletion on behavior following traumatic brain injury. Brain Res 509:41-46.

REGIONAL BLOOD TO BRAIN TRANSPORT OF LACTATE

Joseph C. LaManna, J. Frederick Harrington, Lisa M. Vendel, Kamal Abi-Saleh, W. David Lust, and Sami I. Harik

Departments of Neurology and Neurosurgery, University Hospitals of Cleveland and Case Western Reserve University School of Medicine, Cleveland, Ohio 44106

INTRODUCTION

Brain lactate is elevated in certain pathophysiological conditions such as hypoxic or ischemic brain injury, and plasma lactate levels are elevated during conditions of systemic metabolic stress. In either case, the transport of lactate at the blood-brain barrier (BBB) may play an important role in determining eventual tissue survival. BBB lactate transport is thought to occur via a facilitative, non-concentrative diffusion mechanism (Nemoto et al., 1974; Oldendorf, 1972; Oldendorf, 1973; Gjedde et al., 1975) which is linked to the monocarboxylic acid transporter that also transports pyruvate (Oldendorf, 1973; Pardridge and Oldendorf, 1977). Nevertheless, it is unclear from the previous reports whether the capacity of this system to transport lactate would be functionally meaningful. Here, we measured the rate of lactate influx into brain at plasma concentrations encountered during pathological conditions.

MATERIALS AND METHODS

A dozen male Wistar rats (260-420 g) were kept in the Animal Resource Center with free access to food and water for at least one week before being studied. On the day of the study, the rats were initially anesthetized with halothane and tracheostomized. Surgical preparation was completed under halothane:nitrous oxide:oxygen (2:70:28) anesthesia. Cannulae were inserted into a femoral artery and vein, and into the right atrium of the heart via the external jugular vein. The skin was infiltrated with local anesthetic solution and then closed with wound clips. Halothane was discontinued and curare administered 30 minutes before the final experimental procedure. Rectal temperature was monitored and kept near 37 °C by a feedback controlled infrared lamp. Arterial blood samples were used to determine blood gases, pH, hematocrit, plasma glucose and lactate concentrations. Blood pressure was also monitored. The physiological variables (mean ± sem, n = 11) at the time of the experiment were: body weight, 305 ± 14 g; hematocrit, 44 ± 1; plasma glucose, 13.3 ± 1.0 mM; p_aO_2, 106 ± 9 torr; p_aCO_2, 39 ± 2 torr; pH, 7.39 ± 0.02.

In order to study brain uptake of L-lactate for a range of plasma L-lactate concentrations (1 - 10 mM), rats were given a constant intravenous infusion of 2.5, 5, or 10

The Role of Neurotransmitters in Brain Injury, Edited by
M. Globus and W.D. Dietrich, Plenum Press, New York, 1992

μmoles of L-lactate/gram body weight over 10 minutes. These infusion protocols were based on preliminary studies which showed that the plasma concentration of L-lactate could be elevated to a relatively stable and predictable level by this procedure. Three rats were studied without infusion of L-lactate. The mean (± sem) endogenous plasma lactate in these 3 rats was 1.58 ± 0.09 mM.

Regional blood-to-brain L-lactate transport and regional blood flow were determined by the double label, single pass, atrial bolus injection method (LaManna and Harik, 1985; 1986), with [³H]MPTP as the blood flow indicator (Riachi et al., 1989). The arterial cannula was connected to a syringe fitted to a pump calibrated to withdraw blood at a rate of 1.60 ml/min. The withdrawal pump was started and within seconds a 150 μl bolus of physiological buffer solution was injected into the right atrium. The buffer solution contained 10 - 20 μCi of 1-[methyl-³H]MPTP and 10 - 50 μCi of L-[¹⁴C(U)]lactate in a 10 mM HEPES buffer containing 10 mM glucose and an amount of unlabeled L-lactate to match the plasma lactate concentration. 1-[methyl-³H]MPTP (72.6 Ci/mmol; New England Nuclear), and L-[¹⁴C]lactate (90 Ci/mol; ICN Biomedicals) were evaporated to dryness under reduced pressure, just before use, to remove volatile contaminants. Ten seconds after the atrial injection, rats were decapitated, the withdrawal pump stopped simultaneously and the arterial cannula removed. The withdrawn blood was then transferred to a tared vial, weighed and aliquots measured for radioisotope content. The brain was rapidly removed and bilateral samples from the frontal cortex, parietal cortex, hippocampus, cerebellum and striatum were weighed and their radioisotope content determined. A sample of blood oozing from the foramen magnum was collected in heparinized tubes, and an aliquot of the plasma was counted to estimate the residual vascular radioactive content at decapitation (Sage et al., 1981) as detailed below.

The regional intravascular compartment volumes (V_{pl}), estimated under similar conditions of anesthesia using [¹⁴C]sucrose, were taken from a previous study (Crumrine and LaManna, 1991). These were 13.21, 14.26, 12.73, 15.12, and 10.39 μl plasma/g tissue for the frontal and parietal cortex, hippocampus, cerebellum and striatum. Radioactive counts due to the L-lactate tracer in tissue samples were estimated by multiplying the dpm/μl of plasma from the mixed venous sample by the regional intravascular compartment volumes to give dpm/g, and multiplying this by the weight in grams of each sample to yield the dpm/sample due to intravascular counts. Corrections were then made by subtracting this number from the total dpm of the sample. Regional blood flow (BF) was calculated by the indicator fractionation method (Sage et al., 1981) as previously described (Harik and LaManna, 1988) except that [³H]MPTP was used as the tracer (Riachi et al., 1989): BF (ml/100g/min) = $(F_s \cdot {}^3H_{br})/({}^3H_s \cdot Wt_{br}) \cdot 100$; where F_s is the withdrawal rate of the syringe (ml/min), ${}^3H_{br}$ is the radioactive content (dpm) of the brain sample, ${}^3H_s$ is the radioactive content (dpm) of the withdrawn blood, and the brain sample weight (Wt_{br}) is in grams. Regional permeability-surface area (P*S) products were calculated from the regional extraction fraction and the regional plasma flow according to Renkin (1959) and Crone (1965): P*S (ml/100g/min) = $-F_{pl} \cdot \ln(1-E)$; where F_{pl} is plasma flow (BF · [1 - HCT]) in ml/100g/min, and E is the extraction fraction (Pardridge and Oldendorf, 1975): E = $[({}^{14}C/{}^3H)_{brain} / ({}^{14}C/{}^3H)_{syringe}]$. Unidirectional influx (J) of L-lactate was calculated in each sample as described by Gjedde (1983): J (μmol/100g/min) = P*S · L_{pl}; where L_{pl} is the plasma lactate concentration (μmol/ml).

RESULTS

Infusion of L-lactate had no apparent effect on regional blood flow (Figure 1). Regional blood flow values in the five regions studied in these paralyzed rats (Table 1) were higher than

Table 1. Regional Blood Flow and L-lactate P*S Product

	Regional Blood Flow ml/100g/min	P*S Product ml/100g/min
Cerebral Cortex		
Frontal	286 ± 37	9.2 ± 1.1
Parietal	261 ± 26	8.8 ± 1.0
Hippocampus	153 ± 15	7.6 ± 0.8
Cerebellum	128 ± 9	8.7 ± 1.1
Striatum	210 ± 25	7.3 ± 0.8

Values are means ± sem from 11 rats. Except for the frontal and parietal cortex where blood flows were similar, all regions were significantly different from each other (p < 0.05; paired t-test, 2-tailed, corrected for multiple comparisons). P*S products of the cerebral cortex and cerebellum were different from those of the hippocampus and striatum (p < 0.05).

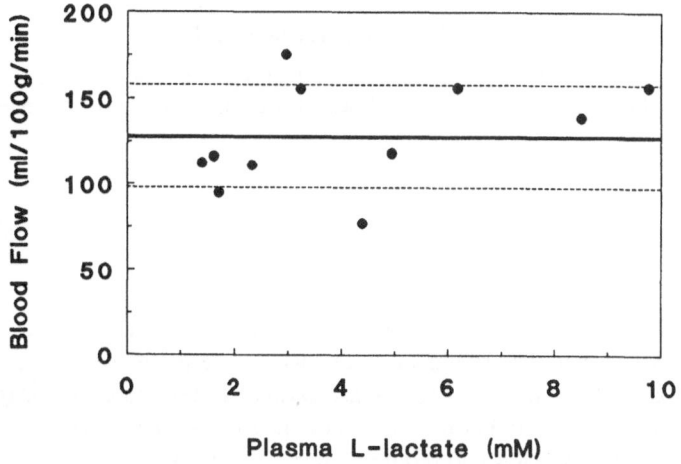

Figure 1. Cerebellar blood flow as a function of plasma L-lactate concentration. The solid horizontal line indicates the mean blood flow and the dashed lines are placed at ± 1 sd of the mean.

in those previously reported for awake, restrained rats and much higher than those of anesthetized rats (LaManna and Harik, 1986). The P*S product for L-lactate (Table 1) was about half that previously reported for D-glucose in awake, restrained rats (LaManna and Harik, 1986). The regional distribution of P*S was also similar to that for D-glucose where cerebral and cerebellar P*S products were similar and about 20% larger than those of the hippocampus and striatum, independent of regional blood flow differences.

For each region, transport constants were estimated from the influx vs plasma lactate curves. Figure 2 shows the graph for the parietal cortex influx curve. These data show that blood-to-brain L-lactate influx increases with increasing plasma lactate concentrations. Assuming that this transport system is saturable and obeys the Michaelis-Menten relationship, we analyzed the data for each region by non-linear regression procedures to obtain the best fit

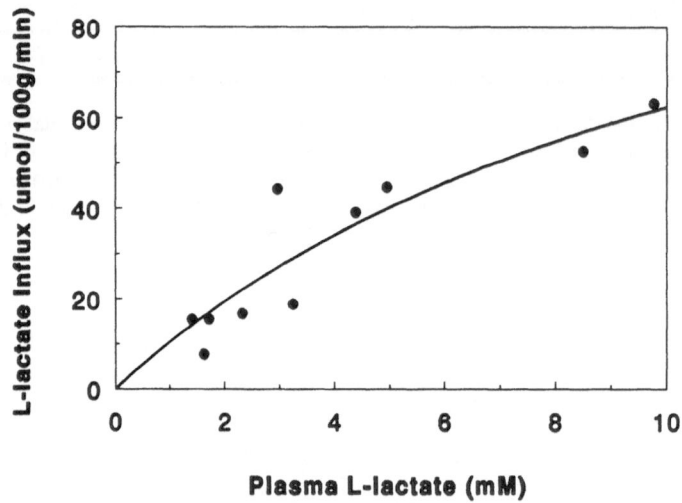

Figure 2. L-lactate influx into the parietal caerebral cortex as a function of plasma lactate concentration. The curve is the non-linear, least squares fit to the Michaelis-Menten equation.

to the equation: $J = (T_{max} \cdot L_{pl})/(K_t + L_{pl})$, where T_{max} is the transport maximum and K_t is the transport affinity constant. The results of these analyses are presented in Table 2. The K_t ranged between 6 and 12 mM, and the T_{max} ranged between 90 and 140 μmol/100g/min, which indicates a rather high capacity but low efficiency transport system (without correction for stereospecific transport, which would result in lower T_{max} estimates).

DISCUSSION

The data reported above demonstrate a basically low efficiency but high capacity transport system for L-lactate. The high capacity is based on the rather poor saturability of the system. Thus, at 10 mM plasma lactate concentration, which is considered high even under pathophysiological conditions, the blood-to-brain lactate influx is about 60 μmol/100g/min.

Table 2. Regional L-lactate Blood-Brain Barrier Transport Constants

	T_{max} (μmol/100g/min)	K_t (mM)	r^2
Cerebral Cortex			
Frontal	90	6.2	.70
Parietal	140	12.4	.83
Hippocampus	112	11.3	.86
Cerebellum	128	11.0	.80
Striatum	102	9.7	.78

Values are from non-linear, least squares fit to the Michaelis-Menten equation:
Influx = $(T_{max} \cdot$ [Plasma lactate])/ K_t + [Plasma lactate])

This contrasts with the T_{max} of about double that, and which is therefore not physiologically attainable. This low efficiency is also evident from the high K_t of 6-12 mM, especially when it is known that plasma lactate levels are normally about 1 mM. It is easier to understand a role for this transporter in the efflux of lactate from brain-to-blood because brain tissue lactate levels can be elevated to much higher levels than plasma levels during metabolic insults such as ischemia.

The transport constants we report here are higher than reported by others for adult rats. Based on the brain uptake index (BUI) method, the K_t was reported at about 1 mM and the T_{max} at about 12 μmol/100g/min (Pardridge and Oldendorf, 1977; Cremer et al., 1979). Because our method did not allow us to study transport at plasma lactate concentrations below 1 mM, it is possible that we have overestimated the actual K_t values. In addition, our analysis did not consider the effects of non-stereospecific diffusion which might have contributed to the higher influx rates we recorded. Besides various technical drawbacks inherent in the BUI method, another plausible explanation for the higher estimates we report could be the effects of anesthesia which lowers both blood flow and D-glucose transport (LaManna and Harik, 1986). Based on BUI values of about 10%, L-lactate influx would be at the reported T_{max} when blood flow was 100 ml/100g/min and plasma lactate was 1 mM.

In conclusion, it appears that the moncarboxylic acid transporter which is credited with the facilitated, bidirectional and non-concentrative transport of a variety of substrates such as pyruvate, has a relatively low affinity for lactate as evidenced by a high K_t. Nevertheless, there appears to be a substantial T_{max} which would not be reached even at plasma lactate levels under the most severe of pathological conditions. This transport system can provide a mechanism for the exit of lactate from the brain when it reaches high tissue levels; and it could provide a pathway for substrate availability to the brain during lactic acidemia.

REFERENCES

Cremer, J.E., Cunningham, V.J., Pardridge, W.M., Braun, L.D., and Oldendorf, W.H., 1979, Kinetics of blood-brain barrier transport of pyruvate, lactate and glucose in suckling, weanling and adult rats, J. Neurochem., 33: 439-445.

Crone, C., 1965, Facilitated transfer of glucose from blood into brain tissue, J. Physiol. (Lond.), 181: 103-113.

Crumrine, R.C. and LaManna, J.C., 1991, Regional cerebral metabolites, blood flow, plasma volume and mean transit time in total cerebral ischemia in the rat, J. Cereb. Blood Flow Metab., 11: 272-282.

Gjedde, A., 1983, Modulation of substrate transport to the brain, Acta Neurol. Scand., 67: 3-25.

Gjedde, A., Andersson, J., and Eklöf, B., 1975, Brain uptake of lactate, antipyrene, water and ethanol, Acta Physiol. Scand., 93: 145-149.

Harik, S.I. and LaManna, J.C., 1988, Vascular perfusion and blood-brain glucose transport in acute and chronic hyperglycemia, J. Neurochem., 51: 1924-1929.

LaManna, J.C. and Harik, S.I., 1985, Regional comparisons of brain glucose influx, Br. Res., 326: 299-305.

LaManna, J.C. and Harik, S.I., 1986, Regional studies of blood-brain barrier transport of glucose and leucine in awake and anesthetized rats, J. Cereb. Blood Flow Metab., 6: 717-723.

Nemoto, E.M., Hoff, J.T., and Severinghaus, J.W., 1974, Lactate uptake and metabolism by brain during hyperlactatemia and hypoglycemia, Stroke, 5: 48-53.

Oldendorf, W.H., 1972, Blood brain barrier permeability to lactate, Europ. Neurol., 6: 49-55.

Oldendorf, W.H., 1973, Carrier-mediated blood-brain barrier transport of short-chain monocarboxylic organic acids, Am. J. Physiol., 224: 1450-1453.

Pardridge, W.M. and Oldendorf, W.H., 1975, Kinetics of blood-brain barrier transport of hexoses, Biochim. Biophys. Acta, 382: 377-392.

Pardridge, W.M. and Oldendorf, W.H., 1977, Transport of metabolic substrates through the blood-brain barrier, J. Neurochem., 28: 5-12.

Renkin, E.M., 1959, Transport of potassium-42 from blood to tissue in isolated mammalian skeletal muscle, Am. J. Physiol., 197: 1205-1210.

Riachi, N.J., LaManna, J.C., and Harik, S.I., 1989, Entry of 1-methyl-4-phenyl-1,2,3,6-tetrahydropyridine into the rat brain, J. Pharmacol. Exp. Ther., 249: 744-748.

Sage, J.I., Van Uitert, R.L., and Duffy, T.E., 1981, Simultaneous measurement of cerebral blood flow and unidirectional movement of substances across the blood-brain barrier: theory, method, and application to leucine, J. Neurochem., 36: 1731-1738.

DELAYED BLOOD-BRAIN BARRIER OPENING WAS INDUCED NOT BY PRESYNAPTIC BUT BY POSTSYNAPTIC ISCHEMIC NEURONAL DAMAGES

Masayasu Matsumoto, Kazuo Kitagawa, Masafumi Tagaya, Toshiho Ohtsuki, Ryuji Hata, Nobuo Handa, Kazufumi Kimura* and Takenobu Kamada

First Department of Internal Medicine, and *Biomedical Research Center, Osaka University Medical School, 1-1-50, Fukushima, Fukushima-ku, Osaka 553, Japan

INTRODUCTION

The precise characteristics of blood-brain barrier opening in selective damages of pre- or postsynaptic sides of neurons following ischemia has not yet been fully elucidated. In this study, we tried to clarify this in three different conditions (i.e., infarction, selective loss of pre- and postsynaptic sides of neurons) following ischemia, by evaluating extravasated albumin and ischemic damages of pre- and postsynaptic sides of neurons with a method of immunohistochemistry. The pre- and postsynaptic damages were separately evaluated by using synapsin I and microtubule-associated protein 2 (MAP2) as a marker protein, respectively (Kitagawa et al., 1989b, 1991b).

MATERIALS AND METHODS

Adult Mongolian gerbils (Meriones unguiculatus) of both sexes, weighing 60-80 g, were used in the present study. For the investigation of selective postsynaptic damages and both pre- and postsynaptic damages, both common carotid arteries were exposed under light ether inhalation and clipped for 5 (n = 8) or 15 min (n = 8), respectively. At postischemic periods of 1 and 4 days, four gerbils each were killed under deep pentobarbital anesthesia by transcardiac perfusion with saline by 200 ml of 0.2% picric acid-2% paraformaldehyde solution (acetated buffer, pH 6.0). Fixed brains were removed, postfixed in the same solution at 4°C for 5 hours and processed for immunohistochemistry as reported previously (Matsumoto et al., 1987). Cell nuclei were visualized by counterstaining with Harris' hematoxylin. Alternative sections were stained with hematoxylin-eosin (HE). The rabbit antiserum for rat serum albumin (1:400), was purchased from Cappel, Inc. Both antisera for MAP2 from the mouse brain (1:400) and for synapsin I from the bovine brain (1:400) were raised in rabbits (Niinobe et al., 1988; Okabe and Sobue, 1987).

For the investigation of selective presynaptic damages, the right common carotid artery of the preselected stroke-prone gerbils, which should develop severe stroke symptoms after unilateral carotid occlusion (Matsumoto et al., 1988, 1990; Kitagawa et al., 1989a, 1991a), was exposed under light ether inhalation and occluded for 30 min with clips. After confirmation of the development of the typical signs of cerebral ischemia during ischemic period for 30 min, the clip was removed, and at the postischemic period of 3 and 7 days, 4 gerbils each were killed and their brains were processed for immunohistochemistry as described above.

RESULTS

At 1 day recirculation after 5-min forebrain ischemia, loss of reaction for MAP2 was visible in the subiculum-CA1, CA2 and CA4 of the

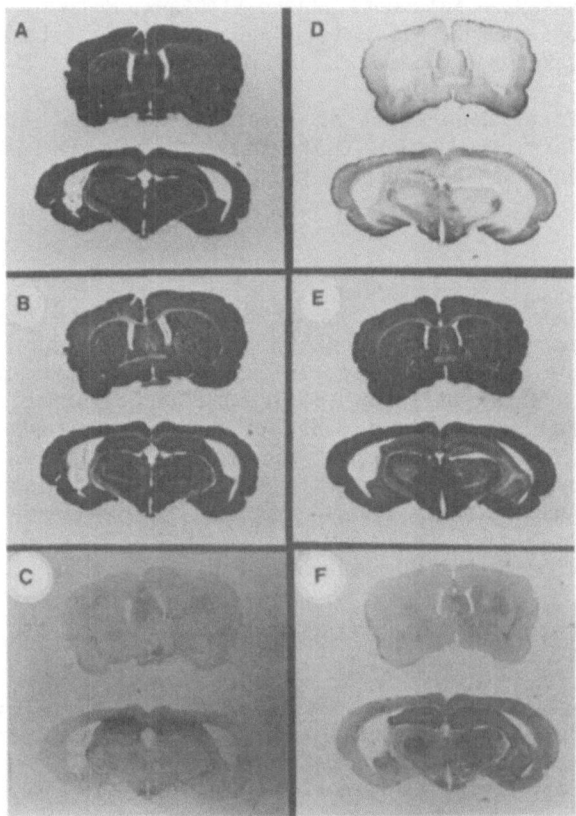

Fig. 1. Representative photographs of immunostaining for MAP2 (A and D), synapsin I (B and E) and albumin (C and F) in the whole gerbil brain at 4 day recirculation after 5-min (A, B and C) or 15-min (D, E and F) forebrain ischemia. For explanation, see text.

hippocampus and faint immunostaining for albumin was visualized only in dead neurons of these brain areas though it was not detected in the CA1 region of the hippocampus with clear immunostaining for MAP2. On the other hand, at 4 day recirculation (Fig. 1A-C), CA1 neurons of the hippocampus were almost destroyed, and clear loss of immunoreaction for MAP2 was observed in the whole CA1 area with massive albumin extravasation, while synapsin I immunostaining was fairly preserved.

At 1 and 4 day recirculation after 15-min forebrain ischemia, HE staining exhibited overt infarction with the destruction of all tissue components in the thalamus and the hippocampus. These areas showed massive extravasation of serum albumin already at 1 day recirculation and loss of immunoreaction for not only MAP2 but synapsin I (Fig 1D-E). On the other hand, in the caudoputamen, while neuronal cells were almost dead and MAP2 immunostaining was completely lost, the immunoreaction for synapsin I was fairly preserved and faint albumin immunoreactivity was also detected in the neuropil and neuronal perikarya. This pattern of immunostaining for MAP2 and synapsin I observed in the caudoputamen was quite similar to that in the CA1 area with 4 day recirculation after 5-min forebrain ischemia.

In unilateral cerebral ischemia, all preselected gerbils exhibited typical severe symptoms during 30 minute ischemia and overt infarction in the whole ischemic hemisphere at 3 and 7 day recirculation. As shown in Fig. 2, in the inner 1/3 of the molecular layer of the dentate gyrus in the nonischemic hippocampus, immunoreaction for synapsin I was lost while MAP2 immunostaining was preserved at 3 and 7 day recirculation. Albumin immunostaining was not detected in these areas.

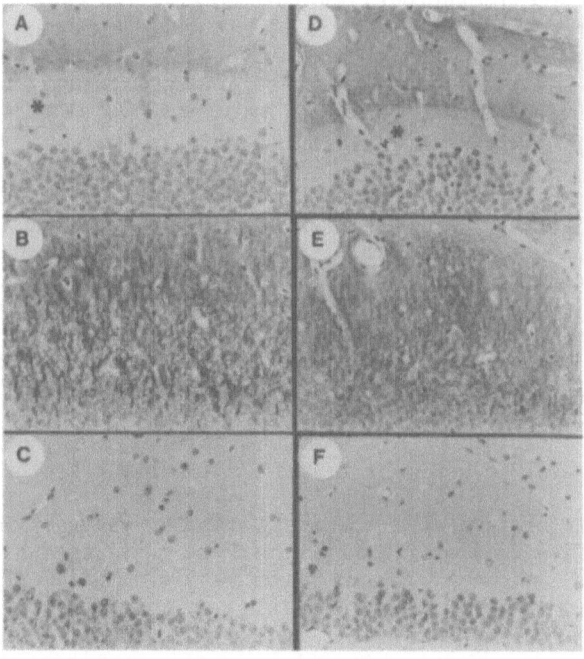

Fig. 2. Representative immunohistochemical reaction for synapsin I (A and D), MAP2 (B and E) and albumin (C and F) in the dentate gyrus of the hippocampus of contralateral nonischemic hemisphere at 3 day (A, B and C) and 7 day (D, E and F) recirculation after 30-min unilateral cerebral ischemia. As denoted by the asterisks in A and D, loss of immunoreaction for synapsin I was visible in inner one third of the molecular layer of the nonischemic dentate gyrus, where immunoreaction for MAP2 was fairly preserved (B and E) and albumin immunoreactivity was not detected (C and F).

DISCUSSIONS

Our recent study demonstrated the immunohistochemical reaction for MAP2 and synapsin I was useful for evaluating the damage of post- and presynaptic sides of neurons, respectively, in light microscopic examination (Kitagawa et al., 1989b, 1991b). The selective neuronal death was recently reported to be selective damage of postsynaptic sides of neurons with the maintenance of presynaptic terminals with electron microscopic observation in CA1 area of hippocampus in gerbils (Kirino, 1990), therefore, in our present study we defined it as a selective loss of MAP2, which is localized in postsynaptic sides (Kitagawa et al., 1989b), with the preservation of immunoreaction for synapsin I, which is localized in presynaptic terminals (Hirokawa et al., 1989). Because the deafferentation following ischemia was reported by Kirino et al. (1990) with electron microscopic observation in the hippocampus of nonischemic hemisphere, we used gerbil model of unilateral cerebral ischemia and defined selective loss of presynaptic terminals as loss of immunoreaction for synapsin I with intact MAP2 immunostaining. On the other hand, in the area of infarction, which was easily detected as a destruction of tissue integrity with HE staining and consistently observed in the thalamus after 15-min bilateral ischemia in gerbils, massive extravasation of serum albumin was visible early after ischemia and was regarded as a consequence of ischemic endothelial necrosis or damage.

In the CA1 area of the hippocampus, where selective neuronal death was consistently observed after 5-minute forebrain ischemia, appearance of neuronal cells and immunoreaction for MAP2 were fairly preserved and albumin immunoreactivity was not visible at 1 day recirculation except for a few dead neurons in the subiculum-CA1 and the CA4. Four days after ischemia, when neuronal cells were destroyed and immunoreaction for MAP2 was completely lost in the whole CA1 area, massive albumin extravasation was clearly visualized (Fig. 1). The loss of immunoreaction for MAP2 and preservation of that for synapsin I was in agreement with the electron microscopic findings exhibiting selective damage of postsynaptic sides (Kirino et al., 1990). Because endothelial cells and glia are intact in morphological aspects in selective neuronal death (Kirino et al., 1982), the present results strongly suggested that albumin extravasation was facilitated by the stimulation of the endothelial or glial cells with the unknown factors derived from destruction of the CA1 neurons, rather than that ischemia in itself directly affected the endothelial or glial cells and caused leakage of serum albumin. The same results of immunostaining were obtained in the caudoputamen in the animals, which received 15-min forebrain ischemia. This indicates that preservation of presynaptic terminals was observed not only in the CA1 but other areas after cerebral ischemia.

If albumin extravasates secondary to the destruction of neuronal cells, it is quite interesting to address the question; which subcellular compartments of neurons, pre- or postsynaptic side, contribute to the extravasation of serum albumin? To answer this question, we used a reproducible gerbil model of unilateral cerebral ischemia, which can consistently produce overt infarction in the ischemic hemisphere and exhibit deafferentation due to Wallerian degeneration in the dentate gyrus of the nonischemic hippocampus (Kirino et al., 1980). In the present study, the selective loss of immunoreaction for synapsin I was clearly visualized at 3 day and 7 day recirculation in the dentate gyrus of nonischemic hippocampus, and no immunoreactivity for albumin was detected in this area exhibiting deafferentation due to Wallerian degeneration (Fig. 2). The difference of albumin extravasation between the damage of pre- and postsynaptic sides of neurons could be explained by the elements included in each side. For example, the ribosomes, nucleic acids and lysosomes are much more abundant in the perikarya and the dendrites than in the presynaptic terminals. As for glial cells, several factors such as hormones, second messengers, polyamines and macromolecules are known to be astrocytic mitogenic and morphogenic factors (Morrison et al., 1985; Nieto-Sampedro et al., 1985). If the stimulants, such as several proteases as included in the lysosomes, leak out from postsynaptic components, this may cause inflammation, stimulate endothelial and/or glial cells, and facilitate the transport of materials across the BBB.

In summary, extravasation of serum albumin was characterized in three different conditions of ischemic brain damage, including cerebral infarction, selective neuronal death and selective loss of presynaptic terminals. In the cerebral infarction, showing destruction of all components (i.e., neurons, glial cells and endothelium), serum albumin massively extravasated early after ischemia, probably due to the ischemic endothelial necrosis. However, in the selective neuronal death, it was not before postsynaptic structures were destroyed that albumin extravasation became apparent. In the selective loss of presynaptic sides, albumin extravasation was not visible. Because endothelium and glial cells were not directly damaged in morphological aspects in selective damage of both pre- and postsynaptic sides, it was thought that extravasation was facilitated by the stimulation of endothelial cells and/or glial cells with unknown factors, which were induced by the destruction of not pre- but postsynaptic elements.

ACKNOWLEDGMENTS

We wish to express our thanks to Mr. M. Tadachi, K. Wakitani and T. Okegawa (Research Institute, Ono Pharmaceutical, Osaka) for careful management of animals, Miss S. Goi, Mr. A. Naito and H. Yoshimura for their technical assistance and Miss K. Moriguchi and M. Sakai for their secretarial assistance. The present study was supported in part by Research Grant for Cardiovascular Diseases (2A-2) from the Ministry of Health and Welfare and a Grant-in-aid (#03670450) from the Ministry of Education, Science and Culture in Japan.

REFERENCES

1. Hirokawa N, Sobue K, Kanda K, Harada A, Yorifuji H (1989) The cytoskeletal architecture of the presynaptic terminals and molecular structure of synapsin I. J Cell Biol 108: 111-126
2. Kirino T, Sano K (1980) Changes in the contralateral dentate gyrus in mongolian gerbils subjected to unilateral cerebral ischemia. Acta Neuropathol (Berl) 50: 121-129
3. Kirino T (1982) Delayed neuronal death in the gerbil hippocampus following ischemia. Brain Res 239: 57-69
4. Kirino T, Tamura A, Sano K. (1990) Chronic maintenance of presynaptic terminals in gliotic hippocampus following ischemia. Brain Res 510: 17-25
5. Kitagawa K, Matsumoto M, Handa N, Fukunaga R, Ueda H, Isaka Y, Kimura K, Kamada T (1989a) Predicition of stroke-prone gerbils and their cerebral circulation. Brain Res 479: 263-269
6. Kitagawa K, Matsumoto M, Niinobe M, Mikoshiba K, Hata R, Ueda H, Handa N, Fukunaga R, Isaka Y, Kimura K, Kamada T (1989b) Microtubule-associated protein 2 as a sensitive marker for cerebral ischemic damage - immunohistochemical investigation of dendritic damage. Neuroscience 31: 401-411
7. Kitagawa K, Matsumoto M, Tagaya M, Ueda H, Oku N, Kuwabara K, Ohtsuki T, Handa N, Kimura K, Kamada T (1991a) Temporal profile of serum albumin extravasation following cerebral ischemia in a newly established reproducible gerbil model for vasogenic brain edema: A combined immunohistochemical and dye tracer analysis. Acta Neuropathol (Berl) in press

8. Kitagawa K, Matsumoto M, Sobue K, Tagaya M, Okabe T, Niinobe M, Ohtsuki T, Handa N, Kimura K, Mikoshiba K, Kamada T (1991b) The synapsin I brain distribution in ischemia. Neuroscience in press

9. Matsumoto M, Yamamoto K, Homburger HA, Yanagihara T (1987) Early detection of cerebral ischemic damage and repair process in the gerbil by use of an immunohistochemical technique. Mayo Clin Proc 62: 460-472

10. Matsumoto M, Hatakeyama T, Akai F, Brengman JM, Yanagihara T (1988) Predicition of stroke before and after unilateral occlusion of the common carotid artery in gerbils. Stroke 19: 490-497

11. Matsumoto M, Hatakeyama T, Morimoto K, Yanagihara T (1990) Cerebral blood flow and neuronal damage during progressive cerebral ischemia in gerbils. Stroke 21: 1470-1477

12. Morrison RS, de Vellis J, Lee YL, Bradshaw RA, Eng LF (1985) Hormones and growth factors induce the synthesis of glial fibrillary acidic protein in rat brain astrocytes. J Neurosci Res 14: 167-176

13. Nieto-Sampedro M, Saneto RP, de Vellis J, Cotman CW (1985) The control of glial populations in brain: Changes in astrocytic mitogenic and morphogenic factors in response to injury. Brain Res 343: 320-328

14. Niinobe M, Maeda N, Ino H, Mikoshiba K (1988) Characterization of microtubule-associated protein 2 from mouse brain and its localization in the cerebellar cortex. J Neurochem 51 :1132-1139

15. Okabe T, Sobue K (1987) Identification of a new 84/82 KDa calmodulin-binding protein, which also interacts with actin filaments, tubulin and spectrin, as synapsin I. FEBS Lett 213: 184-188

Chapter 7

Neurotransmitter Modulation of Cerebral Blood Flow and Metabolism

NEUROEFFECTOR MECHANISMS DURING FOCAL

AND GLOBAL CEREBRAL ISCHEMIA

Michael A. Moskowitz

Stroke Research Laboratory
Massachusetts General Hospital
Harvard Medical School
Boston, MA 02114

INTRODUCTION

The circle of Willis and its tributaries are innervated by sympathetic, parasympathetic and sensory nerve fibers. Over the past 4 years, we have investigated the importance of the latter two pathways, and our findings will be reviewed briefly. The data suggest that parasympathetic projections become activated during focal ischemia whereupon these fibers may contribute to the development of collateral blood flow. Destroying parasympathetic fibers is associated with a 30% increase in the size of focal infarcts, an observation which was made in 3 different species and under different anesthetic conditions. By contrast, sensory fibers become activated during global but not focal ischemia and contribute to the hyperemia during the reperfusion phase. Readers are referred to references (1-4,6,24) for more detailed descriptions).

THE ROLE OF PARASYMPATHETIC FIBERS IN CEREBRAL ISCHEMIA

Lesioning studies in rats have helped to clarify the role of parasympathetic projections in the control of the rat cerebral circulation during ischemia. Such studies were made possible by recent anatomical investigations showing that fibers that arise from the sphenopalatine ganglion and structures within the nasal cavity traverse the ethmoidal foramen to reach the cranial cavity (5).

Sectioning the parasympathetic but not the sensory innervation to the circle of Willis increases the volume of infarction by approximately 33% (from 155 to 198 mm^3) in a tandem occlusion model of MCA-CCA ligation in the SHR (6).

Similar findings were obtained in the normotensive Long
Evans rat following distal MCA and temporary common carotid
artery occlusion (1). Huang recently noted larger striatal
infarcts after proximal MCA occlusion in Sprague Dawley rats
(unpublished observations), and larger cortical infarcts
after tandem occlusion in Fisher-344 after chronic
parasympathetic sectioning (Moskowitz, unpublished
observation). Larger infarcts were only recorded in those
animals demonstrating $\geq 40\%$ (estimate) decrease in
immunoreactive VIP-containing fibers within the ipsilateral
MCA. Kano determined that infarct size increased whether or
not animals were anesthetized with halothane, chloral
hydrate or xylazine and ketamine, or whether animals were
ventilating spontaneously or artificially (1). We conclude
from these studies that parasympathetic fibers modulate
cerebral infarction size in the rat regardless of
anesthetic, site of vascular occlusion (i.e., proximal or
distal MCA), or strain.

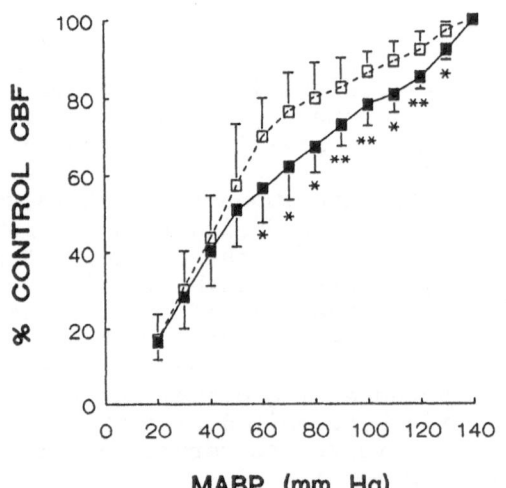

Fig. 1. Blood flow responses (LDF) during
controlled exsanguination within the cortical
barrel fields of SHR following chronic unilateral
selective parasympathetic sectioning. Greater
reductions in flow-velocity were detected on the
denervated side (darkened squares) as perfusion
pressures were lowered between 130-60 mm Hg. *p
<0.05; ** p<0.01 as compared to control side.

In an attempt to explain these observations, flow-
velocity was measured during controlled hemorrhagic
hypotension in SHR 10-14 days after chronic unilateral
sectioning of parasympathetic and/or sensory fibers
innervating pial vessels (5). rCBF was measured in the
cortical barrel fields bilaterally by laser-doppler blood

flowmetry. During hemorrhagic hypotension (140-20 mm Hg), decreases in rCBF were equivalent on the two sides in shams, after selective sensory denervation, or in sectioned animals exhibiting small decreases ($\leq$30%) in immunoreactive VIP-containing fibers within ipsilateral MCA. During controlled hemorrhage, greater reductions in rCBF were present on the parasympathetically-denervated side. These differences were small (5-18%) and were present only between 130 to 60 mm Hg MABP (n=9), and not at lower perfusion pressures. The percent change in vascular resistance was attenuated on the denervated side between 130-60 mm Hg, as well. The blood flow response to hypercapnia was unimpaired, however.

We suggest that the greater reductions (albeit modest reductions) in flow-velocity observed in parasympathetically-denervated SHR during reduced perfusion pressures may explain the larger infarction volumes measured after chronic nerve sectioning. It is possible that the differences in rCBF may have been more profound had the drop in perfusion pressure been abrupt rather than gradual (to similate the conditions in MCA occlusion), or had the surgical sectioning been more effective [only 50% (estimate) of the VIP-containing fibers could be depleted in the ipsilateral MCA after surgical denervation]. VIP-projections from multiple ganglia probably accounts for the subtotal VIP depletion (e.g., the otic (7) as well as carotid miniganglion (8)). Bilateral projections (unpublished observations) also contributed to the the subtotal result in SHR after denervation.

With these important caveats in mind, our working hypothesis is that VIP-containing parasympathetic fibers may promote collateral blood flow in regions of reduced perfusion pressure such as in the peri-infarct zone.

Important blood flow effects have been measured after stimulating parasympathetic projections to the circle of Willis. Electrical stimulation of sphenopalatine ganglia (rat, cat) or its efferent fibers (rat) increases brain blood flow in rat (9-11) and cat (12) as assessed by helium clearance, laser flow probe and ^{14}C-iodoantipyrene methods. Increases were predominantly ipsilateral. Sphenopalatine projections may enhance blood flow by dilating vessels directly because tissue PaO_2 did not fall nor did glucose metabolism increase during electrical stimulation. Blood flow responses appear less clear follwoing electrical stimulation of the gerater sperficial petrosal nerve. The GSPN carries preganglionic parasympathetic nerve fibers to the sphenopalatine ganglia via the facial nere and provides the major (but not exclusive) input to the sphenopalatine ganglia. Because stimulating the GSPN causes less consistent increases in blood flow, it would appear that stimulating postganglionic structures (or activating endogenous mechanisms) might be more suitable for testing and defining the importance of neurogenic pathways to the cerebral circulation.

One or more of several possible mechanisms might activate parasympathetic fibers during focal ischemia including via trigeminal-GSPN connections within brain stem, activation of descending multisynaptic pathways from cortex to brain stem, or local generator mechanisms at the level of the vessel wall. These three possibilities can be tested experimentally.

The effects of parasympathetic denervation on cortical blood flow during focal ischemia may be relevant to conditions associated with autonomic parasympathetic neuropathy such as diabetes mellitus, and may provide one explanation for the propensity of these patients to develop large cerebral infarctions. The findings may also suggest a mechanism by which the levels of cerebral blood flow increase following the administration of drugs during focal ischemia.

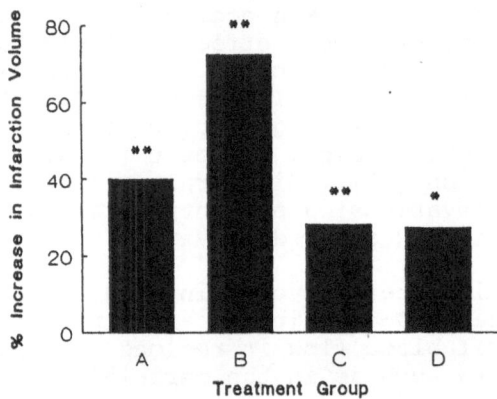

Fig.2. Increase in Infarction Volume (%) after MCA Occlusion in (A)Long Evans-Spontaneous ventilation, (B)Long Evans-Assisted Ventilation (AV), C.SHR-Tandem Occlusion(AV), D.SD-Tamura Model(AV) after Chronic Parasympathetic Sectioning. *p<0.05; **p<0.01.

THE ROLE OF SENSORY FIBERS IN CEREBRAL ISCHEMIA

An important role for sensory fibers in the blood flow responses during postocclusive hyperemia has been obtained by my laboratory and those of my collaborator, Dr. Hermes A. Kontos. Following unilateral trigeminal ganglionectomy in cats, or the topical application of capsaicin to a solitary branch of the MCA, postocclusive hyperemia is attenuated by 40-60% in cortical grey matter (3,4). Chronic trigeminal ganglionectomy markedly reduces the immunohistochemical staining of CGRP in the ipsilateral circle of Willis, and depletes vessels of substance P (4).

Studies using a bilateral closed cranial window technique in cats have shown that superfusion of vessels with alkaline CSF or with norepinephrine constricts pial arterioles to a greater degree in the denervated hemisphere following chronic trigeminal section (2,13). The dilatory response to acute severe hypertension or seizures is also reduced, as is the accompaning extravasation of radiolabeled albumin. The same effect is not observed after trigeminal rhizotomy (14). Because sectioning of the trigeminal root blocks orthodromic transmission but does not cause Wallerian degeneration of cell bodies or peripheral axons, the increases in CBF associated with reperfusion after cerebral ischemia must be mediated by axon reflex-like mechanisms. Sensory nerves do not however, mediate all hyperperfusion states, since the cerebrovascular response to hypercapnia is not impaired.

It is presumed that sensory nerves mediate the increases in CBF by the release of vasoactive neuropeptides. After acute or chronic ischemia of the lower limb, venous levels of SP, but not VIP, increase significantly (15). During ischemia, molecules such as adenosine, lactate, potassium and arachidonate metabolites accumulate in brain, and either stimulate or potentiate the activation of C-fibers. Bradykinin is synthesized in vessel walls during severe hypertension, and both stimulate C-fibers and cause neurogenic inflammation. Potassium may be the coupling mechanism for the activation of perivascular nerves during seizures. Clearly, other mechanisms must also be involved because hyperemia is reduced by only around 50% after sensory denervation. Nevertheless, if hyperemia is responsible for increasing the risk of cerebral edema, hemorrhage, or seizures during reperfusion following global ischemia then strategies aimed at blocking this axon reflex may be of therapeutic importance.

REFERENCES

1.   Kano M, Moskowitz MA, Yokota M., 1991, Parasympathetic denervation of rat pial vessels significantly increases infarction volume following middle cerebral artery occlusion. J Cerebral Blood Flow & Metab 11:628-637.
2.   Moskowitz MA, Wei E, Saito K, Kontos HA., 1988, Trigeminalectomy modifies pial arteriolar responses to hypertension or norepinephrine. Am J Physiol 255:H1-6.
3.   Moskowitz MA, Sakas DE, Wei EP, Kano M, Buzzi, MG, Ogilvy C, Kontos HE., 1989, Postocclusive hyperemia in feline cortical grey matter is mediated by trigeminal sensory axons. Am J Physiol. 257; H1736-1739.
4.   Mcfarlane R, Tasdemiroglu E, Moskowitz MA, Uemura Y, Wei EP, Kontos HA, 1991, Chronic trigeminal ganglionectomy or topical capsaicin application to pial vessels attenuates postocclusive cortical hyperemia but does not influence postischemic hypoperfusion. J Cerebral Blood Flow & Metab 11:261-271.
5.   Suzuki, N., Hardebo JE. and Owman Ch, 1988, Origins and pathways of cerebrovascular vasoactive intestinal
6.   Koketsu N, Moskowitz MA, Yokota M, Shimizu T. Decreased pressure-dependent flow-velocity in cerebral cortex during hemorrhagic hypotension accompanies larger cortical infarcts in SHR after chronic parasympathetic sectioning. Submitted.

7.   Walters BB, Gillespie SA, Moskowitz MA, 1986,
Cerebrovascular porjections from the sphenopalatine and otic
ganglia to the middle cerebral artery of the cat. Stroke 17
(3): 488-494.

8.   Suzuki N, Hardebo JE, Owman Ch, 1989, Origins and
pathways of cerebrovacsulr nerves storing substance P and
calcitonin gene-related peptide in rat. Neuroscience 31:427-
438.

9.   Seylaz J, Hara H, Pinard E, Mraovitch S, MacKenzie ET,
Edvinsson, 1988, Effect of stimulation of the sphenopalatine
ganglion on cortical blood flow in the rat. J Cereb blood
Flow Metab 8:875-878, 1988.

10. Seylaz J, Hara H, Pinard, Sraovitch S, MacKenzie ET,
Edvinsson L, Uddman R., 1989, The cerebrovascular
sphenopalatine system: anatomical and functional aspects.
In:Neurotransmission and Cerebrovascular Function, I. Eds J.
Seylaz and E.T. MacKenzie. Elsevier, Amsterdam, pp317-320.

11. Suzuki N, Hardebo JE, Kahrstrom J, Owman C, 1989,
Electrical stimulation of postganglionic cerebrovascular
parasympathetic fibers enhances cerebral blood flow.
In:Neurotransmission and Cerebrovascular Function, I. Eds J.
Seylaz and E.T. MacKenzie. Elsevier, 1989, Amsterdam, pp
321-324.

12. Goadsby PJ, 1990, Sphenopalatine ganglion stimulation
increases regional cerebral blood flow independent of
glucose utilization in the cat. Brain Res 506:145-148.

13. McCulloch J, Udmann R, Kingman TA, et al, 1986,
Calcitonin-gene related peptide: functional role in
cerebrovascular regulation. Proc Nat Acad Sci USA 83:5731-
5735.

14. Sakas DE, Moskowitz MA, Wei EP, Kontos HA, Kano M,
Ogilvy CS, 1989, Trigeminovascular fibers increase blood
flow in cortical grey matter by axon reflex-like mechanisms
during acute severe hypertension or seizures. Proc Natl Acad
Sci USA 86:1401-1405.

15. Henriksen JH, Bulow JB, Schaffalitzky de Muckadell O et
al, 1986, Do substance P and VIP play a role in the acute
occlusive or chornic ischaemic vasodilation in man? J Clin
Pharmacol 6: 163.

CENTRAL NEURAL MODULATION OF FOCAL CEREBRAL INFARCTION: POSSIBLE RELATIONSHIP TO BRAINSTEM NETWORKS GOVERNING OXYGEN CONSERVING RESPONSES

Donald J. Reis

Division of Neurobiology, Cornell University Medical College
411 East 69th Street, New York, NY 10021, U.S.A.

INTRODUCTION

There is increasing evidence that neurotransmitters released by ischemia may contribute to the magnitude of the resulting brain damage (e.g. ref. 1 as well as other contributions in this volume). The general view is that the release of transmitters during ischemia results from impairment, by hypoxia, of energy-dependent processes in nerve endings required for transmitter inactivation. Normally, the release of neurotransmitters and modulators is governed by nerve impulse activity. This fact raises the question: will normal synaptic release of transmitters in the ischemic zone also modify ischemic damage?

We have recently discovered that electrical stimulation of the cerebellar fastigial nucleus (FN) in rat can reduce by 40-64% the size of focal cerebral infarction produced by occlusion of the middle cerebral artery (MCA)[2,3]. The observations indicate that excitation of neurons at sites remote from the ischemic zone can modify the size of cerebral infarctions. The findings raise questions with respect to the mechanisms which underlie the phenomenon. In this paper I shall review, briefly, the rationale for our studies and indicate that the findings suggest a novel mechanism in brain which may be neuroprotective.

CENTRAL NEURAL REGULATION OF CEREBRAL BLOOD FLOW INDEPENDENT FROM METABOLISM: A POTENTIAL MECHANISM FOR NEUROPROTECTION

The rationale for our studies derives from observations that in focal cerebral ischemia the area of infarction is characterized by a central core in which regional cerebral blood flow (rCBF) and metabolism [as indirectly assessed by measurement of regional cerebral glucose utilization (rCGU)] are profoundly depressed[4,5]. Surrounding the core is a zone, the penumbra, in which rCBF is partly reduced while rCGU is patchily, and markedly, elevated. Thus, in the penumbra, the coupling between rCBF and metabolism is lost with rCGU disproportionally elevated. It is neurons within the penumbra which are salvaged by therapy[6] indicating that following the ischemic insult, these neurons are viable and capable of survival in response to environmental influences. Conceivably, the relative increase in metabolism unsupported by an adequate perfusion contributes to neuronal vulnerability. Hence, we argued that conditions which might selectively increase rCBF or, conversely, reduce metabolism might be neuroprotective.

Traditional views of the physiology of the cerebral circulation would suggest that, with the exception of hypercarbia or hypoxia, rCBF and rCGU are tightly coupled[7]. However, over the past few years it has become evident that neuronal networks entirely contained within the CNS may profoundly increase rCBF without altering rCGU (see ref.

8). One of the cerebral networks from which this primary cerebral vasodilation can be elicited is the FN. Electrical stimulation of the FN will elicit widespread increases in rCBF unassociated in most areas with changes in rCGU[9]. Within the cerebral cortex, the increase in rCBF occurs after a brief latency and long outlast the stimulus[10] suggesting that the effect may depend upon the release of a diffusible mediator(s). The vasodilation depends upon the activity of local neurons in the target excited over a polysynaptic and atropine-sensitive pathway from brainstem[11]. It depends upon the integrity of neurons in the rostral ventrolateral reticular nucleus (RVL)[12] of the medulla, a region from which electrical and chemical stimulation will also increase CBF but not rCGU[13,14]. The observations that excitation of FN or RVL could increase rCBF without rCGU raised the questions: would stimulation of the FN favor neuronal recovery following focal cerebral ischemia? If so, is improvement related to increased rCBF as depicted in the model of Fig. 1A? Or, is the increase independent of changes in flow as shown in Fig. 1B?

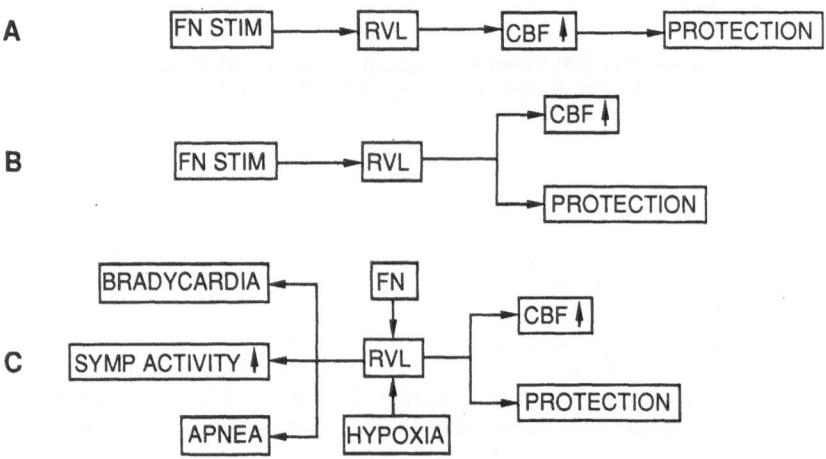

Figure 1.   Models of relationship of excitation of the cerebellar fastigial nucleus (FN) with RVL, elevations of rCBF and neuroprotection. A. Linear model where neuroprotection is a consequence of neurogenically mediated increase in rCBF. B. Model in which elevations of rCBF and neuroprotection are independent effects. C. A model illustrating the hypothesis that stimulation of RVL area by hypoxia not only initiates many of the autonomic patterns of the oxygen-conserving or diving reflex but also is responsible for the elevations in rCBF and, independently, neuroprotection.

REDUCTION OF FOCAL CEREBRAL INFARCTION BY CEREBELLAR STIMULATION

In our studies[2,3,15], rats were anesthetized while recording arterial pressure (AP) and monitoring blood gases. The MCA was ligated unilaterally and the animal retained under anesthesia for one hour. At the end of the hour, wounds were closed, anesthesia discontinued, and animals returned to their cages. In some rats, a stimulating electrode was inserted into the FN or, as control, areas of cerebellum not influencing rCBF, and electrical stimuli delivered over an hour. Some rats received atropine (0.5 mg/kg i.v.). Other controls consisted of rats in which rCBF was elevated for one hour by hypercarbia or by elevating AP with phenylephrine so as to increase rCBF (as measured by a laser-doppler flowmeter) in the contralateral hemisphere to the same extent as that established in other studies by FN stimulation[15]. The effects of FN stimulation were also examined in rats of a different strain. Twenty four hours later the rats were anesthetized, killed, the brain removed, sectioned, stained and the size and distribution of the ischemic infarction established. Blood gases and arterial pressures were comparable between all groups.

Table 1.   Effects of Various Treatments on Volume of Infarction
Produced by Occlusion of MCA in Rat (from refs. 2,15)

Group	Infarct Volume (mm^2)	% Control
Control I (15)	186.9 ± 35.2	100
MCA + FN Stim (13)	112.7 ± 47.1	60.3*
MCA + DN (7)	168.7 ± 28.4	90.3[NS]
MCA + Atropine (9)	169.0 ± 64.5	90.4[NS]
MCA + FN + Atropine (7)	163.4 ± 43.1	87.4[NS]
Control II (5)	196.3 ± 14.9	100
CO$_2$ (5)	201.3 ± 45.5	103[NS]
Hypertension	204.0 ± 35.1	104[NS]

* = p<0.01
NS = not significant

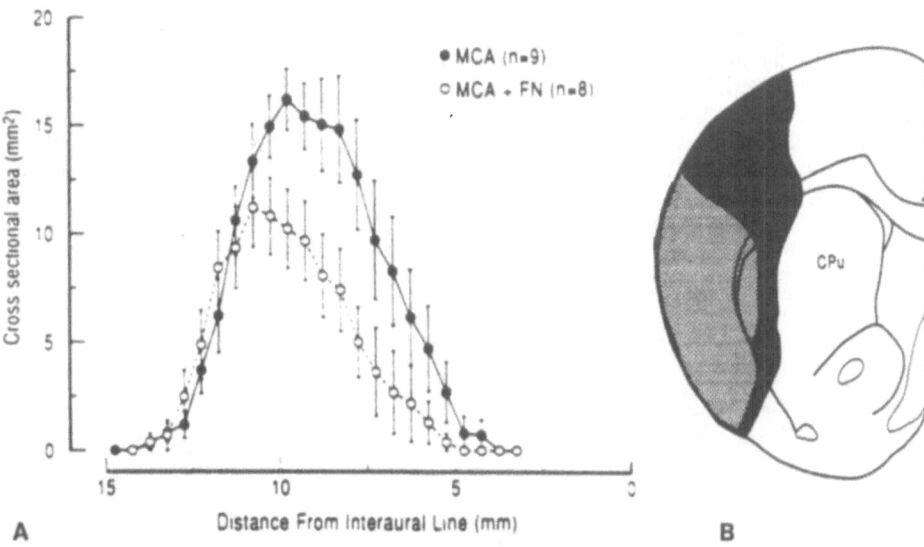

A   B

Figure 2.   Effects of electrical stimulation of the cerebellar fastigial nucleus (FN) on the
distribution of lesions elicited by occlusion of the middle cerebral artery (MCA)
in rat (Sprague-Dawley). A. Area (in mm^2) of lesions at various levels of brain
rostral from intraaural line. Closed circles are controls with volume of 193.7 ±
34.8 mm^3. Open circles represent animals with FN stimulation as described. The
volume is reduced by almost 64% to a mean of 64.5 mm^3 (p<0.05). B. Cross
section of brain at area of greatest salvage. The black and shaded areas together
represent control lesions (summed from 9 animals). The black area represents
the zone which was salvaged by FN stimulation, the shaded area that which was
not (from ref. 2).

As shown in Table 1 and Fig. 2, electrical simulation of the FN reduced the volume and extent of a focal cerebral infarction, not only in SHRs but also rats of the Sprague-Dawley strains. Infarct volumes were reduced between 40-65% with the major area restricted to more posterior portions of the lesion (Fig. 1A,B) corresponding to the penumbra[4] and regions preserved by calcium antagonists[6]. The salvage produced by FN stimulation was not attributable to differences in blood gases or AP between groups nor to strain of rat. Furthermore, the fact that salvage could not be elicited by electrical stimulation of adjacent cerebellar sites and was blocked by atropine (Table 1), indicates that the effects were selective and dependent upon activating the system of FN required for the primary vasodilation. The salvage elicited by FN, however, does not appear attributable to the associated increase in rCBF, since increasing rCBF by hypercarbia or by elevating AP above the autoregulated range in rat had no effect upon the size of the lesion (Table 1).

CONSIDERATION OF THE MECHANISMS WHEREBY CEREBELLAR STIMULATION OFFERS NEUROPROTECTION: RELATIONSHIP TO CENTRAL OXYGEN-PROTECTING REFLEXES

At first blush, it seemed reasonable to assume that the neuroprotective effects of FN stimulation were the result of activating RVL to increase rCBF (Fig. 1A). However, our finding that elevations of rCBF elicited by hypercarbia and/or hypertension were not capable of replicating the effects of FN stimulation (Table 1) suggest that the neuroprotection is not the consequence of increased rCBF. Rather, it may result from coactivation of two independent mechanisms (Fig. 2B): one acting to elevate rCBF, the other neuroprotective. The manner by which the neural signal is transduced in the target is unknown and presently under investigation.

The RVL is a small nuclear region of the medulla which is the principal relay for the increase in rCBF elicited from the FN[12]. With respect to the systemic circulation, the RVL contains sympathoexcitatory reticulospinal neurons innervating preganglionic sympathetic neurons of the spinal cord (see ref. 16). These are tonically active and are responsible for the tonic maintenance of AP as well as mediating the reflex effects upon AP associated with arterial baro- and chemoreflexes, pain, exercise and emotional excitement (see ref. 16 for review). It also may profoundly influence the cerebral circulation: lesions of the RVL abolish the elevation in rCBF elicited by electrical stimulation of the FN, while electrical stimulation of the RVL, like that of FN, increases rCBF globally without increasing rCGU[14,17].

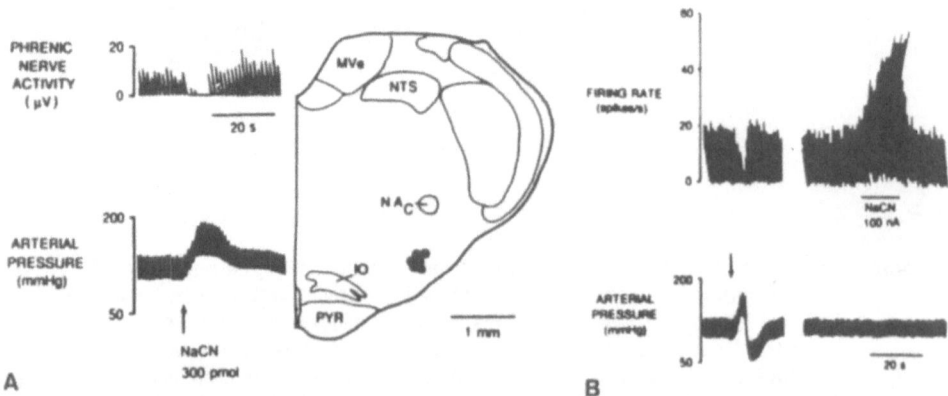

Figure 3. Effects of CN on arterial pressure, respiration and the discharge of a reticulospinal sympathoexcitatory neuron of RVL. A. Microinjection of cyanide into the RVL (containing filled circles) as depicted in drawing of rat brainstem, elevates AP and elicits apnea. B. A RVL neuron which is spontaneously active and characteristically silenced by baroreceptor stimulation with i.v. phenylephrine (at arrow) is potently excited by microiontophoresis of CN (from ref. 24)

The fact that neurons of the RVL also mediate the effects upon the systemic circulation elicited by rendering the brainstem ischemic (the cerebral ischemic response)[18-21] raises the possibility that neurons of the RVL may, in fact, be detecting hypoxia, particularly in view of the intimate association of RVL neurons to local vessels many of which penetrate these neurons[22]. That the RVL may be an "oxygen sensor" has recently been demonstrated by the observations that RVL sympathoexcitatory neurons are excited by brief periods of hypoxia and by microinjection of cyanide (Fig. 3A), but not by hypercarbia[23], and that the same neurons are excited by local microiontophoresis of sodium cyanide (Fig. 3B)[24]. The RVL may also mediate the primary cerebrovascular vasodilation elicited by hypoxia: lesions of the RVL will reduce by almost 50% the elevations in rCBF elicited by hypoxia without affecting resting rCBF[13,14].

The autonomic responses to cerebral ischemia, hypoxia (in the absence of arterial chemoreceptors), or microinjection of cyanide into the RVL mimics the diving or so-called "oxygen conserving reflex" so well developed in diving vertebrates but also present in man[25]. This pattern of response, sympathetic excitation, bradycardia and apnea, also initiated by hypoxia in the fetus, has been proposed to reflect an extraordinarily potent integrated circulatory response in which blood flow to skeletal muscle and viscera are profoundly reduced while blood flow to brain and heart are maintained (see ref. 25). The function of the response, therefore, is to provide protection to brain at a time when oxygen availability is severely compromised. Conceivably, another component of the response would be activation of cerebral mechanisms which can afford cellular protection against hypoxia in addition to enhancing rCBF and orchestrated over pathways emanating from the RVL. Thus, as diagrammed in Fig. 1C, the RVL would be activated by cerebral hypoxia, by direct chemical or electrical stimulation, or indirectly by excitation of the FN to drive the neuroprotecive mechanism. The presence of the neuroprotective network has only been suggested by our experiments. Clearly, further work is required to determine the validity of the hypothesis and to establish the underlying neural, cellular, and molecular mechanisms.

SUMMARY AND CONCLUSIONS

1. Electrical stimulation of the cerebellar fastigial nucleus (FN) will reduce by 40-64% the size of a focal ischemic infarction produced by occlusion of the middle cerebral artery in rat, an effect not replicated by increasing rCBF.

2. The actions of FN on rCBF and other autonomic functions depend upon the integrity of the RVL, a medullary nuclear region which acts as an " oxygen sensor" activating autonomic and cerebrovascular responses to cerebral hypoxia to mimic the diving (oxygen-conserving) reflex.

3. Conceivably, neuroprotection may be an independent mechanism within the diving response and which is activated from RVL in response to FN stimulation.

4. Excitation of intrinsic neural mechanisms may offer protection of brain in stroke.

REFERENCES

1. M.Y. Globus, R. Busto, E. Martinez, I. Valdes, W.D. Dietrich, and M.D. Ginsberg, Comparative effect of transient global ischemia on extracellular levels of glutamate, glycine, and gamma-aminobutyric acid in vulnerable and nonvulnerable brain regions in the rat, J. Neurochem. 57:470 (1991).
2. D.J. Reis, S.B. Berger, M.D. Underwood, and M. Khayata, Electrical stimulation of cerebellar fastigial nucleus reduces ischemic infarction elicited by middle cerebral artery occlusion in rat, J. Cereb. Blood Flow and Metab. 11:810 (1991).
3. S.B. Berger, D.B. Ballon, M. Graham, M.D. Underwood, M. Khayata, R.D. Leggiero, J.A. Koutcher, and D.J. Reis, Magnetic resonance imaging demonstrates that electrical stimulation of cerebellar fastigial nucleus reduces cerebral infarction in rats, Stroke, 21(Sup 3):172 (1990).
4. M. Nedergaard, A. Gjedde, and N.H. Diemer, Focal ischemia of the rat brain: autoradiographic determination of cerebral glucose utilization, glucose content, and blood flow, J. Cereb. Blood Flow Metab. 6:414 (1986).

5. M.D. Ginsberg, Local metabolic responses to cerebral ischemia, <u>Cerebrovasc. Brain Metab. Rev.</u> 2:58 (1990).

6. M.D. Ginsberg, B. Lin, E. Morikawa, W.D. Dietrich, R. Busto, and M.Y. Globus, Calcium antagonists in the treatment of experimental cerebral ischemia, <u>Arzneimittelforschung</u> 41:334 (1991).

7. D.D. Heistad and H.A. Kontos, Cerebral circulation, <u>in</u>: "Handbook of Physiology: The Cardiovascular System," Vol. III, F. Abboud and J. Shephard, eds., Am. Physiol. Soc., Bethesda (1983).

8. D.J. Reis and C. Iadecola, Central neurogenic regulation of cerebral blood flow, <u>in</u>: "Neurotransmission and Cerebrovascular Function II," J. Seylaz and R. Sercombe, eds., Elsevier, Amsterdam, (1989).

9. M. Nakai, C. Iadecola, D. Ruggiero, L. Tucker, and D.J. Reis, Electrical stimulation of the cerebellar fastigial nucleus increases cerebral cortical blood flow without changes in focal metabolism: Evidence for an intrinsic system in brain for primary vasodilation, <u>Brain Res.</u> 260:35 (1983).

10. C. Iadecola and D.J. Reis, Continuous monitoring of cerebrocortical blood flow during stimulation of the cerebellar fastigial nucleus: A study by laser-doppler flowmetry, <u>J. Cereb. Blood Flow Metab.</u> 10:608 (1990).

11. S.P. Americ, C. Iadecola, M.D. Underwood, and D.J. Reis, Local cholinergic mechanisms participate in the increase in cortical cerebral blood flow elicited by electrical stimulation of the fastigial nucleus in rat, <u>Brain Res.</u> 411:212 (1987).

12. K. Chida, C. Iadecola, and D.J. Reis, Lesions of rostral ventrolateral medulla abolish some cardio-and cerebrovascular components of the cerebellar fastigial pressor and depressor responses, <u>Brain Res.</u> 508:93 (1990).

13. M.D. Underwood, C. Iadecola, and D.J. Reis, Neurons in C1 area of rostral ventrolateral medulla mediate global cerebrovascular responses to hypoxia but not hypercarbia, <u>J. Cereb. Blood Flow Metab.</u> 7(Suppl 1):S226 (1987).

14. M.D. Underwood, "Control of the Cerebral Circulation and Metabolism by the Rostral Ventrolateral Medulla: Possible Role in the Cerebrovascular Response to Hypoxia," Ph.D. Dissertation, Cornell University, New York (1988).

15. K. Maiese, L. Pek, S.B. Berger, and D.J. Reis, Reduction in ischemic infarctions produced by occlusion of rat middle cerebral artery by treatment with agents acting at imidazole receptors, <u>J. Cereb. Blood Flow Metab.</u>, in press (1991).

16. D.J. Reis, S. Morrison, and D.A. Ruggiero, The C1 area of the brainstem in tonic and reflex control of blood pressure, <u>Hypertension</u> 11(Suppl.):I8 (1988).

17. M.D. Underwood, C. Iadecola, A. Sved, and D.J. Reis, Stimulation of C1 area neurons globally increases regional cerebral blood flow but not metabolism, <u>J. Cereb. Blood Flow Metab.</u>, in press (1991).

18. M. Kumada, R.A.L. Dampney, and D.J. Reis, Profound hypotension and abolition of the vasomotor component of the cerebral ischemic response produced by restricted lesions of medulla oblongata: relationship to the so-called tonic vasomotor center, <u>Circ. Res.</u> 45:63 (1979).

19. R.A.L. Dampney, M. Kumada, and D.J. Reis, Central neural mechanisms of the cerebral ischemic response: characterization, effect of brainstem and cranial nerve transections, and simulation by electrical stimulation of restricted regions of medulla oblongata in rabbit, <u>Circ. Res.</u> 45:48 (1979).

20. R.A.L. Dampney and E.A. Moon, Role of ventrolateral medulla in vasomotor response to cerebral ischemia, <u>Am. J. Physiol.</u> 239:H349 (1980).

21. D.L. Brown and P.G. Guyenet, Electrophysiological study of cardiovascular neurons in the rostral ventrolateral medulla in rats, <u>Circ. Res.</u> 56:359 (1985).

22. T.A. Milner, V.M. Pickel, S.F. Morrison, and D.J. Reis, Adrenergic neurons in the rostral ventrolateral medulla: ultrastructure and synaptic relations with other transmitter-identified neurons, <u>in</u>: "Prog. Brain Res.", 81, J. Ciriello, C. Polosa, and M. Caverson, eds., Elsevier, Amsterdam (1989).

23. M-K. Sun and D.J. Reis, Evidence that stimulation of oxygen sensors in ventral medulla initiates the cerebral ischemic response, <u>Soc. Neurosci. Abstr.</u> 17:612, 1991.

24. M-K. Sun, I.T. Jeske, and D.J. Reis, Cyanide excites medullary sympathoexcitatory neurons in rats, <u>Am. J. Physiol.</u>, in press (1991).

25. A.S. Blix and B. Folkow, Cardiovascular adjustments to diving in mammals and birds, <u>in</u>: "Handbook of Physiology. Section 2: The Cardiovascular System," J.T Shephard and F.M. Abboud, eds., Am. Physiol. Soc., Bethesda (1983).

EFFECTS OF NMDA ANTAGONISTS AND TEMPERATURE ON REGIONAL CEREBRAL BLOOD FLOW

Eng H. Lo, Gary K. Steinberg

Department of Neurosurgery
Stanford University School of Medicine
Stanford, CA

INTRODUCTION

A large body of experimental data supports the excitotoxic hypothesis of ischemic brain damage [1,2], where the ischemic insult results in a massive pathological release of excitatory amino acids that eventually leads to a lethal influx of calcium into the neurons. Within the framework of excitotoxic ischemic injury, blockade of NMDA receptors and calcium channels may prevent neuronal damage and death [1]. This therapeutic strategy, therefore, relies primarily on the prevention of pathologic activation of NMDA-gated calcium channels, and does not involve improvements of regional cerebral blood flow (rCBF) per se. Mild to moderate hypothermia also appears to protect against cerebral ischemic injury [3,4]. Changes in temperature of a few degrees can significantly decrease the release of excitotoxic amino acids during the ischemic episode [4]. However, it is possible that NMDA antagonists and hypothermia may also influence blood flow, either directly, through secondary alterations related to cerebral metabolism, or by modifications of normal autoregulatory mechanisms including changes in metabolism-flow coupling. These changes may also vary with the specific type of NMDA antagonist, dose, animal species, the use of anesthesia, and the degree of hypothermia involved.

The aim of this study was to measure the effects of NMDA antagonists and temperature on rCBF in the ischemic and normal rabbit brain. A three vessel occlusion model of focal cerebral ischemia in the rabbit brain was used [5,6]. Somatosensory evoked potentials (SEPs) were used to assess the degree of ischemia acheived. Rectal core and temporalis muscle temperature were monitored. rCBF was measured using radioactive microspheres following standard tachniques [6,7]. Systemic parameters were also monitored, including mean arterial pressure, endtidal CO_2, blood gases, glucose and hematocrit.

Ketamine (KT), dextromethorphan (DM), dextrorphan (DX), and MK-801 (MK) were used. As a comparison, the dihydropyridine calcium channel antagonist nimodipine (NP) was also tested. Normal saline (NS) was used in controls. All drugs were administered iv: KT, DM, and DX were infused as a loading dose of 20 mg/kg over 30 min followed by a maintenance dose of 10mg/kg/hr for another 20 min; MK was infused as 1mg/kg over 30 min and 0.75 mg/kg/hr for 20 min; NP was infused as 60 microg/kg/hr for 50 min; and equivalent volumes of NS and vehicle (VH) were used. Four animals per group were measured.

There were no significant differences in systemic parameters between the various groups before or after drug infusion, although NP and MK tended to slightly decrease MAP in some animals. Significant and region-specific alterations in rCBF were demonstrated in the neocortex, subcortex, and midbrain (Fig 1a-c). However, no robust differences between the 7 groups were found in the medulla, pons, and cerebellum (Fig 1d).

In the neocortex, rCBF (mean+SEM) in the NP group was the highest compared to all groups (p<0.01). NP increased rCBF

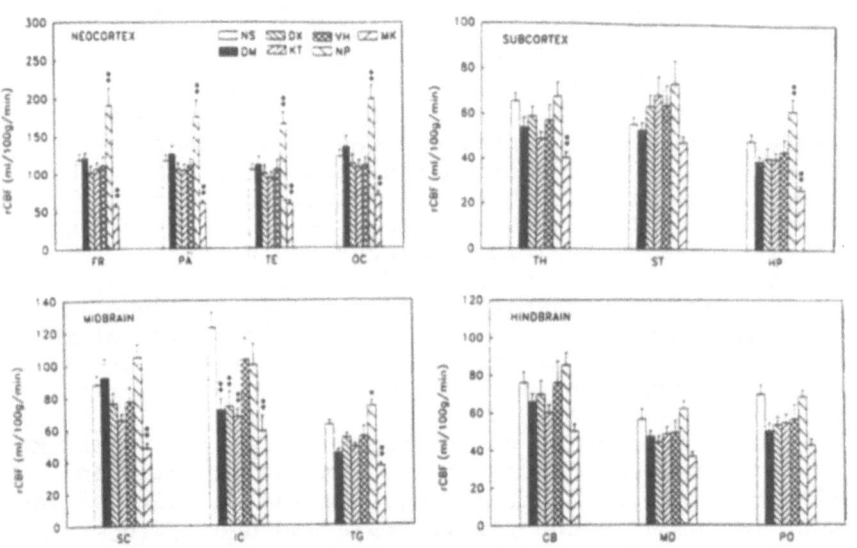

FIGURE 1. rCBF (mean+SEM) in brain regions measured using radioactive microspheres; * p<0.05 and ** p<0.01 for comparison with rCBF in the NS group; n=8 per region per group. A. Neocortex: frontal (FR), parietal (PA), temporal (TE), occipital (OC). B. Subcortex: thalamus (TH), striatum (ST), hippocampus (HP). C. Midbrain: superior colliculus (SC), inferior colliculus (IC), tegmentum (TG). D. Hindbrain: cerebellum (CB), medulla (MD), pons (PO).

in the neocortex by 50-63% compared to the NS group (Fig 1a). rCBF in the MK group was significantly lower than all other groups (p<0.01). MK decreased rCBF by about 43-52% compared to NS (Fig 1a). Cerebral blood flow values in the other groups ranged from 94 ml/100g/min to 135 ml/100g/min.

No significant differences were found in the striatum. In the thalamus, rCBF in the MK group was significantly lower only when compared to the NP and NS groups (p<0.01). In the hippocampus, MK significantly decreased rCBF compared to all other groups (p<0.05 for DM, DX, KT, and VH; p<0.01 for NS and NP). MK decreased flow by 45% compared to NS (Fig 1b). rCBF in the NP group was significantly higher compared to all other groups (p<0.01). NP increased rCBF in this region by 28% compared to NS (Fig 1b). Flow in the remaining groups ranged from 40 ml/100g/min to 44 ml/100g/min.

All the NMDA antagonists tested (KT, DM, DX, MK) were found to significantly decrease rCBF in the inferior colliculus (p<0.01) (Fig 1c). Flows ranged from 59 ml/100g/min to 74 ml/100g/min compared to 100+13 ml/100g/min (NP), 103+14 ml/100g/min (VH), and 123+10 ml/100g/min (NS). There were no significant differences between the NMDA antagonist groups. MK also decreased rCBF in the superior colliculus by 45%, and in the tegmentum by 41% compared to NS (Fig 1c). NP significantly increased rCBF in the tegmentum (p<0.05 with NS; p<0.01 with all other groups) (Fig 1c).

In the hindbrain, the only significant differences were found between the NP and MK groups (p<0.05). NP increased flow and MK decreased flow when compared to NS, but these differences did not reach statistical significance (Fig 1d).

EFFECTS OF NMDA ANTAGONIST DM IN ISCHEMIC BRAIN

rCBF was measured following 1 hr permanent focal ischemia. DM was administered iv 10 min after occlusion as a loading dose of 20 mg/kg over 30 min and maintainence dose of 10 mg/kg/hr (n=7); NS was given as equivalent volumes

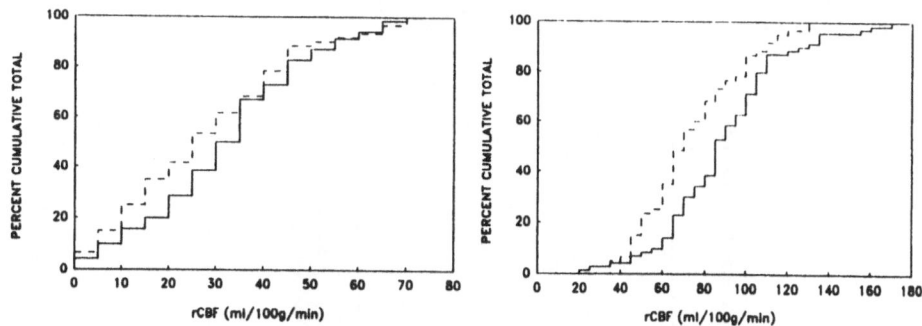

Figure 2. Cumulative probabilty distribution functions demonstrating that rCBF in the DM group (solid line) is higher than rCBF in the NS group (dashed line) in a. the ipsilateral left hemisphere, and b. the contralateral right hemisphere.

(n=7). Arterial pressure was kept at approximately 60-65 mm Hg by varying halothane between 0.4-0.8%.

There were no significant differences in systemic parameters between DM and NS groups. DM did not alter arterial pressure or temporalis muscle temperature. SEPs were abolished within 10 min of arterial occlusion. In both DM and NS groups, there were no decreases in rCBF in cerebellum, but all cortical regions (10 per hemisphere) showed decreased rCBF compared to the contralateral side. There was no difference in the topographic distribution of rCBF perturbations between the two groups; the lowest flows occurred primarily in the ventral regions towards the anterior of the brain, corresponding to the anterior and middle artery distribution.

Frequency distribution analysis was performed by plotting the probability distribution functions for all cortical regions in both groups. In the ipsilateral left hemisphere, the DM group demonstrated an increased distribution of flows (median = 34.5 ml/100g/min) compared to the NS group (median = 27.5 ml/100g/min) (Figure 2a). Most of the increased flow occurred in the 10-30 ml/100g/min range, suggesting that rCBF was improved primarily in the penumbral region. In the contralateral right hemisphere, the DM group also demonstrated a higher distribution of flows (median = 80.5 ml/100g/min) versus the NS group (median = 70.0 ml/100g/min) (Figure 2b). However, the improvement in flow was present across all values of rCBF. Based on these plots, regional means of the frequency distribution were calculated by averaging the rCBF for all regions. It was found that mean rCBF was significantly higher in the DM group for both left (p<0.05) and right hemispheres (p<0.001); rCBF values were increased by 17% and 22% respectively. However, a region-by-region comparison of rCBF between DM-treated and NS-treated animals did not unequivocally demonstrate specific regions of improved flow. This probably reflects the fact that the penumbral regions may vary between animals.

EFFECTS OF TEMPERATURE IN ISCHEMIC BRAIN

rCBF was measured following 4 hrs permanent focal ischemia. Three temperature (temporalis muscle) groups were studied: 37oC, 33oC, and 30oC (n=5 per group). In the

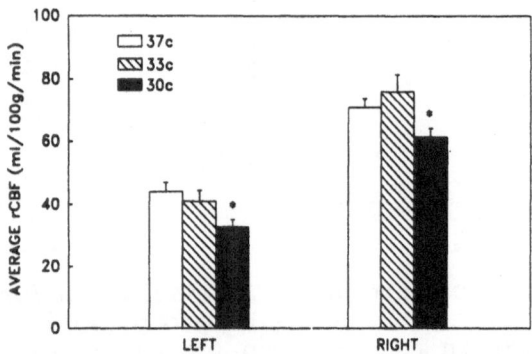

Figure 3. Temperature-dependence of rCBF (mean+SEM) in focal cerebral ischemia. * p<0.05.

hypothermic groups, temperature was lowered immediately after arterial occlusion using a locally-directed fan.

In the lowest temperature group (30oC), rCBF in both occluded (left) and contralateral (right) hemispheres were significantly lower than the other 2 groups (p<0.05) (Fig 3). There were no significant differences between the 33oC group and the normothermic animals. After 4 hrs of permanent ischemia, SEP recovery (expressed as a percentage of pre-occlusion values) was signifcantly improved in the two hypothermic groups (p<0.05). SEP values were 42.3+13% (30oC) and 18.5+6% (33oC) versus 2.2+2% in the normothermic group.

SUMMARY AND CONCLUSIONS

NMDA antagonists have been shown to protect against ischemic neuronal damage in animal models as well as in vitro cell cultures [1,2,5,8]. It is, however, possible that NMDA antagonists may also influence cerebral perfusion patterns, either directly or indirectly, thus modifying their effects on cerebral ischemia [9,10]. Previous results have been varied; some data suggest that NMDA antagonists improve rCBF in the ischemic brain [6,11,12], whereas others demonstrate no improvements in flow [13]. Our data show that although NMDA antagonists do not increase rCBF in the normal rabbit brain, significant improvements in flow, primarily in the penumbra, are present in the ischemic brain.

The neuroprotective effects of hypothermia have been demonstrated to be related to a decrease in efflux of excitotoxic amino acids during the ischemic episode [4]. Our data show that rCBF is not improved by mild to moderate hypothermia. In fact, the lowest temperature group studied actually had decreased blood flow in both ischemic and contralateral hemispheres. These effects may be related to temperature-dependent decreases in cerebral metabolism [14].

In general, it is difficult to seperate primary from secondary effects of rCBF and neuronal protection. Improved neuronal survival due to direct NMDA antagonism may be expressed as increased rCBF due to preserved flow-metabolism coupling. On the other hand, an increase in flow, whether primary or secondary, may additonally contribute to protection against ischemic injury, especially in the non-neuronal cell populations (glial and vascular).

ACKNOWLEDGEMENTS

Research supported by AHA Grant-in-Aid 881069, NIH Grants 2 S07 RR05353-29 and 1 RO1 NS27292-01A2, and the Valerie Bernhardt Cerebral Ischemia Fund.

REFERENCES

1. D. W. Choi, Methods for antagonizing glutamate neurotoxicity, Cerebrovasc Metab Rev 2:105-147 (1990).

2. S. M. Rothman and J. W. Olney, Glutamate and the pathophysiology of hypoxic-ischemic brain damage, Ann Neurol 19:105-111 (1986).

3. R. Busto, W. D. Dietrich, M. Y. T. Globus, I. Valdes, P. Scheinberg and M. D. Ginsberg, Small differences in intraischemic brain temperature critically determine the extent of ischemic neuronal injury, J Cereb Blood Flow Metab 7:729-738 (1987).

4. R. Busto, M. Y. T. Globus, W. D. Dietriech, E. Martinez, I. Valdes and M. D. Ginsberg, Effect of mild hypothermia on ischemia-induced release of neurotransimitters and free fatty acids in rat brain, Stroke 20:904-910 (1989).

5. G. K. Steinberg, J. Saleh, D. Kunis, R. L. DeLaPaz and S. Zarnegar, Protective effect of NMDA antagonists after focal cerebral ischemia in rabbits, Stroke 20:1247-1252 (1989).

6. E. H. Lo and G. K. Steinberg, Effects of dextromethorphan on regional cerebral blood flow in focal cerebral ischemia, J Cereb Blood Flow Metab (in press, 1991).

7. E. H. Lo, G. Sun and G. K. Steinberg, Effects of NMDA and calcium channel antagonists on regional cerebral blood flow, Neurosci Lett (in press, 1991).

8. M. K. Goldberg, P. C. Pham and D. W. Choi, Dextrorphan and dextromethorphan attenuates hypoxic injury in neuronal culture, Neurosci Lett 80:11-15 (1987).

9. A. Kurumaji and J. McCulloch, Effects of MK-801 upon local cerebral glucose utilization in conscious rats and in rats anesthetized with halothane, J Cereb Blood Flow Metab 9:786-794 (1989).

10. D. G. Nehls, C. K. Park, A. G. MacCormack and J. McCulloch, The effects of NMDA receptor blockade with MK801 upon relationship between cerebral blood flow and glucose utilization, Brain Res 511:271-279 (1990).

11. A. M. Buchan, D. Xue, A. Slivka, C. Zhang, J. Hamilton and A. Gelb, MK-801 increases blood flow in a rat model of focal cortical ischemia, Soc Neurosci Abstr 15:804 (1989).

12. G. K. Steinberg, E. H. Lo, D. M. Kunis and G. Grant, Dextromethorphan alters cerebral blood flow and protects against cerebral injury after focal cerebral ischemia, Soc Neurosci Abstr 16: 1278 (1990).

13. C. K. Park, D. G. Nehls, G. M. Teasdale and J. McCulloch, Effect of NMDA antagonist MK-801 on local cerebral blood flow in focal cerebral ischemia in the rat, J Cereb Blood Flow Metab 9:617-622 (1989).

14. D. W. Busija and C. W. Leffler, Hypothermia reduces cerebral metabolic rate and cerebral blood flow in newborn pigs, Am J Physiol 253:H869-873 (1987).

REDUCED LATENCY OF VISUALLY EVOKED POTENTIALS FOLLOWING CORTICAL

INJURY INDICATES SECONDARY GLUTAMATERGIC NEURONAL EXCITATION.

Shichen Xu, Henry G. Wagner, Ferenc Joo, Igor Klatzo,
Robert Cohn*

National Institute of Neurological Disorders and
Stroke, Bethesda, MD20892 *Howard University,
Washington D.C.20060

While searching for physiological evidence of secondary
hyperexcitability following brain injury (Saito et al., 1990), we
observed that the latency of visually evoked responses (VER)s was
reduced following a unilateral cold lesion to the parietal cortex in
the rat. The reduction of latency is interpreted as evidence of a
hyperexcitability developing in the visual pathway following cortex
injury. Many considerations suggest that glutamate plays an
important role in delayed tissue changes following traumatic brain
injuries through actions mediated by NMDA receptors (Choi, 1990). We
thought this reduction of VER latency may be also of neuroexcitatory
nature. This premise was tested by injection of MK-801, a
noncompetitive inhibitor of NMDA receptors, in rats when the latency
of the VER was reduced following a brain injury.

White Sprague Dawly rats, male or female, 100 to 150 days of
age, were used. Electrodes for VER recording were anchored in the
cranium above visual cortex on both sides at least one week before
experiments. The reference electrode was located on the cranial wall
of the right frontal sinus. Under gas anesthesia (1% halothane in
30% O2, 70% nitrous oxide) a cold probe, -100 °C, 4 mm in diameter
was applied to the cranium unilaterally for 40 seconds over the
parietal cortex caudal to the bregma. EEG and body temperature were
monitored and rectal temperature was maintained at 38 °C. At a
frequency of 0.5 Hz, VERs to a stroboscopic xenon flash were
averaged and recorded before and after the cold application. Mk-801
was injected (3mg/Kg, i.p.) 1 or 2 days after induction of cold
lesion when the latency of VER had greatly reduced.

Figure 1 shows the VER on a rat before cold lesion. The VER
showed a dominant negative and dominant positive wave. In most cases
there were some wavelets in front or behind with a large variety in
both latencies and amplitudes. The latency to the peak of dominant
negative wave was found to be 44 ± 3 msec (Mean ± SD) (N=43) before
cold lesion.

Mk-801 was injected 1 or 2 days after cold lesion when the
latency of VER had been reduced . After application of Mk-801, the
reduced latency was prolonged again (Fig. 2). This effect of Mk-801

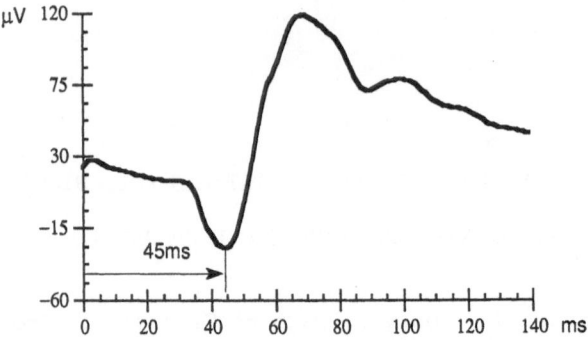

Fig. 1. VER before cold lesion, summed 100 times and averaged. The latency to the dominant negative peak was 45 ms.

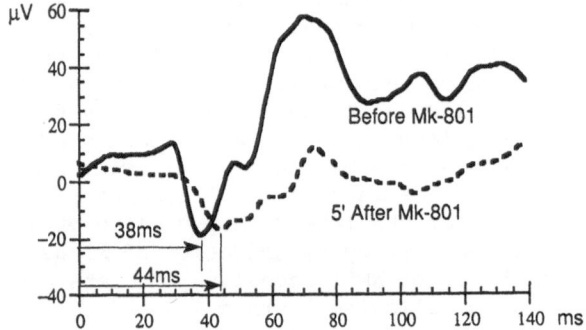

Fig. 2. VER from the same animal as in Fig. 1 but 2 days after cold injury. The latency reduced to 38 ms and became 44 ms 5 minutes after injection of Mk-801.

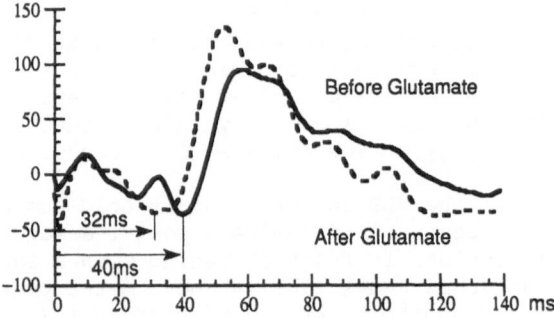

Fig.3. Reduction of latency of VER. Broken line was the VER 20 minutes after application of glutamate, the latency of which was reduced by 8 ms.

appeared within 5 minutes and lasted for about 4 hours. A decrease in EEG and VER amplitude was observed too.

Reduced latency of visually evoked potentials could also be demonstrated by direct application of glutamate on the cerebral cortex. Under pentobarbital anesthesia (90mg/kg, i.p.), an opening through the skull of 2.5 mm diameter was made above the visual cortex. The cortex was exposed by removing dura. The latency of the VER was reduced after the solution of glutamate was applied to the surface of the cortex, see Figure 3.

The glutamatergic nature of hyperexcitation in the vicinity of the cortical lesion is thus supported by its reversibility produced by Mk-801, as well as by effect of direct application of glutamate on the exposed cerebral cortex.

References

Saito N., Chang C.,Kawai K., Joo F., Nowak Jr. T.S., Mies G., Ikeda J., Nagashima G., Ruetzler C., Lohr J., Spatz M. and Klatzo I., 1990, Role of neuroexcitation in development of blood-brain barrier and oededmatous changes following cerebral ischemia and traumatic brain injury, Acta Neurochirugica, Suppl. 51:186.

Choi D.W., 1990, Cerebral hypoxia:some new apporaches and unanswered question, The Journal of Neuroscience, 10:2493.

PRE- OR POSTSYNAPTIC BLOCKING OF GLUTAMATERGIC FUNCTIONING PREVENTS THE INCREASE IN GLUCOSE UTILIZATION FOLLOWING CONCUSSIVE BRAIN INJURY

David A. Hovda, Yoichi Katayama, Atsuo Yoshino, Tatsuro Kawamata and Donald P. Becker

Division of Neurosurgery, UCLA School of Medicine
74-140 CHS, Los Angeles, CA 90024-6901

INTRODUCTION

During the first few minutes following an experimental concussive brain injury in the rat, the cerebral cortex and underlying hippocampus are exposed to an increase in extracellular potassium (1). In addition to this ionic flux, these same regions exhibit an increase in glucose metabolism using [^{14}C]2-deoxy-D-glucose (2DG) autoradiography(2). It has been proposed that any ionic flux due to an traumatic brain injury would result in a stimulation of glycolysis, presumably due to the energy demand of cells to activate ionic pumping mechanisms in their effort to restore ionic balance(3).

In previous work(1,4), it has been demonstrated that following traumatic brain injury there is a marked increase in the concentration of extracellular glutamate. Since the injury-induced ionic flux can be markedly inhibited by blocking the glutamate receptor prior to the insult(1), it was proposed that the release of glutamate may be the mechanism behind the ionic flux. To determine the relationship between the ionic flux and the changes in glucose metabolism these receptors need to be blocked prior to injury and glucose metabolic studies need to be performed immediately after the insult.

Another issue at hand is whether the glutamate increase comes from endogenous (e.g. neuronal) or exogenous (e.g. blood) via an injury induced compromise of the blood-brain barrier (BBB). In our previous work(2) utilizing Evans Blue albumin there was no evidence of a compromised BBB following our fluid percussion injury. However, the Evans Blue method is capable of detecting only gross alterations in BBB permeability, and therefore, we can not rule out more subtle changes which may be detectable using more sensitive procedures.

Given the anatomical organization of the hippocampus, the relatively specific removal of the endogenous source of glutamate to the CA1 region can be achieved by destroying the CA3 cells via kainic acid. Such a procedure, therefore, lends itself to the question of whether exogenous glutamate plays a significant role in the injury-induced metabolic derangement following brain injury.

The Role of Neurotransmitters in Brain Injury, Edited by
M. Globus and W.D. Dietrich, Plenum Press, New York, 1992

METHODS

Microdialysis Experiments

Surgical Preparations

Male Sprague-Dawley rats (Wt=250-300 g) were anesthetized with a mixture of 33% oxygen, 66% nitrous oxide and enflurane (1.5-2.0 ml/min). With the rectal temperature kept at 37.0-38.0°C via a thermostatically controlled heating pad, the femoral vein and artery were catheterized using PE-50 polyethylene tubing. After cannulation, the animals were placed in a stereotaxic frame, the calvarium exposed and a craniotomy was performed for placement of a hollow plastic screw (O.D.=5.0 mm; I.D.=3.5 mm) used to administer a fluid percussion (F-P) injury (see below). The screw was placed epidurally, the center of which was positioned on the midline 2.0 mm anterior to bregma. This position was selected to produce post-traumatic hypermetabolism of the cortex bilaterally and to enable the placement of microdialysis probes in each parietal cortex, Two small burr holes were made posterior to the injury screw (2.5 - 3.5 mm posterior to bregma and 3.0 - 3.5 mm lateral to the midline) and a pair of dialysis probes (CMA/100; Bioanalytical Systems Inc., 2701 Kent Avenue, Purdue Research Park, West Lafayette, IN 47906; O.D.=300 um; effective length = 3.0 mm; cut off = 20,000 MW) were placed vertically into the left and right parietal cortex, 2.5-3.0 mm below the surface of the dura. All surgical wounds were infiltrated with 1% xylocaine.

Microdialysis Procedures

One microdialysis probe was always perfused with Ringer's solution to control for the non-specific effects of microdialysis procedures. The other probe served as a test probe for the administration of drugs. Three drugs [kynurenic acid (KYN: 10.0 mM), 6-cyano-7-nitroquinoxaline-2,3-dine (CNQX: 300 uM, 1.0 and 10.0 mM) or 2-amino-5-phosphonovaleric acid (APV: 100 uM, 1.0 and 10.0 mM)] were tested. All drugs were perfused for 30 min prior to the administration of the F-P pulse. Each dialysate was perfused using a 250 ul Hamilton microsyringe and a microperfusion pump at a rate of 5.0 ul/min. The dialysate was adjusted to pH = 7.4, osmolarity = 308 mOsm and the temperature was maintained at 37.0-38.0°C with a perfusion warmer. Before and after dialysis, the probes were checked microscopically to confirm the integrity of the membrane and the lack of bubbles.

Procedures for Injury Induction

Upon completion of the perfusion, the probes were removed, the injury screw was secured and the burr holes closed using cyanoacrylate and dental acrylic. The concussive brain injury was administered utilizing a fluid percussion described in more detail elsewhere(1,5,6). Briefly, animals were removed from the stereotaxic frame and the injury screw was connected to the saline-filled fluid-percussion device with pressure resistive polyethylene tubing (20 cm). Delivery of gas anesthesia was terminated 60 s before delivery of a transient pressure fluid-pulse to the epidural space. A strain-gage transducer (Stratham PA 85-100) between the injury cylinder and injury screw measured the amplitude (2.3-2.7 atm) and duration (21-23 ms) of the fluid pulse wave. This amplitude produced unconsciousness (judged as the period until occurrence of a toe-pinch withdrawal reflex) lasting up to 5 min after the injury. Although transient apnea was often induced at this injury level, no systemic circulatory collapse was observed. If apnea persisted for more than 10 s respiration was mechanically supported for 30 s with a mixture of oxygen and room air. When respiratory support was required for longer that 30 s the animal was excluded from the study.

Procedure for 2DG Autoradiography

Following the procedure originally described by Sokoloff et al. (1977)(7), 2DG (200 uCi/kg) was slowly injected (i.v.; duration =30 s) followed by the delivery of the F-P injury. The injection of 2DG immediately prior to injury insured that the measurement of glucose utilization would reflect the massive ionic perturbations which occur during the first few minutes following concussion(1). Blood samples were collected periodically through the arterial catheter for measurements of plasma glucose and ^{14}C activity. Forty-five min after the injection, a sample for blood gas analysis was collected and the animals were administered a lethal dose of sodium pentobarbital (100 mg/kg, i.v.). The brains were quickly removed and frozen in powder dry ice. Frozen coronal sections were cut at 20 um in a cryostat at -20°C and mounted onto cover slips. Through the region of the dialysis probes every 10th section was processed for autoradiography with adjacent sections stained for thionin.

Each animal served as his own control given one hemisphere was perfused with the test drug and the other with the vehicle alone. In addition, several different sham control groups were run to control for drug effects on 2DG uptake with and without injury as well as the effect of inserting the microdialysis probe alone. Measurements of local cerebral glucose utilization (lCMRglc; umol/100g/min) were taken at different regions proximal and distal to the site of the microdialysis probe.

CA3 Lesion Experiments

Surgical Preparations

These procedures were identical to that described above except that the craniotomy for the injury screw was positioned 1.0 mm posterior to bregma and 6.0 mm lateral (left) of the midline. This was done to insure that the left hippocampus would exhibit an increase in lCMRglc.

Procedures for Injury Induction

The injury induction was identical to that described above except that the magnitude of the pulse reached between 3.7 and 4.5 atm.

Removal of the CA3 Projection to CA1

Removal of the endogenous innervation of glutamatergic fibers to the CA1 region of the hippocampus was achieved by removal of the CA3 region via intraventricular injection of kainic acid (0.5 ug/ul in phosphate buffer, pH=7.2-7.4) five days prior to the lateral F-P. Control animals received the injection of phosphate buffer alone.

Procedure for 2DG Autoradiography

Using the same procedure as described above the animals were processed for 2DG with every 10th section kept through the region of the dorsal hippocampus. In these animals cortical lCMRglc was calculated for the frontal, parietal and entorhinal cortex. In addition the caudate/putamen, corpus callosum and ventral thalamic nuclei were studied. For the hippocampus, measurements were taken in the dorsal regions of CA1 and CA3 along with the lacunosum-moleculare.

RESULTS

Microdialysis Experiments

In animals who did not receive dialysis of EAA receptor antagonist, the F-P injury resulted in an increase in lCMRglc of between 62.1 to 65.0 representing an increase of 251% above control (p<0.001). However, in regions of the cerebral cortex infiltrated with EAA antagonists, the injury-induced increase in lCMRglc was markedly reduced especially in regions close to the site of dialysis.

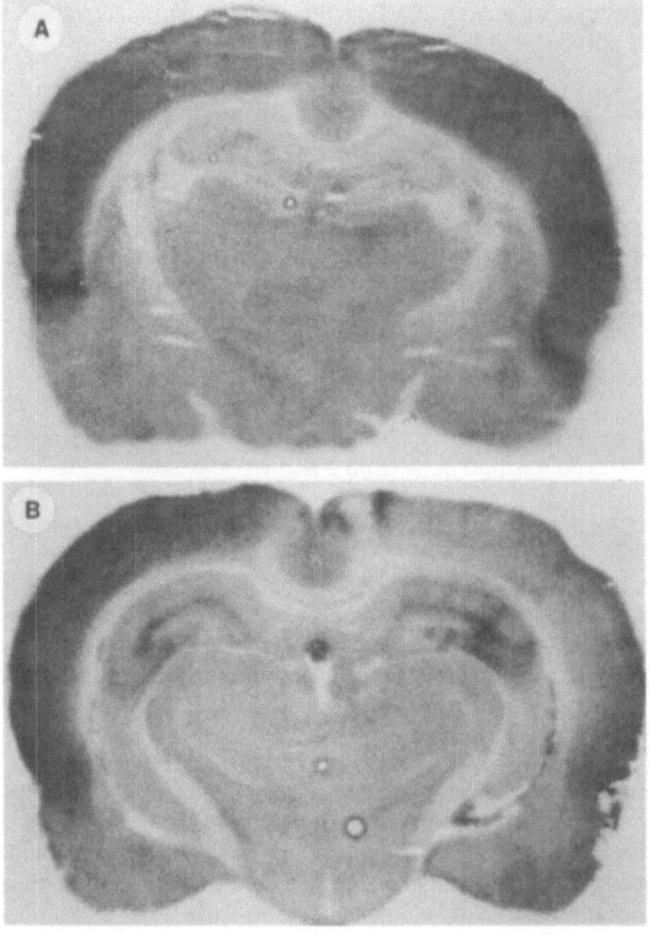

Figure 1. Coronal autoradiographs illustrating the uptake of 2DG seem immediately following a central fluid percussion injury. (A) An animal that did not receive dialysis. (B) An animal that received dialysis of ringers solution in the left cerebral cortex and APV (10 mM) in the right. Note the marked reduction in the increase of 2DG uptake in the right cerebral cortex which had previously been perfused with APV.

In order to evaluate the effects of the EAA antagonists on the increase of 2DG uptake seen following F-P injury, lCMRglc of the drug perfused cerebral cortex were compared to those of the contralateral vehicle perfused cerebral cortex. Although all drugs effectively reduced the level of glucose utilization seen immediately following injury, they exhibited different levels of effectiveness. The administration of KYN resulted in lCMRglc which were 54.4% of control throughout all regions (p<0.01). This represented an effected area of approximately 7 mm in diameter centered around the probe tip. APV, at a concentration of 10 mM and 1 mM, also prevented the post-traumatic metabolic increase resulting in rates which were 59.2% (p<0.05) and 61.9% (p<0.01) of control respectively. The effect of APV was seen over a larger area than that of KYN extending up to 15 mm in diameter (see Fig. 1). The lowest concentration of APV tested (100 uM) showed a mild reduction of lCMRglc affecting only regions closest to the site of dialysis resulting in rates which were 90.3% of control (p<0.05).

Compared to both KYN and APV, CNQX showed milder effects on lCMRglc following F-P injury. Even at the highest concentration (10 mM), CNQX decreased lCMRglc of the injured brain to only 80.9% of control (p<0.01). At 1 mM CNQX had only a mild effect and only in regions closest to the site of dialysis which exhibited levels which were 85.5% of control (p<0.01). At the lowest concentration tested (100 uM) CNQX showed no significant effect on lCMRglc following injury.

CA3 Lesion Experiment

Within the sham-injured groups there was no effect of kainic acid on the lCMRglc for any of the structures measured. Only those animals that were subjected to a F-P injury showed any evidence of metabolic disruption with lCMRglc showing a marked increased in both the cerebral cortex and underlying hippocampus ipsilateral to the injury. The injury-induced increased in lCMRglc within the cerebral cortex was not affected in animals who had received kainic acid injections with both experimental and control animals showing an increase of between 29.4 and 34.8%.

In addition to the cerebral cortex the ipsilateral dorsal hippocampus also increased in lCMRglc immediately after F-P injury, exhibiting a significant increase of 81.5% for the CA1 region (p<0.0001) with a 56% increase within the region of the lacunosum-moleculare (p<0.0001). Finally, the CA3 region also exhibited a marked increase in lCMRglc following injury reaching 106.8% above normal (p<0.0001). However, unlike the cerebral cortex, in animals who had received kainic acid five days prior to F-P the hippocampus did not exhibit any evidence of an increase in lCMRglc (see Table 1).

Table 1. Mean (± standard deviation) local cerebral metabolic rate for glucose (umol/100g/min) of selected regions within the dorsal hippocampus.

	Vehicle-Sham	Kainate-Sham	Vehicle-Injured	Kainate-Sham
Lt. CA1	46.6± 3.1	45.3± 3.9	84.6± 5.0aa,bb	51.4±3.9b,cc
Rt.	47.0± 4.0	45.7± 2.6	56.8± 8.0a,b	47.9±5.9c
Lt. LMol.	71.4±11.8	68.9± 7.8	111.5± 8.0aa,bb	70.7±4.5cc
Rt.	72.8±11.5	68.7± 8.3	85.5±20.3b	65.8±9.0c
Lt. CA3	50.2± 6.5	62.7±12.1	103.8±13.0aa,bb	61.4±5.3c
Rt.	53.3± 7.1	51.1± 4.2	62.8±18.4	56.2±5.9

Abbreviations: Lt: left side (side of injury), Rt: right side (contralateral to side of injury), LMol.: lacunosum-moleculare. Significance level: a p<0.05 and aa p<0.01 compared to vehicle-sham, b p<0.05 and bb p<0.01 compared to kainate-sham, c p<0.05 and cc p<0.01 compared to vehicle-injured.

DISCUSSION

The results of the microdialysis experiments indicate that EAA activated ion channels are involved in the post-traumatic increase in glucose utilization reflecting the energy demand of cells required to drive pumping mechanisms against an ionic perturbation seen immediately following the concussive injury. The differential effects of KYN, APV and CNQX suggest that although all subtypes of the glutamate receptor appear to be involved in this phenomenon, N-methyl-D-aspartate activated channels may play a major role.

The results of the CA3 lesion experiments indicate that the removal of the CA3 projection to CA1 protects the CA1 cells from the metabolic derangement typically seen following injury. This result supports the notion that the glutamate increases following concussive injury is do to its release from neuronal stores. Further work is needed to determine the effects of this injury induced metabolic dysfunction on other cellular functions.

REFERENCES

1. Y. Katayama, D. P. Becker, T. Tamura, and D. A. Hovda, Massive increase in extracellular potassium and the indiscriminate release of glutatme following traumatic brain injury, *J. Neurosurg.* **73**, 889 (1990).

2. A. Yoshino, D. A. Hovda, T. Kawamata, Y. Katayama, and D. P. Becker, Dynamic changes in local cerebral glucose utilization following fluid percussion injury, *Brain Res.* (1991)., in press.

3. B. J. Andersen and A. Marmarou, Isolated simulation of glycolysis following traumatic brain injury, in *Intracranial Pressure VII*, J. T. Hoff and A. L. Betz, Eds. (Springer-Verlag, Berlin, 1989), p. 575.

4. A. I. Faden, P. Demediuk, S. S. Panter, and R. Vink, The role of excitatory amino acids and NMDA receptors in traumatic brain injury, *Science* **244**, 798 (1989).

5. C. E. Dixon, B. G. Lyeth, J. T. Povlishock, R. L. Findling, R. J. Hamm, R. J. Marmarou, A. Young, and R. L. Hayes, A fluid percussion model of experimental brain injury in the rat, *J. Neurosurg.* **67**, 110 (1987).

6. T. K. McIntosh, R. Vink, L. Noble, I. Yamakami, S. Fernyak, S. Soares, and A. L. Faden, Traumatic brain injury in the rat: Characterization of a lateral fluid-percussion model. *Neuroscience* **28**, 233 (1989).

7. L. Sokoloff, M. Reivich, C. Kennedy, M. H. Des Rosiers, C. S. Patlak, K. D. Pettigrew, O. Sakurada, M. Shinohara, The [^{14}C]-deoxyglocse method for the measurement of local cerebral glucose utilization: Theory, procedure and normal values in the conscious and anesthetized albino rat, *J. Neurochem.* **28**, 897 (1977).

FOCAL COLD INJURY ALTERS THE PHARMACODYNAMICS OF

PENTOBARBITAL AND THE METABOLIC RESPONSE TO ANESTHETICS:

EVIDENCE FOR FUNCTIONAL CEREBRAL DEPRESSION

David Archer

Research Laboratory, Department of Anaesthesia, Foothills Hospital
The University of Calgary, Calgary, Alberta

INTRODUCTION

Superficial focal cold injury to the cerebral cortex has been previously shown to be associated with widespread depression of glucose utilization (CMR_{glu}), which is maximal 3 days after the injury (Pappius, 1981). These authors (Pappius, 1981, Pappius and Wolfe, 1983, Pappius et al, 1988) have suggested that the decrease in CMR_{glu} represents functional cerebral depression, and have presented evidence to support a role for cortical monoamines (serotonin and noradrenaline) in this phenomenon. Ginsberg et al, in a model of focal thrombotic infarction, have also observed widespread metabolic effects and have interpreted these findings to represent functional diaschisis (Ginsberg et al, 1989).

Anesthetics such as isoflurane and pentobarbital are thought to reduce cerebral metabolism through reduction of neuronal function (Michenfelder, 1988). Recent investigations (Archer et al, 1990, 1991) in our laboratory have sought evidence to support the functional nature of the depression of CMR_{glu} in the cold injury model of brain injury 1) by examining the effects of the injury on the cerebral metabolic response to anesthetics, and 2) by determining the effects of the injury on the pharmacodynamics of pentobarbital. We proposed that to be consistent with functional cerebral depression, the decrease in CMR_{glu} during anesthesia would be attenuated in brain regions depressed by the cold injury. Furthermore, if the animals were functionally depressed by the cold injury, then lower brain concentrations of pentobarbital would be required to produce "surgical anesthesia" (non-responsiveness to noxious stimulus) in injured than in normal animals. Since pretreatment with p-chlorophenylalanine(PCPA) prevents the depression of CMR_{glu} (Pappius et al 1988), we speculated that any decrease in anesthetic requirements observed in injured animals would be prevented by PCPA pretreatment.

METHODS

General Procedure

Freezing lesions were made as previously described (Pappius, 1981). Briefly, under general anesthesia (halothane 2% in oxygen), a craniectomy was performed and a 3 mm diameter probe was applied to the intact dura for five seconds. In separate groups of injured animals determinations of CMR_{glu}, anesthetic requirements, and cortical indoleamines were performed 3 days after the lesion was made.

Parachlorophenylalanine (PCPA) (methyl ester, Sigma Chemical Co., St. Louis, MO) in a dose of 200 mg/kg was given by intraperitoneal injection 24 hours before the freezing lesion in injured animals, four days before study in uninjured controls.

Determination of Concentrations of Indolamines

Three days after cold injury, the animals were anesthetized with intraperitoneal pentobarbital (30 mg/kg) and decapitated. Following rapid removal of the brain the frontoparietal cortex was dissected free and frozen. Serotonin, (5-HT) and 5-hydroxyindoleacetic acid, (5-HIAA), concentrations in the brain homogenates were determined by high pressure liquid chromatography (HPLC) (Pappius et al, 1988).

Determination of CMR_{glu}

CMR_{glu} was determined using the ^{14}C-deoxyglucose method of Sokoloff (Sokoloff et al, 1977). A four hour period with partial restraint was provided to allow recovery from general anesthesia used to insert femoral arterial and venous catheters. For the anesthetized animals, anesthesia to the end-point of non-response to tail clamp was achieved with either intermittent intravenous boluses of pentobarbital (pentobarbital group) or increasing inspired concentrations of isoflurane in oxygen (isoflurane group). A bolus of 30 Ci of [14 C]DG was injected into the venous catheter over 30 seconds (2-deoxy-D-(1-^{14}C) glucose (^{14}C]-DG); specific activity 50-56 mCi/mmol, New England Nuclear). Timed arterial sampling for determination of plasma ^{14}C]-DG and glucose concentrations commenced 15 seconds after the start of the injection and was continued for 45 minutes during stable anesthesia. The animals were decapitated and the brains rapidly removed and frozen to -50 to -60 °C. Autoradiographs, including ^{14}C-methylmethacrylate calibration standards (New England Nuclear) were prepared from 20 micron thick dried brain sections. Densitometry was performed in 28 brain structures on the autoradiographs with a Photovolt Densitometer (Model 52, Photovolt Corporation, New York, NY).

Calculations were performed on a PDP-12 Computer (Digital Equipment Corp, Maynard, MA). The results, summarized in Table 1, were analyzed by one way analysis of variance (ANOVA).

Determination of Anesthetic Requirements for Pentobarbital

The anesthetic requirements ([Pentobarbital]$_{brain}$ associated with non-response to tail clamp) were determined for normal and lesioned animals (3 days post-injury) with and without PCPA pretreatment. Anesthesia was induced with a continuous intravenous infusion of sodium pentobarbital (1 $mg^{-1}.kg^{-1}.min^{-1}$). At the first failure to respond to tail clamp, the animal was decapitated and the brain removed. The brain was rapidly frozen, homogenized and the pentobarbital extracted with n-butyl chloride(Kelner and Bailey, 1983). The extracts were analyzed by HPLC. Results were analyzed with two-way ANOVA.

RESULTS AND DISCUSSION

Cortical Indoleamines following Cold Injury

Figure 1 summarizes cortical concentrations (mean values $\pm$ standard deviations) of serotonin and its first metabolite, 5-HIAA. In the untreated animals, the effect of the lesion was to increase serotonin levels on the lesioned side and to increase 5-HIAA bilaterally, suggesting a bilateral increase in serotonin turnover. PCPA pretreatment

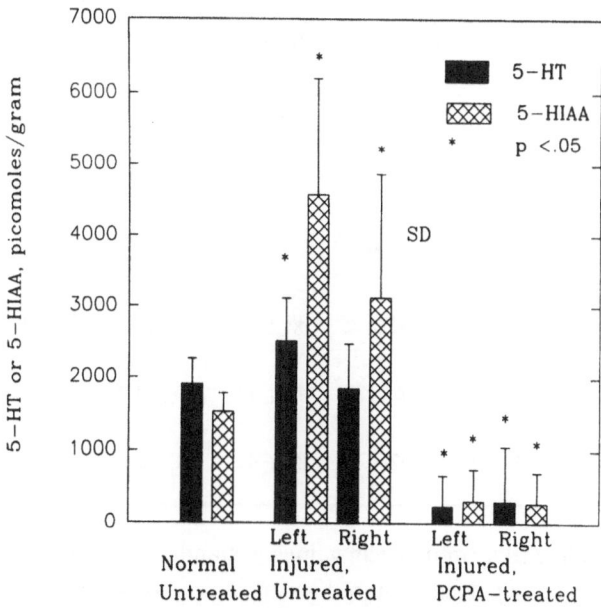

Figure 1. Cortical concentration of Serotonin (5HT) and 5-Hydroxyindoleacetic Acid (5HIAA) in normal, injured and p-chlorophenylalanine (PCPA) treated animals (Figure from Archer et al, 1991. with permission).

reduced both monoamines to less than 20% of normal values. These results confirm previous studies which have shown an increase in cortical serotonin turnover following cold injury, and establish that the dose of PCPA used in this study was effective in preventing 5-HT biosynthesis.

Effect of Pentobarbital and Isoflurane on CMR$_{glu}$

The study (Archer et al, 1990) demonstrated the effect of the cold injury on CMR$_{glu}$ in awake rats and the effect of anesthesia with pentobarbital or isoflurane on CMR$_{glu}$ in injured animals. The results, summarized in Table 1, show that the greatest effect of the lesion in the awake injured rats was in cortical regions. There was a significant difference between the ipsilateral (left) and contralateral (right) cortical CMR$_{glu}$ in the awake animals. The effect of anesthesia with both pentobarbital and isoflurane was to reduce CMR$_{glu}$ and also to abolish the difference in CMR$_{glu}$ between the two hemispheres. Reduction of cortical CMR$_{glu}$ was less on the left side than on the right side (p<.05). These results are consistent with the hypothesis that the depression of CMR$_{glu}$ by the injury is functional, and that in functionally depressed brain the reduction of CMR$_{glu}$ by anesthetics was limited by the decrease in CMR$_{glu}$ associated with the injury.

TABLE 1. CMR$_{glu}$ Following Cortical Cold Injury

	Condition					
	Normal	Injured		Injured Pentobarb		Injured Isoflurane
# Of Animals	(15)	(15)		(10)		(9)
Brain Region	L	R	L	R	L	R
Cortical	122 ± 22** 59 ± 17*	104 ± 25	35 ± 10#	45 ± 12#	36 ± 10#	41 ± 14#
Subcortical	88 ± 23 69 ± 24	76 ± 27	45 ± 14#	46 ± 14#	44 ± 18#	46 ± 18#
Brainstem	122 ± 43** 78 ± 38	102 ± 39	86 ± 41	81 ± 23	85 ± 30	80 ± 42

CRM$_{glu}$, micromoles/100g/min, Mean values $\pm$ standard deviation
* p< .05 - compared to right side, injured group; ** p<.05 - compared to left side, injured group.
# p< .05 - compared to Injured Values in corresponding hemisphere
(Data from Archer et al, 1990)

Effect of Cold Injury on Pentobarbital Pharmacodynamics

Figure 2 shows the brain pentobarbital concentrations (mean values ± standard deviation) at surgical anesthesia in normal and injured animals, with and without PCPA pretreatment. Two way analysis of variance of the data confirmed that the cold lesion was associated with a 30% decrease in $[Pentobarbital]_{brain}$ (p=.005), there was no effect of PCPA pretreatment on $[Pentobarbital]_{brain}$, and that PCPA pretreatment abolished the difference in $[Pentobarbital]_{brain}$ between the normal and the injured animals. Evidence has been presented that the pharmacokinetics of pentobarbital

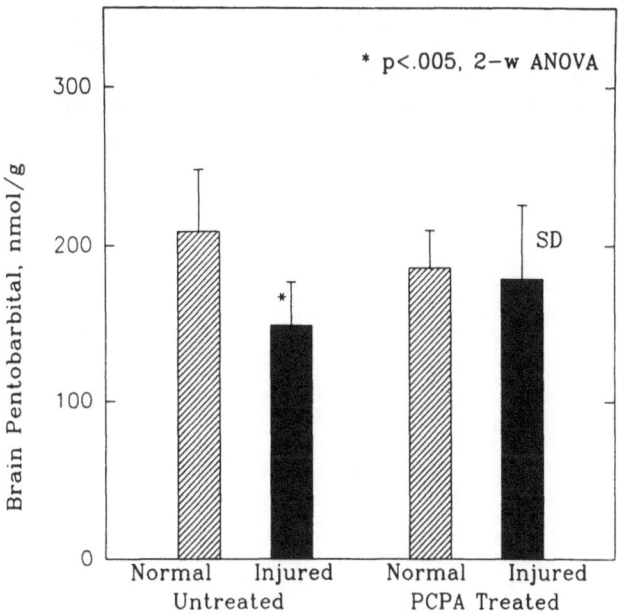

Figure 2. Concentrations of pentobarbital, mean values ± SD in brain samples taken at first failure to respond to tail clamp. (Figure from Archer et al, 1991, with permission)

were not altered by either the cold injury or the PCPA treatment (Archer et al, 1991). These results are consistent with the hypothesis that the cold injury produced a functional change in the nervous system which altered the pharmacodynamics of pentobarbital. The observation that these effects of the cold injury could be abolished by PCPA pretreatment is consistent with previous suggestions that serotonin may play a role in the functional cerebral depression that follows cold injury (Pappius et al, 1988).

REFERENCES

Archer, D.P., Elphinstone, M.G., Pappius, H.M., 1990, The effect of pentobarbital and isoflurane on glucose metabolism in thermally injured rat brain, <u>J Cereb Blood Flow Metabol</u>, 10:624-630.

Archer. D.P., Priddy, R.E., Tang, T.K.K., Sabourin, M.A., Samanani, N., 1991, The influence of cryogenic brain injury on the pharmacodynamics of pentobarbital: Evidence for a serotonergic mechanism, <u>Anesthesiology</u>, 75: In press.

Ginsberg, M.D., Castella, Y., Dietrich, W.D., Watson, B.D., Busto, R., 1989, Acute thrombotic infarction suppresses metabolic activation of ipsilateral somatosensory cortex: Evidence for functional Diasthesis, <u>J Cereb Blood Flow Metabol</u>, 9:329-341.

Kelner, M., Bailey, D.N., 1983, Reversed-phase liquid-chromatographic simultaneous analysis for thiopental and pentobarbital in serum. <u>Clin Chem</u>, 29:1097-1100.

Michenfelder, J.D., 1988, "Anesthesia and the Brain", Churchill Livingstone, New York, pp 36-41.

Pappius, H.M., 1981, Local cerebral glucose utilization in thermally injured rat brain, <u>Ann Neurol</u>, 9:484-491.

Pappius, H.M., Dadoun, R., McHugh, M., 1988, The effect of p-chlorophenylalanine on cerebral metabolism and biogenic amine content of traumatized brain, <u>J Cereb Blood Flow Metabol</u>, 8:324-334.

Pappius, H.M., Wolfe, L.S., 1983, Functional disturbances in brain following injury: Search for underlying mechanisms, <u>Neurochem Res</u>, 8:63-72.

CEREBRAL BLOOD FLOW AND GLUCOSE METABOLISM FOLLOWING EXPERIMENTAL

HEAD INJURY

Katsuji Shima, Anthony Marmarou*, Naoki Koshimae,
Hitoshi Umezawa, Naohide Wako and Hiroo Chigasaki

Department of Neurosurgery, National Defense Medical College,
Tokorozawa, Saitama, Japan, and *Division of Neurosurgery,
Medical College of Virginia, Richmond, V.A., USA

INTRODUCTION

We have as yet very little information as to the changes of local cerebral blood flow (lCBF) and glucose metabolism (lCGU) in the early posttraumatic period in closed head injury. Because of the difficulty in measuring
the CBF soon after head injury, the reported alterations on lCBF after the
injury reveal wide variations, ranging from very low to pronounced hyperemia.
The objective of the present study was designed to detect the changes in
lCBF, lCGU and neuronal damage during the acute phase of concussive closed
head injury in the rat.

MATERIALS AND METHODS

Male Sprague-Dawley rats weighing 300-450 g were anesthetized with
1.5-2 % halothane in O_2 delivered by a closely fitting face mask. After
the reflection of scalp, a stainless steel disc, 10 mm in diameter, was
secured on the skull of the midway area between the bregma and lambda.
Following catheterization the animals were immobilized by a loose-fitting
plaster cast around the hindquaters to facilitate blood sampling for
autoradiographic studies. The physiological condition of each rat was
assessed by measurements of rectal temperature, blood pressure, hematocrit
and blood gases.

After discontinuing halothane inspiration, the animal was placed on a
foam bed. The head injury was delivered by dropping a brass cylinder (450 g)
from the plexiglass tubing positioned above the steel disc on the skull.
Two evels of the injury were selected by varying the height of the weight
drop; moderate injury from 1 meter and severe injury from 2 meters.

Following head injury, the animal was restrained and the autoradiographic studies were done as follows. Fifteen min and 4 h after the injury,
100 microcuries/kg of ^{14}C-iodoantipyrine (^{14}C-IAP) was infused for 30 sec
and arterial blood samples were taken every 5 sec to assess 14C activity.
At 30 sec the rat was decapitated. The brains were rapidly removed, frozen,
cut into 20 micron-thick sections, and subjected to autoradiographs. LCBF
was calculated according to Sakurada et al(1978). LCGU was measured using
the technique designed by Sokoloff et al(1977). Hundred microcuries/kg
^{14}C-2-deoxyglucose(^{14}C-DG) was injected intravenously. Timed blood samples

were taken for analysis of ^{14}C activity and glucose concentration. The animals were decapitated 40 min or 4 h after trauma, respectively 30 min or 45 min after the injection of ^{14}C-DG. Autoradiographs were prepared in the same manner as for lCBF, and lCGU was calculated using the lumped constant for the normal rat determined by Sokoloff et al(1977). LCGU was assessed only in the animals subjected to moderate injury. Posttraumatic neuronal damage was determined by ^{45}Ca autoradiography. Rats subjected to trauma were injected with 100 microcuries/100g ^{45}Ca either 10 min, 1 day, 3 days, 7 days and 10 days after trauma. The ^{45}Ca was allowed to circulate for 4 h, and the brains were removed and autoradiographed in the same manner as for lCBF.

Traumatic changes of blood-brain barrier(BBB) permeability were assessed by intravenous injection of Evans blue (2 % in saline) at 5 min pre- or post-injury.

RESULTS

Eleven of 29 (37 %) animals injured at the high level deteriorated within a few minutes with apnea and rapidly decreasing blood pressure, and died. All animals at low-injury level survived. Subarachnoid hemorrhage around brain stem was observed in 38(6/16) % of moderate injury and in 79 (23/29) % of severe injury. Macroscopic intraparenchymal hemorrhage was not found in any animals of both injury groups. There was no visible extravasation of Evans blue in the brains of all animals measured at 15 min or 4 h post-injury. The abnormal accumulation of ^{45}Ca was not seen by 72 h after moderate injury and 4 h after severe injury. The mean arterial blood pressure (MABP) rapidly increased by 29 % in moderate injury, 60 % in survivors of severe injury and 77 % in nonsurvivors of severe injury over each baseline value, and returned to baseline values within 30 sec in both levels of injury.

Compared with mean values of sham-operated control rats (n=3), moderate injury resulted in mean lCBF increases of 44 % in substantia nigra (p<0.05), and 10-15 % in hypothalamus, septal nucleus and pontine gray, while there were decreases of lCBF in many others (Table 1). In moderate injury group, there was a tendency for variedly changed lCBF to return to control values by 4 h after the injury. In severe injury, lCBF was markedly decreased throughout widespread areas, and none of the areas measured showed any increases in lCBF.

At 40 min post-injury, lCGU showed a significant decrease in all structures examined and the percent changes from control values was more than those of lCBF (Table 2). At 4 h post-injury, there was an insignificant recovery of 5-35 % in lCGU in all regions studies excluding hippocampus and mamillary body.

DISCUSSION

Several investigations of cerebral blood flow and metabolism following experimental head injury have been carried out. Nilsson et al(1989), using the impact acceleration model in rats similar to the present study, found an increase in lCBF in the midbrain and in the central regions of the hemisphere, but not in the superficial gray matter immediate and 30 min after the trauma. Langfitt et al(1966) found the secondary increase in intracranial pressure following an initial transient rise coincident with the impact. They hypothesized that the secondary rise was due to cerebrovascular dilatation and a consequent rise in cerebrovascular volume. The structures with an increase in lCBF following concussive brain injury located in

Table 1. Percent changes in lCBF after head injury compared with control values

| Representative Structures | Moderate Injury | | Severe Injury | |
	15 min posttrauma (n=4)	4 h posttrauma (n=5)	15 min posttrauma (n=6)	4 h posttrauma (n=2)
Frontal Cortex	-27.4 %	- 6.0 %	-29.9 %	-35.6 %
Med. Geniculate Body	-13.5	-23.5	-46.2#	-45.5
Thalamus(Ventral N.)	- 6.3	-31.8*	-31.7	-46.4
Hypothalamus	15.2	-19.1	-35.2#	-29.0
Amygdala	-16.6	-36.5*	-34.2	-35.7
Septal N.	9.7	0.0	- 8.3	-16.9
Hippocampus(Ammon)	-19.1*	-23.3	-29.3	-39.2
Substantia Nigra	44.4*	16.9	-15.7#	-16.2
Pontine Gray	10.0	- 4.9	-23.1#	-34.7

*=p<0.05 vs sham control(n=3); #=p<0.05 vs moderate injury

Table 2. Percent changes in lCGU after head injury compared with control values

| Representative Structures | Moderate Injury | |
	40 min posttrauma (n=4)	4 h posttrauma (n=3)
Frontal Cortex	-44.9**	-10.1 %
Med. Geniculate Body	-37.1*	- 6.4
Thalamus(Ventral N.)	-33.8	-24.5
Hypothalamus	-31.9	5.0
Amygdala	-39.8*	-14.5
Septal N.	- 3.6	19.8
Hippocampus(Ammon)	-37.5	-22.1
Substantia Nigra	-25.6	6.9
Pontine Gray	-44.3*	-15.4

*=p<0.05, **=p<0.01 vs sham control(n=5)

the brain stem and diencephalon correspond to the areas to be considered as the neural center of vasomotor activity. Our results show a marked reduction of CBF and depressive metabolism in severe closed head injury. The result is more consistent with the clinical experience.

REFERENCES

Dienel, G. A., 1984, Regional accumulation of calcium in postischemic rat brain, J Neurochem, 43:913.
Langfitt, T. W., Tannanbaum, H. M., and Kassell, N. F., 1966, The etiology of acute brain swelling following experimental head injury, J Neurosurg, 24:47.

Nilsson, B., Dick, A., Ecklöf, B, Jagodzinski, Z., and Pontén, U., 1975,
    Regional blood flow in the brain ahnd in the cervical cord in experi
    mental head trauma, in:"Cerebral circulation and metabolism," T. W.
    Langfitt, L. C. McHenry Jr., M. Reivich, H. Wollman, eds.,Springer,
    Berlin, Heiderberg, New York.
Sakurada, O., Kennedy, C., Jehle, J., Brown, J. P., Carbin, G. L., and
    Sokoloff, L., 1978, Measurement of local cerebral blood flow with
    [^{14}C]-iodoantipyrine, Am J Physiol, 234:H59.
Sokoloff, L., Reivich, M., Kennedy, C., DesRosiers, M., H., Patlak, C$_{14}$S.,
    Pettigrew, K. D., Sakurada, O., and Shinihara, M., 1977, The [^{14}C]
    deoxyglucose method for the measurement of local cerebral glucose
    utilization: Theory, procedure, and normal values in the conscious
    and anesthetized albino rat, J Neurochem, 28:897.

REMOTE EFFECTS OF SMALL DEEP HEMISPHERIC INFARCTION ON THE CORTICAL

BLOOD FLOW -CT AND 133Xe INHALATION STUDY

A. Kushi[1], K. Yamaguchi[1], N. Katsuyama[1], M. Nakano[1],
J. Mukawa[2], Y. Hirata[3], M. Itoh[4], T. Fujiwara[5], S. Yoshioka[5],
K. Yamada[5], and T. Matsuzawa[5]

1 Department of Radiology, 2 Department of Neurosurgery, and
3 1st Department of Anatomy, Faculty of Medicine, University
of the Ryukyus. 4 Nuclear Medicine Division, Cyclotron
Radioisotope Center, and 5 Department of Radiology, Research
Institute for TB and Cancer, Tohoku University, Japan.

INTRODUCTION

Many reports have discussed the remote effects of unilateral ischemic
infarction on the cortical circulation. However in these studies, the subject
was dealt with during acute phase of stroke and the infarct has been too
large to analyze the circulation in the distant related area[1,2,3]. We
investigated the remote effect of a small deep ischemic lesion on the
cortical circulation in 15 patients with a lacunar infarct during the
chronic phase.

SUBJECT AND METHOD

We included 15 patients in whom computed tomography (CT) confirmed the
presence of a single lacunar infarct in the unilateral putamen (right 3
cases, left 5 cases) or internal capsule (right 4 cases, left 3 cases). The
size of the infarcts was estimated by multiplying the length by the width of
the hypodense lesion in a slice. This product was multiplied by the
depth, summed across the several slices showing the lesion. The volume of all
infarcts were less than 1cm^3. CT scanning was performed using third
generation scanners during the chronic phase.

Table 1 summarizes the patient data and clinical findings. Fifteen
patients aged 45-74 years old who suffered from dizziness, headache, or mild
hemiparesis. Eleven had hypertension and two had diabetes. The clinical
evaluation in the chronic stage took place when rCBF was measured.

We also measured rCBF in 18 controls aged 43-81 years old who came from
out-patient clinics in the hospital. They had normal CT appearances, no brain
disorders, hypertension, or diabetes by history. All patients and controls
were right-handed.

Regional cerebral blood flow was measured using the two dimensional
^{133}Xe inhalation technique with 14 scintillation detectors for each side of
the head, with collimator tubes perpendicular to the brain surface.

The Role of Neurotransmitters in Brain Injury, Edited by
M. Globus and W.D. Dietrich, Plenum Press, New York, 1992

Table 1. Clinical summary of the patients

CASE No.	AGE (yr)	SEX	NEUROLOGIC MANIFESTATIONS AT THE TIME OF THE rCBF STUDY	FOCUS*	SIZE (cm3)	INTERVAL+ (days)
1	63	F	Headache	R.putamen	0.1	old
2	74	F	Vertigo	R.putamen	0.1	old
3	63	F	Headache,Vertigo	R.putamen	0.5	old
4	57	F	Vertigo	L.putamen	0.1	old
5	74	F	Dysarthria	L.putamen	0.2	355
6	47	M	Headache	L.putamen	0.4	old
7	49	M	R. slight paresis of lower limb	L.putamen	0.5	old
8	53	M	R. hemiparesis	L.putamen	0.5	27
9	68	F	L. slight hemiparesis	R.I.C.(p)	0.1	118
10	69	M	L. hemiparesis	R.I.C.(p)	0.4	32
11	55	F	L. slight hemiparesis	R.I.C.(p)	0.6	39
12	61	F	L. hemiparesis	R.I.C.(p)	1.0	1010
13	73	F	Vertigo	L.I.C.(p)	0.1	old
14	56	F	R. slight hemiparesis	L.I.C.(p)	0.2	23
15	44	F	R.slight hemiparesis	L.I.C.(p)	0.4	24

*Location of the lacunar infarct
+Interval from the onset to the rCBF measurment
Abbreviations:R=right;L=left;I.C.=internal capsule;p=posterior
limb;old=Unknown onset of the stroke,but it was recognized as a chronic
phase by its series of CT studies.

Channel(CH)1,8,CH2,9,CH3,10,CH4,11,CH5,12,CH6,13 and CH7,14 are corresponding to the frontal pole,prefrontal area,central sulcus,anterior part of the lateral sulcus,posterior part of the lateral sulcus,parietal lobe and temporooccipital region,respectively(Figure 1). During the measurements the subjects were lying in a quiet room,resting,with their eyes covered and their ears plugged. Arterial 133Xe concentrations were estimated from recordings of end-tidal radioactivity of 133Xe gas during the investigations and the error caused by the recirculation of 133Xe was corrected. Mean arterial blood pressure was measured by ausculation after rCBF was measured.

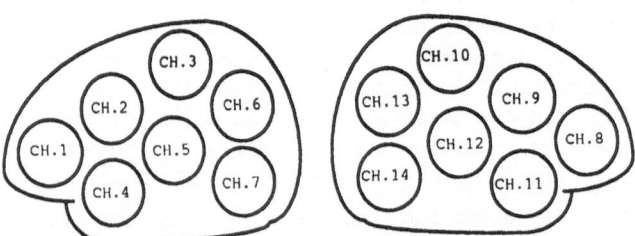

L. Hemisphere   R. Hemisphere

Figure 1. Location of the detectors

The rCBF values were caluculated as the ISI following the method of Obrist et al[4]. The absolute rCBF values were not evaluated because of the variability with age and other physiological conditions[5,6]. The relative flow in each region was calculated from the following equation:

Relative flow(%)=100x(rCBF-mCBF)/mCBF
mCBF:the mean rCBFs in all regions

We analyzed the result of rCBF measurements as to global or focal changes in the patients with 8 putaminal or 7 capsular infarct and compared with those of the controls. In that case,we compared the rCBF of the patients with the mean rCBFs in symmetrical regions of bilateral hemiheres of controls. We also investigated the patients with a right or left putaminal (or capsular) infarct. Statistical analyses were performed by paired or unpaired t test.

RESULT

Controls

rCBF in the two hemispheres was almost identical,except for anterior part of the lateral sulcus(p<0.05). rCBFs of the anterior part of the brain demonstrated high values compared with posterior part(Table 2).

Patients with lacunar infarction

Hypoperfusion area

Reductions of the rCBFs around the bilateral central sulci were present with the patients with a putaminal infarct(P<0.01,Table 3). The patients with a capsular infarct demonstrated significant hypoperfusion at the bilateral parietal lobes(P<0.01,Table 4).

Table 2. Regional cerebral blood flows of the controls

LOCATION	LEFT HEMISPHERE mean ± SD	RIGHT HEMISPHERE mean ± SD
FRONTAL POLE	8.0 ± 7.2	6.8 ± 5.2
PREFRONTAL AREA	0.9 ± 5.3	1.8 ± 7.3
CENTRAL SULCUS	-2.8 ± 5.2	-5.9 ± 5.1
LATERAL SULCUS (A)	1.2 ± 5.1	4.4 ± 6.0
LATERAL SULCUS (P)	0.9 ± 4.8	2.0 ± 3.3
PARIETAL LOBE	-3.7 ± 6.0	-6.0 ± 4.2
TEMPOROOCCIPITAL REGION	-5.1 ± 4.3	-2.6 ± 8.1

Abbreviations:A=anterior part;P=posterior part

Table 3. rCBFs of the patients with a putaminal infarct

LOCATION	IPSILATERAL HEMISPHERE mean ± SD	CONTRALATERAL HEMISPHERE mean ± SD
FRONTAL POLE	12.8 ± 8.5	11.9 ± 9.6
PREFRONTAL AREA	4.0 ± 9.0	-0,9 ± 4.3
CENTRAL SULCUS	-11.9 ± 5.0 ⇊	-12.0 ± 6.1 ⇊
LATERAL SULCUS (A)	8.5 ± 9.4	8.3 ± 10.7
LATERAL SULCUS (P)	6.3 ± 8.0 ↑	1.3 ± 9.3
PARIETAL LOBE	-8.9 ± 4.9 ↓	-8.4 ± 5.8
TEMPOROOCCIPITAL REGION	-6.0 ± 7.3	-4.9 ± 8.1

Abbreviations: Significant high (↑) or low (↓) values compared with those of the controls. P<0.01(Double arrows),P<0.05(Single arrow)

Hyperperfusion area

The patients with a putaminal infarct demonstrated hyperperfusion at the ipsilateral posterior part of the lateral sulcus(P<0.05,Table 3). The patients with a capsular infarct presented hyperperfusion at the ipsilateral frontal pole and contralateral prefrontal area(P<0.05,0.01,Table 4).

Table 5 summerized rCBFs of the patients with a right or left lacunar infarct.

Table 4. rCBFs of the patients with a capsular infarct

LOCATION	IPSILATERAL HEMISPHERE mean ± SD	CONTRALATERAL HEMISPHERE mean ± SD
FRONTAL POLE	13.3 ± 8.6 ↑	13.3 ± 14.9
PREFRONTAL AREA	3.1 ± 3.8	7.5 ± 5.0 ⇈
CENTRAL SULCUS	-8.5 ± 7.4	-5.0 ± 6.4
LATERAL SULCUS (A)	1.5 ± 12.2	1.7 ± 8.2
LATERAL SULCUS (P)	2.7 ± 6.0	5.2 ± 9.5
PARIETAL LOBE	-12.6 ± 7.6 ⇊	-10.7 ± 6.3 ⇊
TEMPOROOCCIPITAL REGION	-4.0 ± 10.5	-7.6 ± 5.7

Abbreviations: Significant high (↑) or low (↓) values compared with those of the controls. P<0.01(Double arrows),P<0.05(Single arrow)

Table 5. Significant high or low rCBFs of the patients with a right or left lacunar infarct

Focus	Ipsilateral hemisphere		Contralateral hemisphere	
	High value	Low value	High value	Low value
R.putamen	Frontal pole** Lateral sulcus (A)*	None	Lateral sulcus (A)*	Central sulcus* Lateral sulcus (P)*
L.putamen	Lateral sulcus (P)*	Central sulcus**	None	Central sulcus*
R.internal capsule(p)	Frontal pole*	Parietal lobe*	Prefrontal area*	None
L.internal capsule(p)	None	None	None	Parietal lobe* T.O.R.**

Abbreviations:R=right;L=left;A=anterior part;P=posterior part;T.O.R.= Temporooccipital region
*P<0.05,**P<0.01 (unpaired t-test) Significant difference between the controls and the patients.

DISCUSSION

Hypoperfusion area

Diaschisis has been described in acute and chronic stroke[7]. The patients with a putaminal infarct and the patients with a capsular infarct,each demonstrated reductions of rCBF at the bilateral central sulci and parietal lobe. There are many neuronal connection between putamen or internal capsule and cerebral cortex. Our results indicate the irreversible ischemic damage of the connecting fibers that pass through the putamen or internal capsule may induce neuronal loss or antegrade and retrograde degeneration in the related cortical areas.

Hyperperfusion area

Hyperfrontal distribution of the rCBF is characteristic phenomenon of the resting normal human brain[8]. It is also reported that stress induced increase of hyperfrontality. The patients with a capsular infarct demonstrated hyperperfusion at the frontal part of the brain. This group may have been under more stress than the others. However,it is interesting that the patients with a right hemispheric lacunar infarct alone demonstrated this phenomenon in our study.

The patients with a putaminal infarct presented hyperperfused areas at the ipsilateral posterior part of the lateral sulcus. This increas in rCBF, may have been caused by damage of the inhibitory circuits or transneural activation. To examine these mechanisms,we need to determine whether an increase in blood flow matches that in metabolism.

CONCLUSION

Small lesions including lacunar infarcts effect the circulation of the distant cortical areas. These effects depend on the location of a lacunar infarct. Furthermore,we observed not only cortical hypoperfusion but also cortical hyperperfusion. Our results suggest that these phenomena may result from transneural depression or activation.

REFERENCES

1.Lavy,S.,Melamed,E., and Portnoy,Z.,The effect of cerebral infarction on the regional cerebral blood flow of the contralateral hemisphere.,Stroke 6:160 (1975)
2.Orgogozo,J.M.,Larsen,B.,Skyhøj,T.,Skinhøj,E., and Lassen,N.A., Evidence of cortical disconnection in deep hemispheric strokes as revealed by rCBF.,Acuta Neurol Scand 60:258 (1979)
3.Skynøj Olsen,T.,Larsen,B.,Bech Skriver,E.,Herning,M.,Enevoldsen,E., and Lassen,N.A., Focal cerebral hyperemia in acute stroke. Incidence,pathophysiology and clinical siginificance.,Stroke 12:598 (1981)
4.Obrist,W.D.,Thompson,H.K.,Wang,H.S., and Wilkinson,W.E., Regional cerebral blood flow estimated by 133Xenon inhalation.,Stroke 6:245 (1975)
5.Frackowisk,R.S.J.,Lenzi,G.L.,Jones,T., and Heather,J.D., Quantitative measurments of regional cerebral blood flow and oxygen metabolism in man using 150 and positoron emission tomography:Theory,procedure,and normal values., J Comput Assist Tomogr 4:727 (1980)
6.Powers,W.J.,Wayne-Martin,W.R.,Herscovitch,P.,Raichle,M.E., and Grubb,R.L.:Extracranial-intracranial bypass surgery:Hemodynamic and metabolic effects.,Neurology 34:1168 (1984)
7.Hanyu,H.,Arai,H.,Kobayashi,Y.,Hatano,N.,Katsunuma,H., and Suzuki,T.,Remote Effects in Cerebral Infárction:123I-IMP SPECT Study, Jpn J Ncul Med 27:629 (1990)
8.Ingvar,D.H.,"Hyperfrontal" distribution of the cerebral gray matter flow in resting wakefullness;On the functional anatomy of the concious state., Acuta Neurol Scand 60:12 (1979)

WHICH HAS A MORE SIGNIFICANT ROLE, LOCUS CERULEUS OR NUCLEUS TRACTUS SOLITARII, IN HEMISPHERIC VASOMOTOR RESPONSES?

Yuichi Maruki, Kunio Shimazu, Takeshi Ohkubo, Hotaek Kim, Hideyoshi Sugimoto, Yoshio Asano, Yoshihiko Nakazato, Masahiko Sawada, Katsuhiko Hamaguchi
Department of Neurology, Saitama Medical School
Saitama Japan

INTRODUCTION

Locus ceruleus (LC) and nucleus tractus solitarii (NTS) are important brainstem nuclei in the regulation of cerebrovascular tone. Their functional significance in regulation of vasomotor responses has not tobe completedly elucidated. The aim of the present study was to investigate which nucleus had more important roles in the hemispheric vasomotor responses.

MATERIAL AND METHODS

Twenty-six *macaca fuscata* were stereotaxically operated under ketamine-hydrochloride anesthesia for producing right LC lesion (n=14) or right NTS lesion (n=12) by an electrical coagulation method (100V,0.8mA, 60sec.) as we previously reported[1]. One week after the lesion, the animals were anesthetized with ∂-chloralose(25mg/kg,ip) and urethane(500mg/kg,ip). For artificial ventilation, a bolus injection of pancuronium bromide (0.1mg/kg) followed by continuous administration (0.01mg/kg/hour) throughout the experiment was conducted. Endotidal PCO_2 and rectal temperature were maintained within physiological range. Catheters were inserted to femoral artery and vein in order to measure and control blood pressure. Bilateral internal carotid arteries were separated for measuring blood flow.

MEASUREMENT

Bilateral internal carotid arterial blood flow (ICBF) was continuously measured by using zero adjusted electromagnetic flowmeters. Blood pressure and pulse rate were continuously recorded through the catheter inserted in femoral artery. Arterial blood gases and pH were also measured. Cerebral vasomotor responses were

Table 1 Physiological parameters at measuring the steady state ICBF. In any parameters, there were no differences between LC and NTS lesion groups.

Lesion	LC(n=14)	NTS(n=12)
Body Weight (kg)	5.9±1.4	6.5 ± 1.7
Rectal Temperature(°C)	36.9±0.7	37.3±0.2
MABP (mmHg)	111.8±13.0	120.8 ± 14.2
apH	7.40±0.05	7.40± 0.04
PaO2 (mmHg)	79.5±18.6	88.5 ± 10.5
PaCO2 (mmHg)	36.9±0.7	37.3± 0.2 (mean±SD)

quantitatively tested by means of controlled changes in PaCO2, % Chemical Vasomotor Index (%CVI:ΔICBF/ΔPaCO2/steady state ICBF), and by means of controlled changes in blood pressure, % Autoregulation Index (%AI: ΔICBF/ΔMABP/steady state ICBF).

RESULTS

1)Physiological parameters: In physiological parameters at measuring the steady state ICBF, there were no significant differences between LC and NTS lesion groups. (Table1)

2)Steady state ICBF: In LC lesion, ICBF of lesion side (27.1±9.6ml/min/100g, mean±SD) was significantly higher than that of contralateral side (22.9±7.8)(p<0.05).   On the contrary, there was no significant asymmetry with the NTS lesion.(Fig 1).

3)Vasomotor response to changes in PaCO2
a)Induced hypocapnia (hyperventilation) : %CVI of LC lesion side (2.5±1.0/mmHg) was significantly lower than that of the contralateral side(3.0±0.9).   However, the NTS lesion did not effect the vasomotor reactivity to hypocapnia(Fig 2,3).

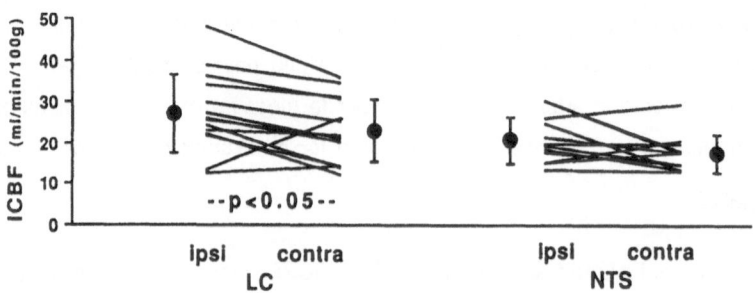

Fig.1 Steady state ICBF: In the LC lesion group, steady state ICBF of ipsilateral side of the lesion was significantly higher than contralateral side.   There were, however, no significant differences in steady state ICBF between the NTS lesion groups.

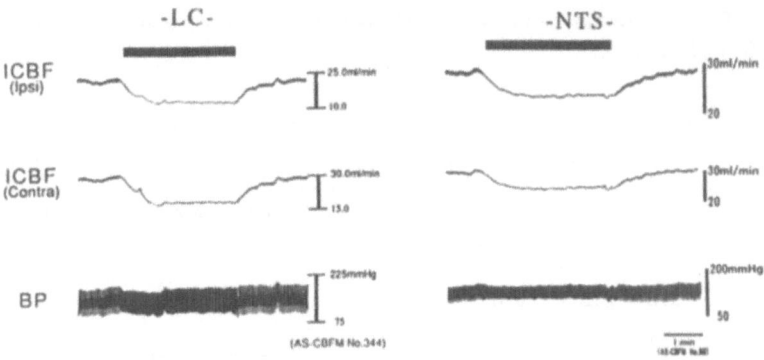

Fig.2 Actual recordings during induced hypocapnia: In the LC lesion, ICBF of contralateral side of the lesion decreased more than ICBF of ipsilateral side. In the NTS lesion, reactivity to induced hypocapnia were almost equal between the two sides.

b)Induced hypercapnia (CO_2 inhalation): In the LC lesion group, %CVI of ipsilateral and contralateral sides were 7.1±5.0, 6.5±1.4, respectively. In the NTS lesion group, %CVI of ipsilateral and contralateral sides were 5.3±2.2, 5.4±2.3, respectively. There were no significant lateralities of %CVI in either LC or NTS lesion.

4) Vasomotor responses to changes in blood pressure: No significant differences were obtained between ipsi- and contralateral side of either LC or NTS lesion (Table2).

DISCUSSION

We have examined regulatory mechanisms of the cerebral blood flow and cerebral vasomotor responses in the carotid arterial system and vertebral arterial system

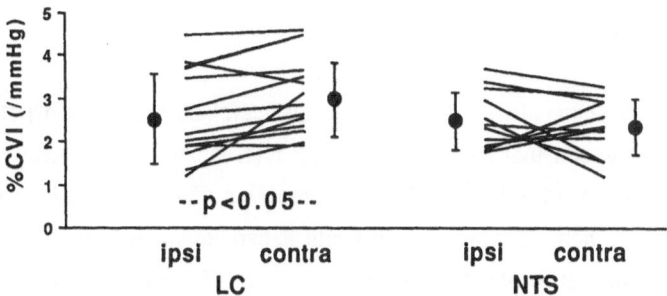

Fig3 %CVI of induced hypocapnia: In the LC lesion group, % CVI of ipsilateral side was significantly lower than that of contralateral side. However, there was no significant asymmetry in the NTS lesion.

Table 2. There were no significant asymmetries in % Autoregulation Index between either LC or NTS lesion.

| LESION | LC | | NTS | |
	Ipsilateral	Contralateral	Ipsilateral	Contralateral
Induced Hypotension	0.76±0.36	0.64±0.45	0.61±0.42	0.77±0.57
Induced hypertension	0.92±0.50	0.82±0.53	1.25±0.73	1.12±0.99

The locus ceruleus and the nucleus tractus solitarii have been recently demonstrated as important brainstem nuclei through the modulation of neurotransmitters. These nuclei have been suggested to modulate the cerebral hemodynamics of the carotid and vertebral arterial systems. In fact, we have recently reported that the NTS has a more important role on regulating the vertebral blood flow than the LC[3]. In the present study we compared the differences between LC and NTS lesions on the carotid arterial system. These results suggest that a unilateral LC lesion not only affects the resting tone but also the vasomotor response during hypocapnia. Morphological studies have demonstrated that the LC has direct ipsilateral noradrenergic innervation to cerebral intraparenchymal vessels[4,5,6]. There are, however, few reports about the asymmetry of responses between ICBF to the stimulation or lesion of the LC[1,7]. The main reason why asymmetrical responses could not be demonstrated might be due to the acute experiments. Our previous study showed significant differences between acute and chronic studies after the LC lesion[8]. In the acute study, when we electrically stimulated the LC, blood pressure increases were too great to examine the cerebral hemodynamics accurately. In the present study, we measured cerebral hemodynamics at least one week after the lesion and documented asymmetrical patterns of ICBF and cerebral vasomotor responses.

The NTS has some efferent fibers to the ipsilateral LC[9] and the hypothalamus[10]. There are some reports that the acute unilateral NTS lesion or stimulation significantly changed cerebral blood flow[11,12]. The NTS receives efferent fibers from organs that control the cardiac hemodynamics. In acute experiments, it is therefore difficult to exclude the hemodynamic effects from these organs. The present study showed that chronic unilateral NTS lesion did not effect cerebral hemodynamics.

We conclude that the unilateral LC has vasoconstrictive effect on the ipsilateral hemisphere and its effect is stronger than the NTS.

REFERENCE

1. K. Shimazu, T. Ohkubo, H. Kim et al, Role of locus ceruleus in hemispheric and vertebral blood flow and their vasomotor responses in "Neurotransmission and Cerebrovascular function I" J.Saylaz and E.T. Mackenzie, ed., Elsevier Science Publisher B.V. : 209,(1989).

2. Y.Maruki, Comparison of cerebral vasomotor Responses between Internal Carotid Artery and Vertebral Artery. J. Saitama Med. School, 11, 331(1984).

3. K.Shimazu, T.Ohkubo, Y. Maruki et al, Differences between Locus ceruleus and Nucleus Tractus solitalii on vertebrobasilar vasomotor responses. J. Cerb Blood Flow Metab. 11 (supple 2): 691(1991).

4. B.K.Hartman, D.Zide and S.Undenfriend, The use of Dopamine ß-hydrooxylase as a maker for the central noradrenergic nervous system in rat brain. Proc. Natl. Acad. Sci U.S.A.. 69: 2722(1972).

5. L.Edvinsson, M. Lindvall, K.C. Nielsen et al, Are brain vessels innervated also by central(non-sympathetic) adrenergic neurons? Brain Res., 476: 71(1989).

6. T.Itakura, K.Yamamoto, M.Tohyama et al, Central dural innervation of arterioles and capillaries in the brain. Stroke. 8: 360(1977).

7. P.J.Goadsby, G.A.Lambert and J.W.Lance, Differential effects on the internal and external carotid circulation of the monkey evoked by locus coeruleus stimulation. Brain Res., 249: 247(1982).

8. Y.Nakazato, K. Shimazu, T. Ohkubo et al, Changes in hemispheric vasomotor responses a after unilateral lesion of locus ceruleus comparison between acute and chronic studies. J. Cerb Blood Flow Metab, 11 (supple 2): 692 (1991).

9. J.M. Cedarbaum, G.K. Aghajanian, Afferent projection to the rat locus coeruleus as determined by a retrograde tracing technique. J Comp Neurol, 190: 1 (1978).

10. C.A. Ross, D.A. Ruggiero, D.J.Reis, Projection from the nucleus tractus solitarii to the rostral ventrolateral medulla. J Comp Neurol 242: 511(1985).

11. M. Nakai, An increase in cerebral blood flow elicited by electrical stimulation of the solitary nucleus in rats with cervical cordotomy and vagotomy. Jpn J Physiol 35: 57 (1985).

12. T. Ishitsuka, C. Iadecola, M.D. Undewood et al, Lesions of nucleus tractus solitarii globally impair cerebrovascular autoregulation. Am J Physiol 251: H269 (1986).

Chapter 8

Neurotransmitters and Free Radical Mediated Injury

REGIONAL RECOVERY OF EICOSANOIDS AFTER FOREBRAIN ISCHEMIA BY

MICRODIALYSIS IN RATS

PM Patel, JC Drummond, MD Mitchell, TL Yaksh,
DJ Cole

Departments of Anesthesiology, VA Medical
Center and University of California, San Diego;
Department of Obstetrics, University of Utah,
Salt Lake City, Utah; Loma Linda University,
Loma Linda, CA.

INTRODUCTION

The production of prostaglandins in the post-ischemic brain may affect the evolution of ischemic neuronal injury. In particular, it has been suggested that the relative balance between prostaglandins with opposing effects on the vasculature might influence post-ischemic cerebral perfusion, and thereby alter the the the extent of neuronal injury[1, 2]. An exacerbation of neuronal injury would be predicted if prostanoids with vasoconstrictive properties are produced in excess of those that produce vasodilatation. A recent investigation in our laboratory had demonstrated that thromboxane B_2 (a stable metabolite of thromboxane A_2) was produced in greater quantities than 6-keto-prostaglandin $PGF1a$ (a stable metabolite of prostacyclin) in both the caudate nucleus (CN) and dorsal hippocampus (HPC) in the rat[3]. However, the ratio of TxB_2 to 6-keto-PGF_{1a} was substantially greater in the CN. Therefore, a greater susceptibility of the CN to ischemic damage would be predicted. This is not thought to be the case, and accordingly, additional factors must also play a role. These factors might include the production of other eicosanoids with vasoactive effects.

The present study was therefore undertaken to evaluate the relative regional production of the prostanoids 6-keto-PGF_{1a}, TxB_2, PGD_2, PGE_2, PGF_{2a} and the leukotrienes LTC_4 and LTB_4 in the caudate nucleus and dorsal hippocampus in a rat model of forebrain ischemia. The eicosanoid concentrations were measured in the microdialysate, collected after the implantation of microdialysis probes in the caudate nucleus and dorsal hippocampus, by radioimmunoassay. This approach permitted a regional evaluation of the post-ischemic eicosanoid production in the brain. In addition, an assessment of time related changes in the concentrations of eicosanoids in the extracellular space of these two structures was made possible.

MATERIALS AND METHODS

Eighteen spontaneously hypertensive rats of the same age and weight (275 - 325 grams) were anesthetized with isoflurane, intubated orotracheally and mechanically ventilated. Two microdialysis probes were inserted into the caudate nucleus bilaterally to a depth of 7.0 mm. Two additional probes were inserted into the dorsal hippocampus bilaterally to a depth of 5.8 mm. The probes were perfused with mock CSF at a rate of 2 μl/min. Two hours after probe insertion, the animals were randomly alloted to one of three groups (n=6 for each group). The control group animals were left undisturbed for the duration of the experiment. The two experimental groups were subjected to temporary incomplete forebrain ischemia (bilateral carotid artery occlusion with simultaneous hypotension) according to the technique of Smith et al [4] for intervals of 8 and 20 minutes. The microdialysate was collected in 30 minute epochs for 1 hour during the pre-ischemic interval and for 3 hours during reperfusion. During the ischemic interval, microdialysate was collected for 8 or 20 minutes, depending upon the duration of carotid occlusion.

Within each group, the microdialysate from corresponding 30 minute epochs (120 μl per epoch) in each structure was pooled. This yielded a total of 720 μl of microdialysate for a given epoch per structure. The pooled microdialysate was then divided into seven aliquots of 100 μl each. The levels of TxB_2, 6-keto-PGF_{1a}, PGF_{2a}, PGD_2, PGE_2, LTC_4 and LTB_4 in aliquot were measured by radioimmunoassay. This approach permitted only a single measurement of the eicosanoid level for each time point in any given group.

RESULTS

The eicosanoid concentrations in the microdialysate from the animals in the control group are presented in Figures 1. Significant quantities of prostanoids were not detected in the microdialysate from the CN. By contrast, substantial levels of PGF_{2a} and PGE_2 were detected in the microdialysate from the HPC. The PGF_{2a} levels declined gradually until two hours after probe insertion. Thereafter, the PGF_{2a} levels remained stable (approximately 750 pg/ml) for the duration of the study. PGE_2 levels increased gradually and continuously throughout the study, reaching peak levels of 1339 pg/ml.

In the animals that were rendered ischemic, baseline levels of the eicosanoids were not different from those of the control group. Upon reperfusion after forebrain ischemia, the levels of all prostanoids measured increased significantly in the CN and HPC in both ischemic groups (Figures 2 and 3). In the CN, the greatest increases were in the levels of TxB_2 and PGF_{2a}, followed by PGE_2, 6-keto-PGF_{1a} and PGD_2. The concentrations of TxB_2 and PGF_{2a} declined gradually during the reperfusion period whereas the levels of PGE_2 and 6-keto-PGF_{1a} increased gradually and continuously during the reperfusion

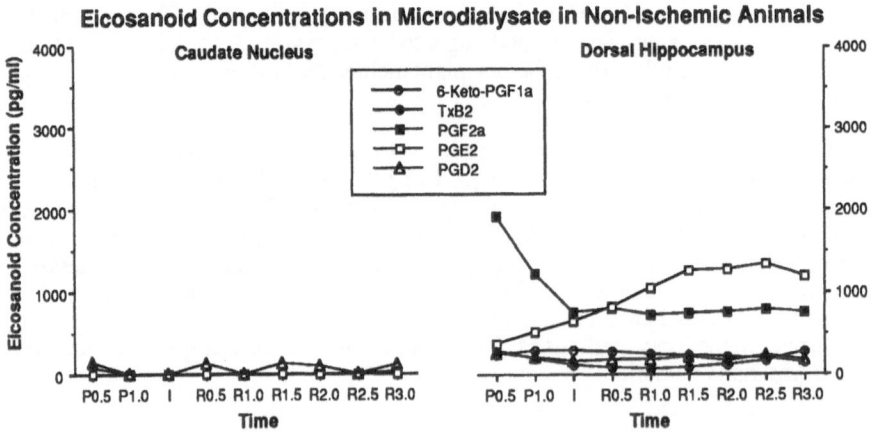

Figure 1. Eicosanoid concentrations in microdialysate from the caudate nucleus (left) and dorsal hippocampus (right) in non-ischemic control animals. Each data point represents an eicosanoid level measured in microdialysate that was obtained by pooling microdialysate from corresponding epochs from six animals.

period. By contrast, in the HPC, the greatest increases were in the levels of PGF_{2a}, followed by TxB_2, PGE_2, 6-keto-PGF_{1a} and PGD_2. Hence, the net production of prostanoids that produce vasoconstriction (TxB_2 and PGF_{2a}) was considerably greater in the HPC. The levels of PGF_{2a} and TxB_2 declined gradually upon reperfusion. The level of PGE_2, however, gradually and continuously increased throughout the study. Leukotrienes B_4 and C_4 were not detected in the microdialysate from either structure.

DISCUSSION

It has been proposed that an imbalance in the production of eicosanoids that have opposing effects on the vasculature might influence the evolution of post-ischemic neuronal injury[1,2]. A recent investigation in our laboratory had demonstrated that the ratio of TxB_2 production to that of 6-keto-PGF_{1a} was considerably greater in the CN than in the HPC, thereby suggesting that the former structure might be more vulnerable to ischemic injury. However, in that investigation, only TxB_2 and 6-keto-PGF_{1a} were measured. The present study was therefore undertaken to evaluate the regional production of several other eicosanoids. The results demonstrate regional heterogeneity in the post-ischemic eicosanoid production in the brain. Specifically, while TxB_2 was present in greater quantities in the CN, PGF_{2a}, a potent vasoconstrictor[5], was present in significantly greater quantities in the HPC. Therefore, in contrast to the inference drawn from the initial investigation, the ratio of the vasoconstricting (TxB_2 and PGF_{2a}) and vasodilating (6-keto-PGF_{1a}) prostanoids was significantly greater (more 'unfavorable') in the hippocampus than in the caudate. This is in keeping with the greater vulnerability to ischemic injury of the former structure[6].

Differences in the recovery of PGE_2 in the microdialysate from the caudate and hippocampus were also evident. Upon probe insertion, significant amounts of PGE_2 were detected in the microdialysate from the hippocampus but not from the caudate. Furthermore the PGE_2 levels gradually increased and remained elevated throughout the duration of the experiment. Surprisingly, forebrain ischemia did not influence the time related changes in the concentrations of PGE_2 recovered from the hippocampus that were seen following probe insertion. The reason such differences between the two structures were seen is not clear. Changing activities of rate limiting enzymes following penetration injury is suggested. The results are, however, consistent with those reported by Kempski

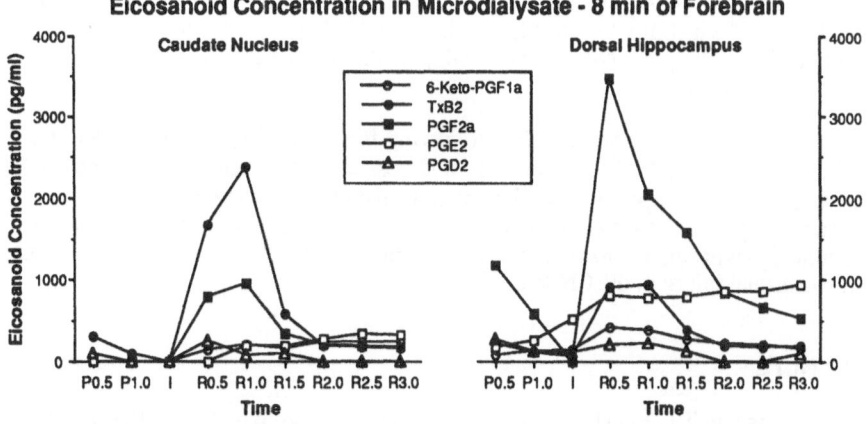

Figure 2. Eicosanoid concentrations in microdialysate from the caudate nucleus (left) and dorsal hippocampus (right) prior to (P), during (I) and after (R) 8 minutes of forebrain ischemia. Each data point represents an eicosanoid level measured in microdialysate that was obtained by pooling microdialysate from corresponding epochs from six animals.

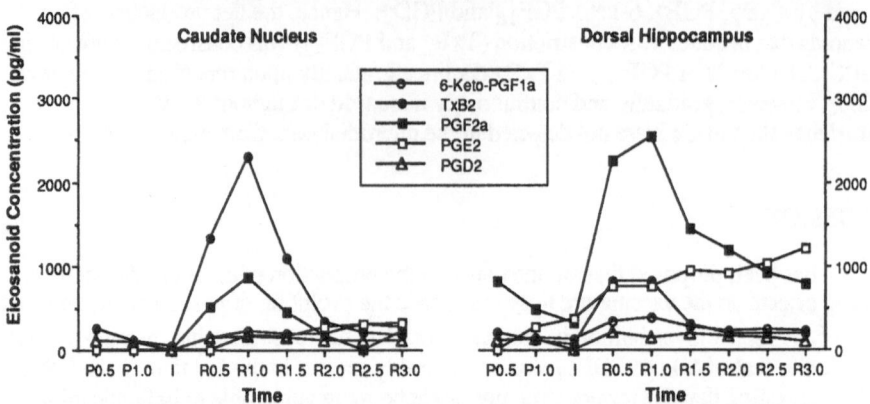

Figure 3. Eicosanoid concentrations in microdialysate from the caudate nucleus (left) and dorsal hippocampus (right) prior to (P), during (I) and after (R) 20 minutes of forebrain ischemia. Each data point represents an eicosanoid level measured in microdialysate that was obtained by pooling microdialysate from corresponding epochs from six animals.

et al[7]. These investigators, using an ex vivo technique, showed that PGE_2 levels peaked 4 hr post ischemia and that the PGE_2 levels in the hippocampus were greater than the levels in the striatum. The recovery of PGD_2 from both structures was not significantly different.

The present observations regarding post-ischemic increases in vasoconstrictor/vasodilator prostanoid ratios are consistent with several previous demonstrations that interventions intended to decrease post-ischemic thromboxane production attenuate the sequelae of ischemia. [8, 9, 10]. Similarly, prostacyclin and its analogues have been shown to improve post-ischemic cerebral blood flow [11] and to reduce post ischemic memory deficit[12]. The available data are therefore consistent with the notion that interventions that reduce the formation of vasoconstricting prostanoids attenuate ischemic neuronal injury.

In contrast to the prostanoids, the leukotrienes B_4 and C_4 were not detected in the microdialysate. The inability to detect leukotrienes in the dialysate may be due to the possibility that they are produced in very small quantities or that they do not gain access to the extracellular space.

In conclusion, our data show that, following forebrain ischemia, TxB_2, PGF_{2a}, PGE_2, PGD_2 and 6-keto-PGF_{1a} are present in increased concentrations in both the caudate nucleus and the dorsal hippocampus. However, the data demonstrate regional heterogeneity in post ischemic eicosanoid production in the brain. It therefore remains to be determined to what extent manipulation of arachidonic acid metabolism influences post ischemic neuronal injury in the CN and HPC.

REFERENCES

1. M.K. Stevens and T.L.Yaksh , Time course of release in vivo of PGE2, PGF2alpha, 6-keto PGF1alpha and TxB2 into the brain extracellular space after 15 min of complete global ischemia in the presence and absence of cyclooxygenase inhibition. J Cereb Blood Flow Metab 8:790-798 (1988).

2. C. Galli , A. Petroni, A. Bertazzo, and S. Sarti , Arachidonic acid and its metabolites during cerebral ischemia and recirculation. Ann NY Acad Sci 559:352-64 (1989).

3. P.M. Patel, J.C. Drummond, M.D. Mitchell, T.L. Yaksh, and D.J. Col, Eicosanoid production in the caudate nucleus and dorsal hippocampus after forebrain ischemia : A microdialysis study. J Cereb Blood Flow Metab (1991).

4. M.L. Smith, G. Bendek, N. Dahlgren, I. Rosen, T. Wieloch, and B.K. Siesjo, Models for studying long-term recovery following forebrain ischemia in the rat. A 2-vessel occlusion model. Acta Neurol Scand 69:385-401 (1984).

5. J.D. Pickard, L.A. MacDonell, E.T. MacKenzie, and A.M. Harper, Prostaglandins-induced effects in the primate cerebral circulation. Eur J Pharmacol 43:343-351(1977).

6. T. Kirino, Delayed neuronal death in the gerbil hippocampus following ischemia. Brain Res 239:257-69 (1982).

7. O. Kempski, E. Shohami, Dv Lubitz, J.M. Hallenbeck, and G. Feuerstein, Postischemic production of eicosanoids in gerbil brain. Stroke 18:111-119 (1987).

8. L.C. Pettigrew, J.C. Grotta, H.M. Rhoades, and K.K. Wu, Effect of thromboxane synthase inhibition on eicosanaoid levels and blood flow in ischemic rat brain. Stroke 20:627-632 (1989).

9. T. Nakagomi, T. Sasaki,T. Kirino, A. Tamura, M. Noguchi, I. Saito, and K. Takakura, Effect of cyclooxygenase and lipoxygenase inhibitors on delayed neuronal death in the gerbil hippocampus. Stroke 20:925-929 (1989).

10. T. Sasaki, T. Nakagomi, T. Kirino, A. Tamura, M. Noguchi, I. Saito, and Takakura K: Indomethacin ameliorates ischemic neuronal damage in the gerbil hippocampal CA1 sector. Stroke 19:1399-1403 (1989).

11. Y. Masuda, Y. Ochi, Y.Ochi, T.Karasawa, N.Hatano, T. Kadokawa, and T. Okegawa T., Protective effect of a new prostacyclin analogue OP-2507 against cerebral anoxia and edema in experimental animals. Eur J Pharmacol 123:335-344 (1986).

12. Borzeix MG, Cahn R, Cahn J: Effects of new chemically and metabolically stable prostacyclin analogues (iloprost and ZK 96480) on early consequences of a transient cerebral oligemia in the rat. Prostaglandins 35:653-664 (1988).

IN VIVO GENERATION OF HYDROXYL RADICALS DURING GLUTAMATE EXPOSURE

Donald P. Boisvert

Room 162 Heritage Research Bldg.
Department of Surgery, University of Alberta
Edmonton, Alberta, Canada   T6G 2S2

INTRODUCTION

Irreversible damage occurs in neurons exposed in vivo or in vitro to abnormally high concentrations of excitatory amino acids (EAAs) such as glutamate, aspartate, and kainate.   These EAAs activate receptors that control calcium influx and most evidence points to excessive intracellular calcium as the primary event leading to cell death.   High intracellular calcium levels produce diverse secondary and tertiary effects; some of these, e.g. protease/lipase activation, are well known, but the relative importance of individual calcium-mediated events in producing neuronal damage is unclear.

It is known that many processes activated by calcium lead directly or indirectly to free radical production[1], and several recent studies[2,3,4,5,6,7] suggest a link between EAA toxicity and oxygen derived free radicals. Alterations in endogenous free radical scavengers[3] or the provision of scavengers exogenously[2,4] have both been effective in altering EAA neuronal toxicity.   The present study assessed the effect of in vivo glutamate exposure on the production of hydroxyl free radicals by striatal neurons.

METHODS

Adult male Wistar rats were used.   The experimental protocol was approved by the University of Alberta Health Sciences Animal Welfare Committee and the treatment of the animals conformed to the Canadian Council on Animal Care Guidelines.

Trans-striatal microdialysis fibre implantation was performed under pentobarbital anesthesia, as previously described.[8]   At 24h after fibre implantation the animals were again anesthetized using isoflurance, 3%, via nose cone.   The trachea was cannulated via a midline incision and the animal placed on mechanical ventilation.   Anesthesia was subsequently maintained with isoflurane, 1.0%.   A femoral artery was cannulated and arterial blood gas determinations were made approximately every 30 min.   Ventilation was adjusted to provide arterial pH, pCO_2 and pO_2 at or near 7.40, 40 mm Hg and >100 mm Hg respectively.   Head temperature was monitored using a needle thermocouple positioned adjacent to the occipital region and maintained between 37 and $38^\circ C$ by means of a small heating pad.

The Role of Neurotransmitters in Brain Injury, Edited by
M. Globus and W.D. Dietrich, Plenum Press, New York, 1992

A microinfusion pump was used to perfuse the microdialysis fibre with Ringer solution (pH~6.5) at a rate of $3\mu l/min$. The samples were measured for their content of uric acid and 2,5-dihydroxybenzoic acid (2,5-DHBA) using high pressure liquid chromatography with electrochemical detection. Each sample was injected via a $20~\mu l$ loop onto a 5C18, 150 x 4.6 mm Nucleosil column. The mobile phase consisted of a 20mM Na citrate, 18mM Na acetate, 4% methanol, and phosphoric acid to adjust the pH to 6.0. The flow rate was 1.6 ml/min. The compounds were detected with the electrode set at +0.70V against an Ag/AgCl reference electrode.

In order to employ 2,5-DHBA levels as an index of hydroxyl radical production [9,10], salicylate 100 mg/kg i.p. was administered 2 h before testing the effects of glutamate or other agents. Previous work in this laboratory has shown that striatal extracellular levels of 2,5-DHBA increase gradually and then plateau at 90-120 min after salicylate administration[11]. Changes in in vivo brain levels of uric acid are known to reflect the activity of xanthine oxidase, which produces the superoxide radical and possibly the hydroxyl radical as byproducts[1].

The experimental protocol consisted of five groups:
Group I,II: Switch to dialysate containing 0.25M or 0.5M glutamate at 2h post-salicylate (n=6/group).
Group III: Deferoxamine (DES) 90 mg/kg i.m. at 6h and 2h before switching to dialysate containing oxypurinol (10^{-3}M), a specific xanthine oxidase inhibitor, and glutamate (0.5M) at 2h post-salicylate (n=6).
Group IV: Switch to dialysate containing oxypurinol (10^{-3}M), a specific xanthine oxidase inhibitor, and glutamate (0.5M) at 2h post-salicylate (n=5).
Group V: A control group in which no glutamate or other compound was used (n=6).

The peak heights in the chromatograms were expressed as a percentage change from baseline, where baseline equals the mean of the three measurements preceding (-20-0 min.) the switch to dialysate containing glutamate.

Statistical comparisons between appropriate sets of groups were made using the student t-test (unpaired).

RESULTS

In rats in which no changes were made in the dialysate over the 60 min test period no changes in uric acid or 2,5-DHBA occurred, as shown in Fig. 1. When the dialysate was switched to one containing 0.25M glutamate rapid increases in extracellular levels of both compounds occurred over 60 min, reaching 250-300% above baseline. Similar results were obtained when 0.5M glutamate was used (see fig. 3).

The effects of pretreatment with DES, a potent iron chelating agent that is known to inhibit iron-dependent hydroxyl radical production, are shown in Fig. 2. The increase in 2,5-DHBA produced by 0.25M glutamate was significantly diminished in the group of rats given DES. In contrast, a small non-significant reduction in the glutamate-induced increase in uric acid was seen.

To assess further whether a relationship exists between glutamate-induced increases in uric acid and 2,5-DHBA, oxypurinol was added to the dialysate. The results are shown in Fig. 3. As expected, the increase in uric acid was completely inhibited. The increase in 2,5-DHBA however, was identical to that obtained in the absence of oxypurinol.

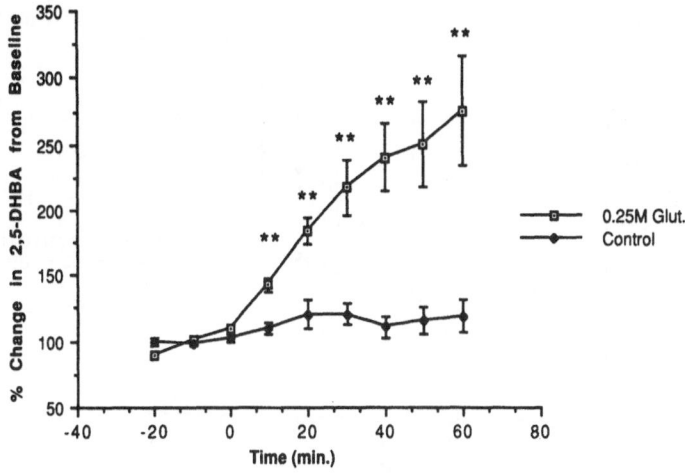

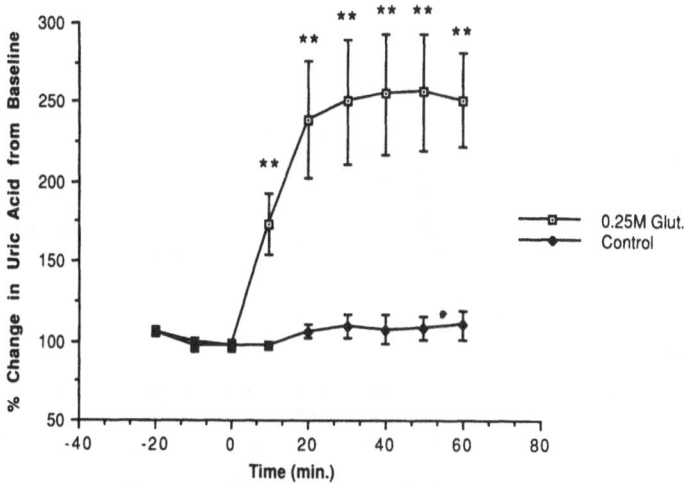

Fig. 1.   Effect of 0.25M glutamate in dialysis fluid on striatal extracellular levels of 2,5-DHBA (top) and uric acid (bottom).   **p<0.01.   Values = mean +/- SEM.

DISCUSSION

Previous investigators [12,13,14] have employed tissue measurement of 2,5-DHBA to assess in vivo hydroxyl radical formation. The use of brain microdialysis provides a clear advantage in that serial changes can be assessed. This technique was used recently to provide in vivo evidence that the 2,5-DHBA does arise from iron-dependent hyroxyl radical formation with subsequent hydroxylation of salicylate [11]. A marked increase in 2,5-DHBA was measured when FeSO4/EDTA (10^{-3}M) was added to the dialysis fluid. This increase has been inhibited by pretreatment with DES (unpublished results).

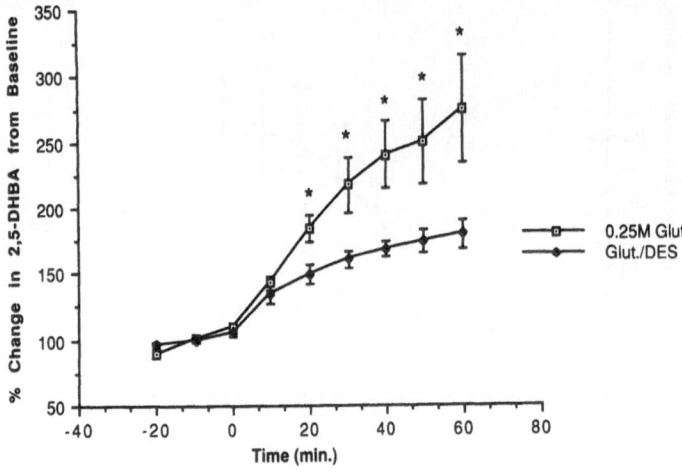

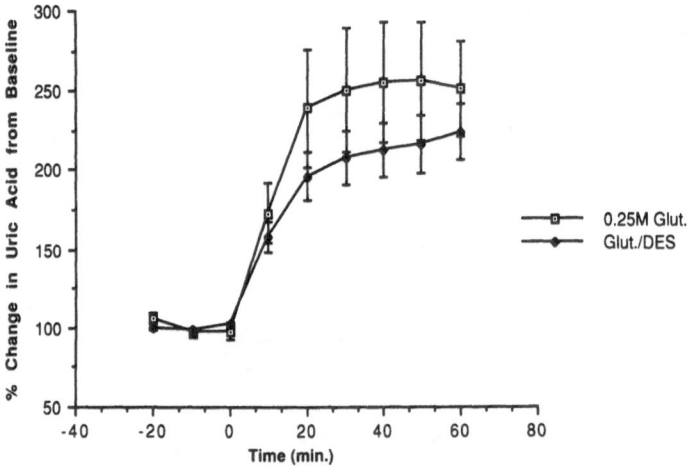

Fig. 2.   Effect of pretreatment with DES on the 2,5-DHBA
(top) and uric acid (bottom) response to 0.25M
glutamate. * p<0.05. Values = mean +/- SEM.

The data reported here demonstrate that in vivo exposure of striatal
neurons to glutamate results in large, rapid increases in 2,5-DHBA and uric
acid in the extracellular space.   A marked inhibition of the increase in 2,5-
DHBA following administration of the iron chelating agent DES is consistent
with the stimulation of hydroxyl radical formation by glutamate, but the
mechanism for this response is unclear.   One plausible suggestion is that
glutamate-induced increases in intracellular calcium activate proteases that
digest metalloproteins, leading to increased availability of iron to catalyze
hydroxyl radical formation[1].   Since activation of phospholipases and release
of arachidonic acid is a consequence of EAAs, and inhibition of lipid
peroxidation with 21-aminosteroids provides some protection against
excitotoxic neuronal injury, others have suggested the arachidonic acid
cascade as a source of free radicals[7].

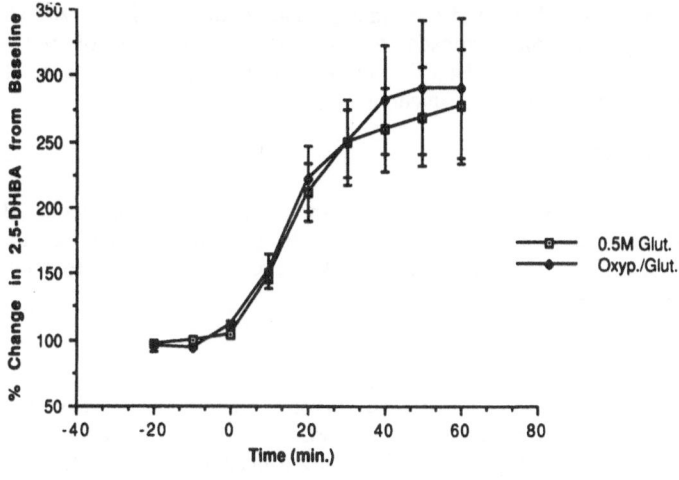

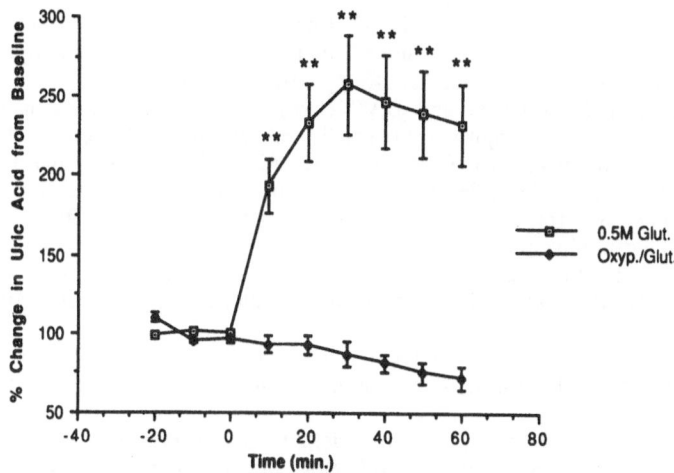

Fig. 3. Effect of oxypurinol (10-3M) in the dialysis fluid on the 2,5-DHBA (top) and uric acid (bottom) response to 0.50M glutamate. **p<0.01. Values = mean +/- SEM.

There is strong evidence that elevated intracellular calcium promotes the conversion of xanthine dehydrogenase to xanthine oxidase[1]. The latter enzyme catalyzes the formation of uric acid from hypoxanthine and xanthine. Thus, it is not surprising that the metabolic stress produced during glutamate exposure resulted in elevated uric acid levels. A result of xanthine oxidase activity however, is the formation of superoxide radical that could act as a source of the more active hydroxyl radical. This does not appear to be the case here, since the inclusion of oxypurinol in the dialysis fluid totally inhibited the uric acid increase during glutamate exposure but had no effect on the increase in 2,5-DHBA.

In summary, this study suggests that hydroxyl free radical production is stimulated in neurons exposed to glutamate. In view of the well known destructive capacity of hydroxyl radicals, this might represent an important route by which EAA toxicity occurs.

In summary, this study suggests that hydroxyl free radical production is stimulated in neurons exposed to glutamate. In view of the well known destructive capacity of hydroxyl radicals, this might represent an important route by which EAA toxicity occurs.

ACKNOWLEDGEMENT

This work was supported by an operating grant from the Medical Research Council of Canada. The author is grateful for the excellent technical assistance provided by Mr. R. Kozak, and for the secretarial assistance provided by Ms. M. Hildebrand.

REFERENCES

1. B. Halliwell and J.M.C Gutteridge, "Free Radicals in Biology and Medicine", Oxford University Press, Oxford (1989).
2. J.A. Dykens, A. Stern, and E. Trenkner, Mechanism of kainate toxicity to cerebellar neurons is analagous to reperfusion tissue injury, J. Neurochem. 49:1222 (1987).
3. P.H. Chan, L. Chu, S.F. Chen, E.J. Carlson, and C.J. Epstein, Reduced neurotoxicity in transgenic mice overexpressing human copper-zinc-superoxide dismutase, stroke 21 (Suppl III): III-80 (1990).
4. H. Monyer, D.M. Hartley, and D.W. Choi, 21-aminosteroids attenuate excitotoxic neuronal injury in cortical cell cultures, Neuron 5:121 (1990).
5. D.E. Fellegrini-Giampietro, G. Cherici, M. Alesiani, V. Carla, and F. Moroni, Excitatory amino acid release from rat hippocampal slices as a consequence of free-radical formation, J. Neurochem. 51:1960 (1988).
6. D.E. Pellegrini-Giampietro, G. Cherici, M. Alesiani, V. Carla, and F. Moroni, Excitatory amino acid release and free radical formation may cooperate in the genesis of ischemia-induced neuronal damage, J. Neurosci. 10:1035 (1990).
7. S.M. Oh, and A.L. Betz, Interaction between free radicals and excitatory amino acids in the formation of ischemic brain edema in rats, stroke 22:915 (1991).
8. G. Damsma, D.P. Boisvert, L.A. Mudrick, D. Wenkstern, and H.C. Fibiger, Effects of transient forebrain ischemia and pargyline on extracellular concentrations of dopamine, serotonin, and their metabolites in the rat striatum as determined by in vivo microdialysis, J. Neurochem. 54:801 (1990).
9. M. Grootveld, and B. Halliwell, Aromatic hydroxylation as a potential measure of hydroxyl-radical formation in vivo, Biochem. J. 237:499 (1986).
10. R.A. Floyd, J.J. Watson, and P.K. Wong, Senstive assay of hydroxyl free radical formation utilizing high pressure liquid chromatography with electrochemical detection of phenol and salicylate hydroxylation products, J. Biochem, Biophys. Methods 10:221 (1984).
11. D.P. Boisvert, In vivo assessment of hydroxyl free radical production, J. Cereb. Blood Flow Metab. 11 (Suppl 2): 5637 (1991).
12. R.A. Floyd, R. Henderson, J.J. Watson, and P.K. Wong, Use of salicylate with high pressure liquid chromatography and electrochemical detection (LCED) as a sensitive measure of hydroxyl free radicals in adriamycin treated rats, J. Free Radic. Biol. Med. 2:13 (1986).
13. W. Cao, J.M. Carney, A. Duchan, R.A. Floyd, and M. Chevion, Oxygen free radical involvement in ischemia and reperfusion injury to brain, Neurosci. Lett. 88:233 (1988).
14. R. Udassin, I. Ariel, Y. Haskel, N. Kitrossky, and M. Chevion, Salicylate as an in vivo free radical trap: studies on ischemic insult to the rat intestine, Free Radic. Biol. Med. 10:1 (1991).

BRAIN EPOXYGENASE METABOLITES OF ARACHIDONIC ACID PRODUCE

OXYGEN RADICALS AND AFFECT IN VIVO PLATELET AGGREGATION

Earl F. Ellis and Shivachar C. Amruthesh

Department of Pharmacology and Toxicology
Medical College of Virginia
Richmond, VA 23298

INTRODUCTION

Cyclooxygenase and lipoxygenase metabolites of arachidonic acid (AA) are known to affect neurotransmitter function and inflammatory processes. Cyclooxygenase metabolism of arachidonic acid has been shown to be associated with oxygen free radical formation [1]. These cyclooxygenase-dependent free radicals are produced after experimental neural trauma or acute hypertension and contribute to endothelial lesions in the cerebral microcirculation [2]. Recently a third pathway of arachidonic acid metabolism has been examined, the so-called third pathway, or epoxygenase pathway [3]. It is known that this pathway produces epoxides of arachidonic acid as well as 19- and 20-hydroxylated derivatives of arachidonic acid. One of the products of the epoxygenase pathway, 5,6-epoxyeicosatrienoic acid (5,6-EET) is known to be metabolized by cyclooxygenase enzyme [4], therefore offering the possibility that third pathway metabolites may also produce oxygen radicals. The goal of the current study was to determine if brain makes epoxyeicosatrienoic acids (EETs) from AA and if so whether they affect cerebral vascular resistance or in vivo platelet aggregation.

METHODS AND MATERIALS

Brain Metabolism of Arachidonic Acid

Blood-free mouse brain slices were incubated with ^{3}H arachidonic acid, the metabolites extracted with ethyl acetate and separated by reverse phase HPLC [5]. Certain products isolated by reverse phase HPLC were further purified by normal phase HPLC. Metabolite confirmation was obtained by comparison of HPLC retention times with the retention time of known standards. Structural confirmation was performed by using GC/MS in the electron impact and negative ion chemical ionization modes.

Studies of In Vivo Cerebral Arterioles

Metabolites isolated by HPLC, as well as authentic synthetic epoxygenase metabolites, were tested for vasoactivity using the acute cranial window technique and in vivo microscopy in anesthetized rabbits [6]. Adult rabbits were anesthetized with chloralose-urethane and supplemented with sodium pentobarbital. The animals were intubated and ventilated to produce normal blood gases. Arterial and venous catheters were inserted for measurement of blood pressure as well as anesthetic injection.

The Role of Neurotransmitters in Brain Injury, Edited by
M. Globus and W.D. Dietrich, Plenum Press, New York, 1992

Studies of In Vivo Platelet Aggregation

A craniectomy was performed on urethane-anesthetized mice, and the light plus dye method of Rosenblum and El-Sabban utilized to induce in vivo platelet aggregation [7]. Fluorescein (0.2 ml) was injected intravenously followed by mercury lamp illumination of the cerebral microcirculation, which produces injury and induces in vivo arteriolar and venular platelet aggregation. The effect of epoxygenase metabolites and indomethacin were determined by measuring the time necessary for the first arteriolar and venular platelet aggregate to occur in the cerebral microcirculation.

RESULTS

Figure 1 shows a typical reverse phase HPLC chromatogram of the profile of metabolites produced by mouse brain slices. We arbitrarily labeled 11 (I-XI) radioactive peaks, many of which co-chromatographed with authentic radioactive standards. In addition to products co-chromatographing with prostaglandins and previously described lipoxygenase products, a number of products co-chromatographed similarly to known, authentic epoxygenase metabolites. Among these were included 5,6-EET and 14,15-EET. Subsequent normal phase analysis of peaks VIII-XI determined that in addition to 5-, 12-, and 15-HETE, the brain also produced products with normal phase HPLC retention times identical to 14,15-EET and 5,6-EET. We next took the products with the mobility of 14,15- and 5,6-EET and derivatized them for examination by gas chromatography/mass spectrometry. Both electron impact and negative ion chemical ionization analysis of these products confirmed the identity as 14,15- and 5,6-EET.

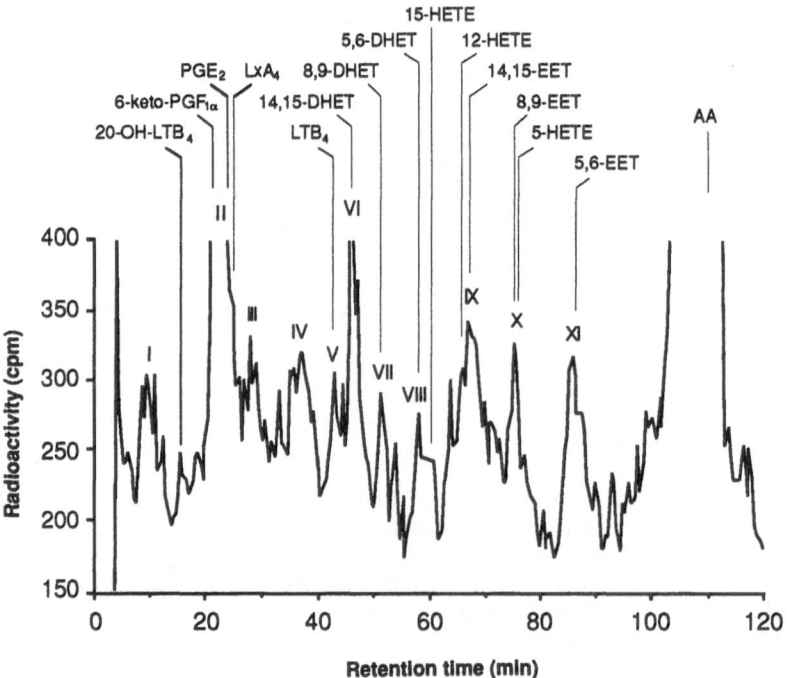

Figure 1. Reverse-phase HPLC elution profile of metabolites (I-XI) formed from [³H]-arachidonic acid by mouse brain slices. Retention time of authentic standards is indicated.

The effects of products isolated from reverse phase HPLC were next tested for their vasoactivity on in vivo cerebral arterioles in rabbits. As shown in figure 2, the product identified as 5,6-EET produced a transient vasodilation. Other products, such as reverse phase HPLC peaks IV and VI had no effects on arteriolar diameter. Additionally, we found that product XI had vasoactivity similar to that produced by authentic 5,6-EET. In other studies (data not shown) we found that 8,9-EET had reduced activity and 11,12- and 14,15-EET has essentially no vasoactivity compared to 5,6-EET [8].

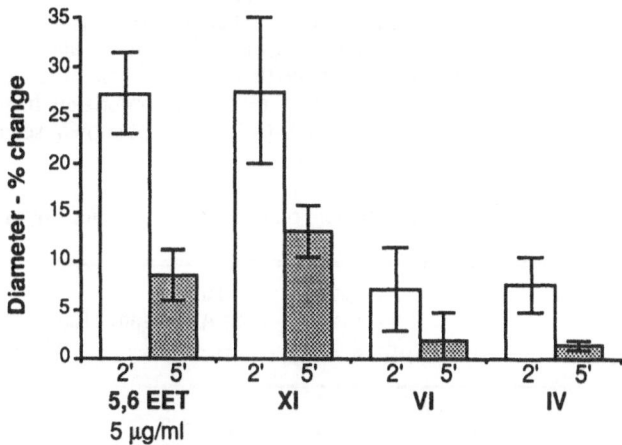

Figure 2. The effect of mouse brain metabolites of arachidonic acid and authentic 5,6-EET on in vivo cerebral arteriolar diameter. Responses at 2 and 5 minutes after application are depicted.

We next examined the mechanism by which 5,6-EET induces cerebral arteriolar dilation. Since it is known that arachidonic acid-induced dilation in the cerebral microcirculation is produced by vasoactive oxygen radicals formed concomitant with cyclooxygenase metabolism of arachidonic acid [8], we hypothesized that a similar mechanism of vasoactivity may be occurring if 5,6-EET is metabolized by brain cyclooxygenase. To test this hypothesis, we first determined the effect of indomethacin on 5,6-EET-induced dilation. As shown in figure 3, indomethacin virtually eliminated 5,6-EET-induced dilation. Next, to determine whether this dilation was due to oxygen radicals, we co-applied with 5,6-EET superoxide dismutase and catalase, two known scavengers of oxygen radicals. As also shown in figure 3, these oxygen radical scavengers dramatically reduced dilation produced by 5,6-EET.

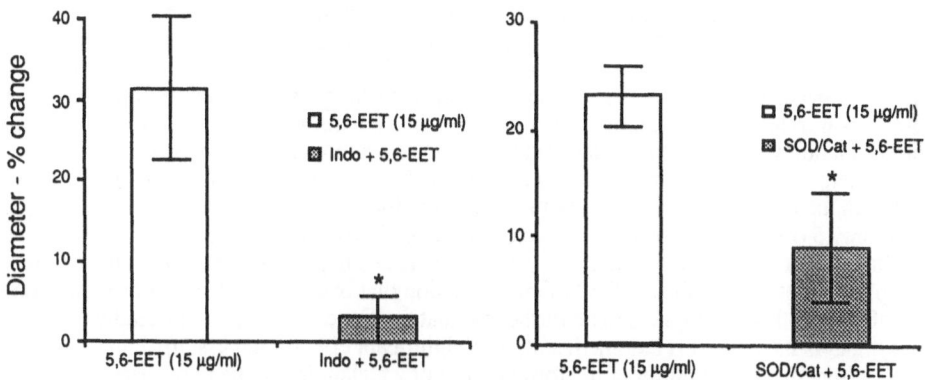

Figure 3. The effect of indomethacin and SOD + catalase on arachidonic acid-induced rabbit cerebral arteriolar dilation. P < 0.05 vs control.

We next tested effects of two epoxides on in vivo platelet aggregation. The epoxides we examined included 14,15-EET, which we had shown to be produced by brain tissue, as well as another epoxide, 8,9-EET. Whether 8,9-EET is produced by brain tissue is at this time not completely determined. In addition, to confirm the validity of our light plus dye method to stimulate platelet aggregation we also examined the effect of indomethacin, a known cyclooxy-genase inhibitor, on in vivo platelet aggregation. As shown in Table 1 14,15-EET was able to significantly reduce the speed at which platelet aggregation occurred in cerebral arterioles. This effect of 14,15-EET was almost as dramatic as that produced by indomethacin. However 8,9-EET had no effect on the rate at which in vivo platelet aggregation occurs. No agents affected venular aggregation. We also determined the effect of indomethacin and the EETs on throm-boxane B_2 formation in mouse serum. We found that indomethacin produced an 80% reduc-tion in thromboxane production while 14,15-EET reduced thromboxane formation by 50%. Thromboxane formation was not affected by 8,9-EET (data not shown, see ref. 10).

Table 1. The effect of EETs and indomethacin on time to first platelet aggregate in cerebral arterioles

Experimental Condition	N	Arteriolar Diameter (μm, $\bar{x} \pm$ SE)	Time to First Aggregate (sec $\pm$ S.E.)	Experimental Control	P value
indomethacin (0.5 mg/kg)	18	35 ± 1	127 ± 7	1.35	< 0.01
control	17	34 ± 1	95 ± 7		
14,15-EET (0.3 mg/kg)	10	42 ± 1	118 ± 4	1.26	< 0.01
control	11	43 ± 1	94 ± 4		
14,15-EET (0.15 mg/kg)	10	37 ± 1	105 ± 4	1.06	NS
control	10	38 ± 1	99 ± 4		
14,15-EET (0.03 mg/kg)	10	39 ± 1	105 ± 5	1.06	NS
control	10	38 ± 1	99 ± 5		
8,9-EET (0.3 mg/kg)	10	39 ± 2	97 ± 6	1.07	NS
control	10	39 ± 2	91 ± 6		

DISCUSSION

The current work shows that whole brain slices are capable of synthesizing 5,6- and 14,15-EET, as well as several other unidentified arachidonic acid metabolites. The elucidation and confirmation of these EETs is based on their chromatographic retention in normal phase and reverse phase HPLC, gas chromatographic retention time and mass spectra produced both by electron impact and negative ion chemical ionization techniques. In addition these two epox-ides appear to have potent biological activities on the cerebral microcirculation. 5,6-EET-in-duced dilation was inhibited by classical inhibitors of cyclooxygenase or oxygen free radical scavengers. The production of oxygen radicals by 5,6-EET but not other epoxides is consis-tent with the previous literature showing that, of the four epoxides produced by the third path-way, only 5,6-EET is a substrate for cyclooxygenase [4]. An important consequence of our finding that 5,6-EET induces dilation by cyclooxygenase-dependent free radical formation is that previous studies of the cerebral microcirculation that based their conclusions simply on the use of indomethacin will have to be further evaluated. For example in many studies investigators have utilized indomethacin, seen significant effects, and assumed that all these effects were due to inhibition of the formation of prostaglandins of the two-series. The current study suggests that an additional mechanism may be inhibited by indomethacin, that being the cyclooxygenase metabolism of P-450 metabolites formed by the brain.

Our studies with 14,15-EET and in vivo platelet aggregation indicate that this brain-pro-duced eicosanoid may modulate platelet aggregation in the cerebral microcirculation. Whether adequate amounts of 14,15-EET can be produced by the brain or brain vasculature to have a physiological effect on in vivo platelet aggregation is unknown. This uncertainty is a conse-quence of our current inability to measure the formation of epoxides from endogenous arachidonic acid in the brain or brain microcirculation. Our ongoing studies are addressing the degree to which endogenous arachidonic acid is metabolized by this third pathway. However our results clearly show that ,at least at pharmacologic doses in vivo, 14,15-EET significantly prolongs the time required for the light plus dye method to induce platelet aggregation. Furthermore, analysis of serum thromboxane showed that 14,15-EET reduced thromboxane formation by 50%. However whether the effect of 14,15-EET on in vivo platelet aggregation is a consequence of thromboxane reduction is uncertain. This stems from the fact that Fitzpatrick et al using 14,15-EET and other EETs in in vitro studies of platelet aggregation and thromboxane formation showed that inhibition of in vitro aggregation by EETs was not always associated with inhibition of thromboxane B_2 formation [9]. In addition, our other ongoing studies indicate that a 50% inhibition of thromboxane B_2 is not likely adequate to affect in vivo aggregation as assessed by the light plus dye technique [11]. Therefore the mechanism by which 14,15-EET inhibits platelet aggregation may not be related to its effect on thromboxane B_2 formation.

In summary, our studies show the existence of a new third pathway of arachidonic acid metabolism in brain tissue. We have only begun to examine the potential physiologic or pathophysiologic consequences of this pathway in the brain. However indications are that these third pathway metabolites may affect cerebrovascular resistance and affect hemostasis in the brain.

ACKNOWLEDGEMENTS

This work was supported by a grant-in-aid and Fellowship from the American Heart Association and National Institute of Neurological Disorders and Stroke grants 27214 and 12587. E. Ellis is the recipient of a Javits Neuroscience Investigator Award. The assistance of Brian Thomas in the GC/NICI/MS analysis is appreciated.

REFERENCES

1. R. Kukreja, H.A. Kontos, M.L. Hess, and E.F. Ellis, PGH synthase and lipoxygenase generate superoxide in the presence of NADH or NADPH, Circ. Res. 59:612-619 (1986).

2. H.A. Kontos, E.P. Wei, E.F. Ellis, W.D. Dietrich, and J.T. Povlishock, Prostaglandins in physiological and in certain pathological responses of the cerebral circulation, Fed. Proc. 40:2326-2330 (1981).

3. F.A. Fitzpatrick and R.C. Murphy, Cytochrome P-450 metabolism of arachidonic acid: formation and biological actions of "epoxygenase"-derived eicosanoids, Pharmacol. Rev. 40:229-241 (1989).

4. E. Oliw, Metabolism of 5,6-oxidoeicosatrienoic acid by ram seminal vesicles: formation of two stereoisomers of prostaglandin I1, J. Biol. Chem. 259:2716-2721 (1984).

5. S.C. Amruthesh and E.F. Ellis, Brain synthesis and cerebrovascular action of epoxygenase metabolites of arachidonic acid, J. Neurochem. (1991).(in press)

6. T. Kamitani, M.H. Little, and E.F. Ellis, Evidence for a possible role of the brain kallikrein-kinin system in the modulation of the cerebral circulation, Circ. Res. 57:545-552 (1985).

7. W.I. Rosenblum and F. El-Sabban, Platelet aggregation in the cerebral microcirculation: effect of aspirin and other agents, Circ. Res. 40:320-328 (1977).

8. E.F. Ellis, R.J. Police, L.M. Yancey, J.S. McKinney, and S.C. Amruthesh, Dilation of cerebral arterioles by cytochrome P-450 metabolites of arachidonic acid, Am. J. Physiol. 259:H1171-H1177 (1990).

9. F.A. Fitzpatrick, M.D. Ennis, M.E. Baze, M.A. Wynalda, J.E. McGee, and W.F. Liggett, Inhibition of cyclooxygenase activity and platelet aggregation by epoxyeicosatrienoic acids, J. Biol. Chem. 261:15334-15338 (1986).

10. M.L. Heizer, J.S. McKinney, and E.F. Ellis, 14,15-Epoxyeicosatrienoic acid inhibits platelet aggregation in cerebral arterioles, Stroke (1991) (in press)

11. M.L. Heizer, J.S. McKinney, and E.F. Ellis, The effect of dietary n-3 fatty acids on platelet aggregation in the cerebral microcirculation, Am. J. Physiol. (1991).(submitted for publication)

AUTHOR INDEX

SUBJECT INDEX*

---

*Page numbers indicate the first page of the paper in which the subject
is discussed.